T0331168

Integrated Circuit Design

This textbook seeks to foster a deep understanding of the field by introducing the industry integrated circuit (IC) design flow and offering tape-out or pseudo tape-out projects for hands-on practice, facilitating project-based learning (PBL) experiences.

Integrated Circuit Design: IC Design Flow and Project-Based Learning aims to equip readers for entry-level roles as IC designers in the industry and as hardware design researchers in academia. The book commences with an overview of the industry IC design flow, with a primary focus on register-transfer level (RTL) design, the automation of simulation and verification, and system-on-chip (SoC) integration. To build connections between RTL design and physical hardware, FPGA (field-programmable gate array) synthesis and implementation is utilized to illustrate the hardware description and performance evaluation. The second objective of this book is to provide readers with practical, hands-on experience through tape-out or pseudo tape-out experiments, labs, and projects. These activities are centered on coding format, industry design rules (synthesizable Verilog designs, clock domain crossing, etc.), and commonly-used bus protocols (arbitration, handshaking, etc.), as well as established design methodologies for widely-adopted hardware components, including counters, timers, finite state machines (FSMs), I2C, single/dual-port and ping-pong buffers/register files, FIFOs, floating-point units (FPUs), numerical hardware (Fourier transform, matrix-matrix multiplication, etc.), direct memory access (DMA), image processing designs, neural networks, and more.

The textbook caters to a diverse readership, including junior and senior undergraduate students, as well as graduate students pursuing degrees in electrical engineering, computer engineering, computer science, and related fields. The target audience is expected to have a basic understanding of Boolean Algebra and Karnaugh Maps, as well as prior familiarity with digital logic components such as AND/OR gates, latches, and flip-flops. The book will also be useful for entry-level RTL designers and verification engineers who are embarking on their journey in application-specific IC (ASIC) and FPGA design industry.

Integrated Circuit Design

IC Design Flow and Project-Based Learning

Xiaokun Yang

CRC Press is an imprint of the
Taylor & Francis Group, an **informa** business

Designed cover image: Shutterstock ©

First edition published 2025
by CRC Press
2385 NW Executive Center Drive, Suite 320, Boca Raton FL 33431

and by CRC Press
4 Park Square, Milton Park, Abingdon, Oxon, OX14 4RN

CRC Press is an imprint of Taylor & Francis Group, LLC

ISBN: 978-1-032-03079-1 (hbk)
ISBN: 978-1-032-03392-1 (pbk)
ISBN: 978-1-003-18708-0 (ebk)

DOI: 10.1201/9781003187080

Typeset in CMR10 Roman
by KnowledgeWorks Global Ltd.

Dedicated to my wife Ruan Zhan
and our daughter Bella Yang

Dedicated to my parents Hong Yang and Guizhi Xu
and my sister Xiaojing Yang

Contents

List of Abbreviations

ABV Assertion-Based Verification

AHB Advanced High-Performance Bus

AMBA Advanced Microcontroller Bus Architecture

APR Automated Placement and Routing

ATPG Auto Test Pattern Generation

ASIC Application-Specific Integrated Circuit

AXI Advanced eXtensible Interface

BFM Bus Functional Model

CDC Clock Domain Crossing

CDV Coverage-Driven Verification

CLB Configurable Logic Block

CMOS Complementary Metal-Oxide-Semiconductor

CRV Constrained-Random Verification

DFT Design for Test

DRC Design Rule Check

DUT Design-Under-Test

DMA Direct Memory Access

ECO Engineering Change Order

EDA Electronic Design Automation

FPGA Field-Programmable Gate Array

GDS Graphic Data System

HDL Hardware Description Language

HCL Hardware Construction Language

HLS High Level Synthesis

IC Integrated Circuit

IDM Integrated Device Manufacturer

IO Input and Output

IP Intellectual Property

LVS Layout Versus Schematic

LUT Look-Up Table

LSB/MSB Least-/Most-Significant Bit

MPW Multi Project Wafer, or Shuttle

MOF Maximum Operational Frequency

MOSFET Metal-Oxide-Semiconductor Field-Effect Transistors

NRE Non-Recurring Engineering

OOP Object-Oriented Programming

Pre/Post-Sim Pre-/Post-Layout Simulation

Post-STA Post-Layout Static Timing Analysis

PCB Printed Circuit Board

RAM Random Access Memory

ROM Read-Only Memory

RTL Register-Transfer Level

SDF Standard Delay Format

SPF Standard Parasitic Format

STA Static Timing Analysis

UVM Universal Verification Methodology

VMM Verification Methodology Manual

WNS Worst Negative Slack

Preface

Purpose of the Book

The primary objective of this book is to equip readers for entry-level roles as integrated circuit (IC) designers in the industry and as hardware design researchers in academia. It seeks to foster a deep understanding of the field by introducing the industry IC design flow and offering tape-out or pseudo tape-out projects for hands-on practice, facilitating project-based learning (PBL) experiences.

Industry IC Design Flow

The IC design industry places significant emphasis on the intricate IC design flow, encompassing a comprehensive suite of electronic design automation (EDA) tools for tasks such as simulation, synthesis, layout, timing analysis, and more, along with established design and verification methodologies. This book commences with an overview of the industry IC design flow, with a primary focus on register-transfer level (RTL) design, the automation of simulation and verification, and system-on-chip (SoC) integration. To build connections between RTL design and physical hardware, field-programmable gate array (FPGA) synthesis and implementation is utilized to illustrate the hardware description and performance evaluation.

Part III of the book includes three appendices/tutorials on commonly used toolsets and industry design and simulation approaches. These encompass the Vim editor, Siemens ModelSim simulator, AMD Vivado, as well as guidance on the structured project directory, fundamental Unix/Linux commands, and the effective automation of simulations using TCL scripts. All three appendices/tutorials are readily accessible through the accompanying Git repository, ensuring that readers can easily utilize supplementary materials and resources to enhance their learning experience.

Project-Based Learning

RTL design is pivotal as it not only describes hardware functionality but also directly impacts chip performance, encompassing aspects such as area, speed, power efficiency, and successful chip fabrication within specified timing constraints. Hence, the second objective of this book is to provide readers with practical, hands-on experience through tape-out or pseudo tape-out experiments, labs, and projects. These activities are centered on coding format, industry design rules (synthesizable Verilog designs, clock domain crossing, etc.), and commonly used bus protocols (arbitration, handshaking, etc.), as well as established design methodologies for widely adopted hardware components, including counters, timers, finite state machines (FSMs), I2C, single/dual-port and ping-pong buffers/register files, FIFOs, floating-point units (FPUs), numerical hardware (Fourier transform, matrix-matrix multiplication, etc.), direct memory access (DMA), image processing designs, neural networks, and more.

The project-based learning unfolds systematically throughout the book. For instance, it commences with the construction of a timer featuring two-level counters in Chapter 6, followed by the design of FSMs in Chapter 7. Subsequently, it progresses to the creation of an I2C design, a widely used serial bus protocol in modern chipsets, achieved through the integration of an FSM and a timer-based datapath in Chapter 8. It is imperative to underscore that most of the design projects showcased in this book are open-source in the accompanying Git repository. This not only enables readers to utilize these projects but also encourages them to expand and experiment with the designs.

The Contents at a Glance

The book is structured into three distinct parts as below.

Part I: Fundamentals of IC Design and Simulation with Verilog HDL (Chapters 1–8)

Part I is dedicated to introducing the industry IC design flow and encompasses fundamental knowledge and skills for RTL design and simulation using Verilog hardware description language (HDL). It provides practical examples aimed at enhancing the learning experience and skill development.

- **Chapter 1** functions as an introduction to the IC industry, offering background information and the essential knowledge required for RTL design with Verilog HDL.
- **Chapter 2** provides a concise overview of the industry IC design flow, encompassing both the front-end (specification, RTL design, verification, SoC integration, etc.) and back-end procedures (synthesis, layout, timing check, etc.), fabrication stages, and packaging and testing.
- **Chapters 3–5** explore the foundational aspects of Verilog, with an emphasis on creating synthesizable designs and establishing a basic simulation environment. The EDA tools employed in these chapters encompass Vim editor and Siemens ModelSim.

PBL 1–9 showcase the design of fundamental combinational and sequential circuits, the hierarchical design methodology, and the creation of automated simulation environments.

- **Chapter 6** centers on the synthesis results using AMD Vivado, establishing a connection between Verilog descriptions and specific hardware.
  PBL 10–16 are provided to facilitate creating a linkage between Verilog code and their described hardware circuits.

- **Chapter 7** covers basic FSM design and simulation, serving as a fundamental timing controller for digital ICs.
  PBL 17 offers an example of utilizing an FSM template for detecting a specific digit sequence.

- **Chapter 8** introduces design integration involving FSMs and datapaths, presenting two design cases: I2C and master-slave bus interfaces. This chapter serves as a versatile template, demonstrating how digital circuits can be integrated with timing controllers and well-structured datapaths.
  PBL 18–21 are provided to demonstrate FSMD designs across various hardware architectures, alongside advanced design methodologies like pipeline and parallel computing.

Part II: Advanced IC Design and Integration (Chapters 9–12)

Part II provides a comprehensive exploration of various design architectures (streaming design, iterative design, pipeline and parallel computing, etc.), timing considerations/constraints employing RTL design, and SoC integration. This part illustrates these concepts through a variety of design examples including numerical hardware accelerators and neural engines. Additionally, Part II explores the prevalent SoC architecture utilized in the industry, focusing particularly on AMBA AXI (Advanced Microcontroller Bus Architecture - Advanced eXtensible Interface), alongside other commonly-used design components located on SoCs such as DMA and image/video processing units.

- **Chapter 9** centers on numerical hardware design and integration, leveraging the FPUs provided by this book. Examples featured in this chapter encompass a floating-point (FP) matrix-matrix adder, *axpy* computation, and a fundamental *ddot* design. These examples represent essential hardware accelerators frequently employed in scientific computing.
  PBL 22–27 demonstrate RTL design and simulation for register files, as well as numerical hardware components like FP multiplication-addition circuits and matrix-matrix multipliers.

- **Chapter 10** explores high-performance design structures with a particular emphasis on streaming and iterative designs. This chapter illustrates

how specialized hardware designs can be employed to improve data processing efficiency and/or providing cost-effective solutions to meet a range of hardware design specifications.
PBL 28–30 showcase streaming and iterative design options tailored for various numerical applications, including iterative *ddot* design and Fourier transform.

- **Chapter 11** presents an introduction to timing constraints, emphasizing their critical importance in high-speed design and their substantial influence on the success or failure of timing analysis and chip fabrication.
PBL 31–41 focus on exploring designs with attainable MOF (maximum operational frequency) and introduce design rules and options, including signal management across asynchronous clock domains and the implementation of high-speed designs using pipeline structures.

- **Chapter 12** is dedicated to SoC integration, particularly emphasizing the AMBA AXI bus, a leading bus protocol in the industry. It introduces various typical design components frequently employed in SoCs, including DMA, image processing units, and neural network engines.
PBL 42–50 exemplify hardware design architectures and methodologies, encompassing topics such as floating-to-fixed point conversion, approximate design, hardware reuse, and more. These are tailored to meet various design specifications, addressing concerns such as accuracy, hardware resource and power constraints, and computational latency bounds. Industry-adopted projects, including DMA arbitration, image processing units, and sigmoid neural networks, are utilized in these projects to provide practical insights and applications.

Part III: Tutorials on EDA Tools and Essential Skills Related to IC Design and Simulation

This part provides tutorials that introduce EDA tools and essential knowledge and skills for IC design and simulation.

- **Appendix/Tutorial A** introduces the structured project directory and foundational Unix/Linux commands, accompanied by an exhaustive tutorial for proficiently utilizing the Vim editor.

- **Appendix/Tutorial B** furnishes a tutorial for using the Siemens ModelSim simulator, covering navigation of the GUI interface and effective automation of simulations through TCL scripts.

- **Appendix/Tutorial C** offers a comprehensive tutorial on using AMD Vivado for FPGA implementations, with a particular focus on RTL analysis for synthesized circuits and performance evaluation. This includes assessing hardware resource utilization, power consumption, and computational speed.

The Intended Audience

The book provides an overview of the IC design flow, with a specific focus on front-end aspects, covering design specification, RTL design, simulation, SoC integration, FPGA synthesis and implementation, and performance evaluation. It caters to a diverse readership, including junior and senior undergraduate students, as well as graduate students pursuing degrees in electrical engineering, computer engineering, computer science, and related fields. The target audience is expected to have a basic understanding of Boolean Algebra and Karnaugh Maps, as well as prior familiarity with digital logic components such as AND/OR gates, latches, and flip-flops.

This book also serves as an invaluable resource for entry-level RTL designers and verification engineers who are embarking on their journey in application-specific IC (ASIC) and FPGA design industry, offering essential knowledge and practical insights for their roles in the field.

How to Use This Book

This book is thoughtfully designed to encompass the material for one or two courses at the junior and senior undergraduate levels, and within graduate studies. To fully derive value from the content, it is recommended that readers have completed a prerequisite course covering topics such as Boolean Algebra, Karnaugh Maps, and fundamental digital logic. Below are two examples of teaching courses using this textbook.

Introduction to IC Design Flow/Digital System Design

This course is designed specifically for junior and senior undergraduate students, as well as those pursuing graduate studies in digital system/IC design and simulation. The course materials covered in Part I and Part III of the book, spanning Chapters 2–8 and Appendixes A–C, provide a solid foundation for instruction. Moreover, a series of PBL examples, numbered 1–21, have been included to complement the training lectures and facilitate practical hands-on labs and projects. Further teaching suggestions are available on the accompanying GitHub repository.

Advanced Digital System Design

This course is specifically tailored for senior undergraduate and graduate students who are pursuing studies in high-performance hardware design and SoC integration. The course materials, primarily covered in Part II and Part III of the book, spanning Chapters 1 and 9–12, along with Appendixes A–C, establish a robust foundation for instruction. For students without prior Verilog coding experience, Chapters 3–5 can be utilized to introduce the fundamental Verilog syntax for RTL design and simulation. Moreover, a series of practical PBL examples, numbered 6–16 and 22–50, can be integrated to complement the training lectures and provide hands-on labs and projects. Further teaching suggestions are available on the accompanying GitHub repository.

Supplements and Resources (Git Repository)

Numerous supplementary resources are accessible on the Git repository: `https://github.com/LBL-ICS/IC-Design/`, encompassing:

- The Verilog designs for experiments, labs, and projects presented in the book.
- Appendices A–C containing tutorials on the recommended project directory, Vim Editor, Siemens ModelSim, AMD Vivado, and other valuable commands and scripts for IC design projects.

Instructors teaching IC design-related courses can also access the following additional resources:

- Lecture slides covering all chapters.
- Extensive teaching suggestions for two courses: "Introduction to IC Design Flow" and "Advanced Digital System Design".
- Verilog design and simulation materials for experiments, labs, and projects outlined in the book.
- Thorough solution manual for exercises/assignments.
- Two examination tests to facilitate course assessment.

Acknowledgments

The author expresses deep gratitude to Mario Vega, his former student and research assistant at the University of Houston Clear Lake, for his invaluable contribution in providing the foundational FPUs generated using Chisel hardware construction language (HCL). Mario, currently a researcher at Lawrence Berkeley National Laboratory, played a pivotal role in enhancing the content of this book.

Furthermore, the author acknowledges the research endeavors at Berkeley Laboratory since 2022 which have greatly expanded the book's scope to include code generator designs using Chisel HCL. Chisel, a domain-specific language developed by the University of California, Berkeley, provides an efficient solution for generating Verilog code for hardware implementations. This advancement has facilitated the integration of FPUs into the book's content in a highly productive manner.

Bio

Dr. Xiaokun Yang currently serves as an associate professor at the College of Science and Engineering at the University of Houston Clear Lake, located in Houston, Texas. Additionally, since 2022, he has held the role of affiliate faculty at Lawrence Berkeley National Laboratory, situated in Berkeley, California. Dr. Yang earned his Ph.D. in the Department of Electrical and Computer Engineering at Florida International University (FIU) in the spring of 2016. He also brings extensive industry experience, having worked as a senior ASIC design and verification engineer at Advanced Micro Devices (AMD) and China Electronic Corporation (CEC) during 2007–2012. Dr. Yang's research interests center on specialized hardware design and acceleration for future high-performance computing, design automation for numerical hardware and machine learning, and advanced high-performance SoC architecture.

Errata

This book has been self-prepared, encompassing all elements including text, tables, figures, code, indexing, and formatting. Acknowledging the possibility of errors, an accompanying Git repository provides an updated errata sheet and serves as a platform for reporting any identified mistakes.

X. Yang
January 2024
Houston, TX

Aid Sheet of Verilog HDL Design and Simulation

TABLE 0.1
Summary of Verilog HDL description.

Blocks		**Statements**	**Basic Operators**	**Tasks and Functions**
assign	**concurrent assign**	Basic Operators	~ (NOT), & (AND), | (OR), ∧ (XOR) + (Adder)	**System Tasks:** $display, $time, $stop, $readmemh, $random **Compiler:** \`define, \`include, \`timescale, \`ifdef-\`endif **Tasks** **Functions**
	conditional assign	?:		
always	**level trigger always**	if-else case-endcase		
	edge trigger always	if-else		
initial		if-else case-endcase for/ while/ repeat/ forever		

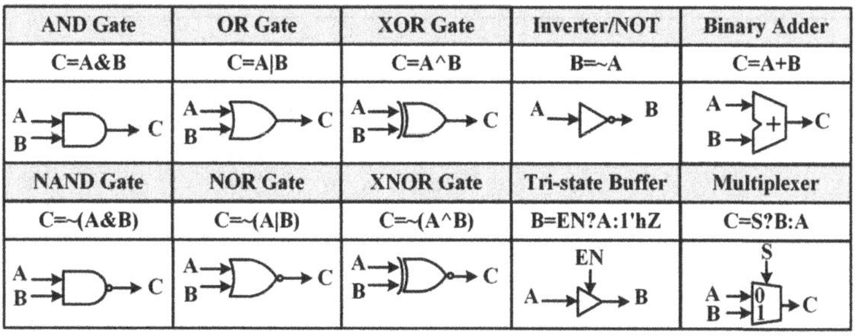

AND Gate	OR Gate	XOR Gate	Inverter/NOT	Binary Adder
C=A&B	C=A|B	C=A^B	B=~A	C=A+B
NAND Gate	**NOR Gate**	**XNOR Gate**	**Tri-state Buffer**	**Multiplexer**
C=~(A&B)	C=~(A|B)	C=~(A^B)	B=EN?A:1'hZ	C=S?B:A

FIGURE 0.1
Combinational circuit design with Verilog HDL.

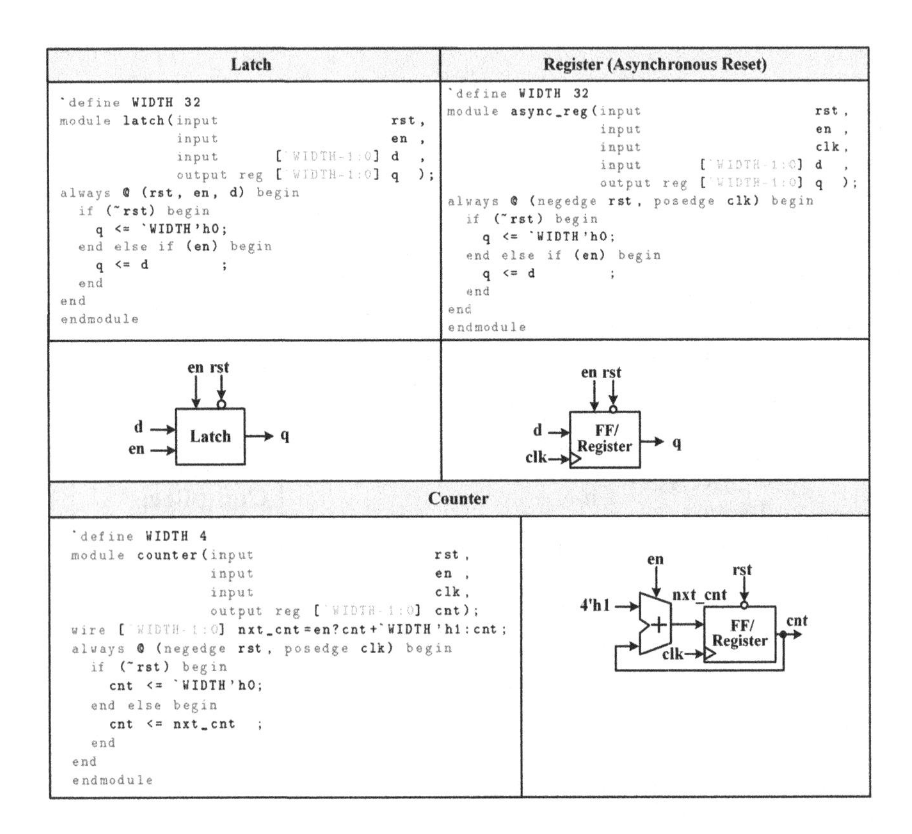

FIGURE 0.2
Sequential circuit design (Latch, Register, and Counter) with Verilog HDL.

Single-Port Register File

```
`define RAM_WIDTH 32
`define RAM_DEPTH 1024
`define RAM_ADDR_WIDTH 10
`define RAM_WE_WIDTH 4
`define WE4 `RAM_WE_WIDTH'hf
`define WE3 `RAM_WE_WIDTH'h7
`define WE2 `RAM_WE_WIDTH'h3
`define WE1 `RAM_WE_WIDTH'h1
module single(input                              clk  ,
              input                              en   ,
              input      [`RAM_WE_WIDTH-1:0]     we   ,
              input      [`RAM_ADDR_WIDTH-1:0]   addr ,
              input      [`RAM_WIDTH-1:0]        din  ,
              output reg [`RAM_WIDTH-1:0]        dout);
reg [`RAM_WIDTH-1:0] ram[0:`RAM_DEPTH-1];

always @(posedge clk) begin
  if (en) begin
    case(we)
      `WE4: ram[addr]<=din;
      `WE3: ram[addr]<={ram[addr]
                        [`RAM_WIDTH-1:`RAM_WIDTH-32*1],
                        din[`RAM_WIDTH-32*1-1:0]};
      `WE2: ram[addr]<={ram[addr]
                        [`RAM_WIDTH-1:`RAM_WIDTH-32*2],
                        din[`RAM_WIDTH-32*2-1:0]};
      `WE1: ram[addr]<={ram[addr]
                        [`RAM_WIDTH-1:`RAM_WIDTH-32*3],
                        din[`RAM_WIDTH-32*3-1:0]};
      default:  ram[addr]<=`RAM_WIDTH'h0;
    endcase
  end else begin
    dout <= ram[addr];
  end
end
endmodule
```

clk, en, we, addr, din → simple_single → dout

Dual-Port Register File

```
module dual(input                              clka ,
            input                              ena  ,
            input      [`RAM_WEA_WIDTH-1:0]    wea  ,
            input      [`RAM_ADDR_WIDTH-1:0]   addra,
            input      [`RAM_WIDTH-1:0]        dina ,
            input                              clkb ,
            input                              enb  ,
            input      [`RAM_ADDR_WIDTH-1:0]   addrb,
            output reg [`RAM_WIDTH-1:0]        doutb);
reg [`RAM_WIDTH-1:0] ram[0:`RAM_DEPTH-1];

always @(posedge clka) begin
  if (ena) begin
    case(wea)
      `WEA4: ram[addra]<=dina;
      `WEA3: ram[addra]<={ram[addra]
                          [`RAM_WIDTH-1:`RAM_WIDTH-32*1],
                          dina[`RAM_WIDTH-32*1-1:0]};
      `WEA2: ram[addra]<={ram[addra]
                          [`RAM_WIDTH-1:`RAM_WIDTH-32*2],
                          dina[`RAM_WIDTH-32*2-1:0]};
      `WEA1: ram[addra]<={ram[addra]
                          [`RAM_WIDTH-1:`RAM_WIDTH-32*3],
                          dina[`RAM_WIDTH-32*3-1:0]};
      default: ram[addra]<=`RAM_WIDTH'h0;
    endcase
  end
end

always @(posedge clkb) begin
  if (enb) begin
      doutb <= ram[addrb];
  end
end
endmodule
```

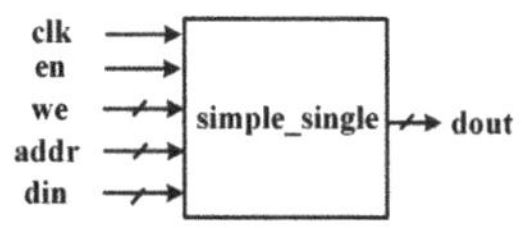

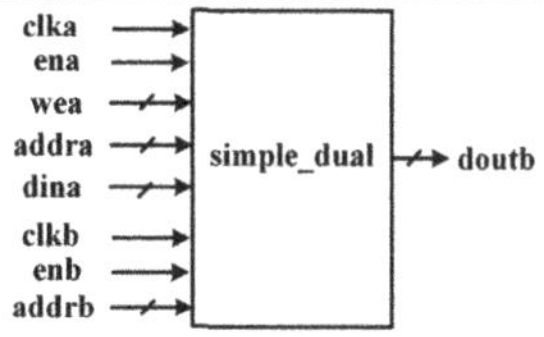

clka, ena, wea, addra, dina, clkb, enb, addrb → simple_dual → doutb

FIGURE 0.3
Sequential circuit design (Register Files) with Verilog HDL.

Part I

Fundamentals of IC Design and Simulation with Verilog HDL

1

Introduction to IC Design

Chapter 1 provides a comprehensive review of the history and roadmap of **integrated circuit (IC)** design industry. It also introduces key aspects of the IC design, encompassing both **field-programmable gate arrays (FPGAs)** and **application-specific integrated circuits (ASICs)**. Additionally, it offers an overview of Verilog design fundamentals, serving as foundational knowledge for understanding **register-transfer level (RTL)** design and simulation.

1.1 History and Road Map of IC Design

1.1.1 A brief history of IC design

The IC design industry has a storied history that spans several decades, exerting a pivotal influence on the evolution of modern electronics since the invention of the first IC.

A. The First IC (1958)

In 1958, Jack S. Kilby achieved a groundbreaking milestone when he introduced the world's first functional IC at Texas Instruments. This initial IC was a revolutionary electronic device that combined multiple electronic components – transistors, resistors, and capacitors – onto a single semiconductor substrate. This invention paved the way for the miniaturization of electronic components and the development of complex ICs that underlie modern electronics. Kilby's extraordinary achievement was later recognized with the Nobel Prize in Physics in 2000.

B. Moore's Law (1965)

Gordon Moore, a co-founder of Intel Corporation, stands as another influential figure in the realm of the IC industry. In 1965, he made a remarkable prediction, now famously known as Moore's Law. Moore's insight proposed that the number of transistors on a chip would approximately double every 18–24 months, resulting in an exponential increase in the processing power of

DOI: 10.1201/9781003187080-1

microchips and the density of transistors over time. This visionary declaration has transcended its era, evolving into a guiding principle for the semiconductor industry, propelling the continuous advancement of IC technology.

> *Moore's Law*
>
> Moore's Law is the observation that the number of transistors in a dense IC roughly doubles every 18–24 months.
>
> *Gordon Moore*, Co-founder of Intel

C. Advancements in the 1970s and 1980s

Looking ahead, the IC industry has witnessed remarkable developments, largely propelled by Moore's Law. These advancements encompass computational performance, operational frequency, power efficiency, and the emergence of multi-core implementations. During the early 1970s and 1980s, the semiconductor industry placed a significant emphasis on increasing the operational frequency of ICs, driven by the growing demands of the personal computer industry.

D. Multi-Core SoCs (1990s and Early 2000s)

In the 1990s, a significant shift occurred as the semiconductor industry recognized the importance of addressing power consumption as a critical design constraint, especially with the proliferation of smartphones and mobile devices. Balancing the enhancement of operational frequency with power efficiency has become a major challenge. Consequently, the industry adopted a new strategy by embracing multi-core architectures within system-on-chips (SoCs). In these multi-core designs, the operating frequency of individual cores was scaled down in inverse proportion to the number of cores integrated on a single IC. Such designs allowed smartphones to operate for extended periods while maintaining high computational performance.

1.1.2 Today's IC industry and technology

In today's ever-evolving landscape, new design requirements for microprocessors continue to propel the ongoing progress of IC design technology. Factors like the advent of 5G technology and the integration of complex machine learning capabilities are adding further complexity to modern IC designs. To illustrate this, Table 1.1 offers a comprehensive comparison of the specifications for Apple Inc.'s A11 through A17 SoCs, spanning the years 2017 to 2023 [5]. The A-series comprises a family of SoCs utilized in the iPhone, specific iPad models, and the Apple TV.

TABLE 1.1
Design and fabrication specification of Apple A11-A17 SoCs.

SoC	**A11**	**A12**	**A13**	**A14**	**A15**	**A16**	**A17**
Year	2017	2018	2019	2020	2021	2022	2023
CPU (Bits, Core)	64,6	64,6	64,6	64,6	64,6	64,6	64,6
Frequency (GHz)	2.38	2.49	2.65	3.10	3.23	3.46	3.78
GPU (Core)	3	4	4	4	5	5	6
Neural Engine (Core, TOPS)	2, 0.6	8, 5	8, –	16, 11	16, 15.8	16, 17	16, 35
Transistor (Billions)	4.3	6.9	8.5	11.8	15	16	19
Technology TSMC (nm)	10	7	7	5	5	4	3
Die size (mm^2)	87.66	83.27	98.48	88	–	–	–

A. Maximum CPU Clock Rate

These SoCs are integral components of Apple's A series, featuring 64-bit six-core CPUs. Each SoC is comprised of two high-performance cores and four energy-efficient cores, with the high-performance cores capable of operating at higher frequencies compared to the energy-efficient cores.

Over the years, we have observed remarkable progress in clock frequencies across A11 to A17 SoCs. The A11 core, introduced in 2017, featured a maximum clock rate of 2.39 GHz, whereas the A17 core, introduced in 2023, achieved an impressive clock speed of 3.78 GHz.

B. GPUs and Neural Engines

Within this group, the A11 SoC is equipped with a minimum of three graphics processing unit (GPU) cores, while the A17 SoC boasts a maximum of six GPU cores. Notably, beginning with the A11 SoC, dedicated neural engines have seamlessly integrated into these chips, greatly enhancing their machine learning capabilities. Among these SoC designs, the A17 truly stands out with an impressive neural processing performance of 11.8 trillion operations per second (TOPS). It's important to clarify that TOPS is a specific measure of neural processing capabilities and should not be confused with floating-point operations per second (FLOPS), which is a metric commonly used to gauge general computing power.

C. Transistor Technology

In the semiconductor industry, aside from clock frequency and multi-core designs, the manufacturing technology and transistor density hold immense significance. The A11 SoC made its debut with TSMC's 10 nm FinFET technology, marking a significant stride in semiconductor miniaturization. Subsequent iterations, the A12 and A13, took advantage of TSMC's 7 nm technology, further enhancing efficiency and performance. The A14 and A15 raised the bar by adopting TSMC's cutting-edge 5 nm technology, a milestone in chip fabrication. Most recently, the A17 has pushed the envelope even further, leveraging TSMC's advanced 3 nm technology. As the technology node size shrinks, the functional density of the chip design increases, resulting in smaller, more efficient chips.

D. Moore's Wall

Moore's Law has served as a guiding principle for scaling transistors and advancing IC performance over several decades. However, it has become increasingly evident that the rate of transistor count increase has started to decelerate in recent years, even though growth continues. This can be clearly observed in the transition from 11.8 billion transistors in A14 SoC in 2020 to an impressive 19 billion transistors in A17 SoC of 2023. The slowdown in transistor count can be attributed to a formidable opponent: the fundamental constraints imposed by the laws of physics.

Today, the miniaturization of transistors has reached a point where they are approaching the size of individual atoms. This presents a substantial challenge to silicon lithography technologies, necessitating increasingly intricate and cost-prohibitive efforts to sustain further miniaturization. In response to these challenges, innovative technologies are necessary to drive computational performance in the post-Moore's Law era.

1.1.3 Beyond Moore's law

Experts anticipate that Moore's Law will reach its physical limitations in the near future. The International Technology Roadmap for Semiconductors (ITRS) 2.0, published in 2015, projected that the reduction of transistor gate length would continue until at least 2028 [8]. Additionally, ASML, a prominent semiconductor equipment supplier, predicted that feature size reduction would encounter limitations by the end of this decade [7].

Despite these projections, the relentless march of technological advancements is expected to continue through innovative approaches. Both the industry and academia will continue to explore novel materials and transistor structures, develop specialized designs tailored for specific applications, delve into heterogeneous integration with different platforms and chiplets, and pioneer next-generation chip architectures, among other strategies. These endeavors are driven by the goal of overcoming physical constraints and relentlessly pushing the boundaries of semiconductor technology, thereby fueling the evolution of high-performance computing.

A. New Materials and New Transistor Structures

From 1975 to 2002, the semiconductor industry primarily relied on geometrical scaling to shrink the size of planar transistors, leading to improved performance. Since 2003, the focus has shifted to horizontal scaling, along with the use of new materials and the exploitation of physical effects to enhance transistor performance. For example, III–V materials have been incorporated into FET structures to improve high-hole mobility, while germanium has been utilized for high electron mobility.

Between 2011 and 2022, MOSFET device architecture has evolved significantly, moving from planar 2D designs to 2.5D FinFETs, and currently to gate-all-around (GAA) MOSFETs, available in nanowire or nanosheet configurations. These GAA MOSFETs are expected to be stacked vertically for 3D VLSI, ushering in the third era in semiconductor technology, expected between 2025 and 2040. It's worth noting that the memory industry has already embraced monolithic 3D integration for devices like NAND flash, reducing the need for further miniaturization while increasing capacity.

B. Specialized Design on Future High-Performance Computing

Given the absence of a new transistor technology poised to replace complementary metal-oxide-semiconductor (CMOS), the semiconductor industry has shifted its focus toward design specialization as a compelling avenue for advancing high-performance computing. A notable example of this approach is exemplified by the purpose-built architectures of the Google tensor processing unit (TPU) and Tesla Dojo D1, which have demonstrated significantly enhanced efficiency when compared to general-purpose CPUs and GPUs.

C. Chiplet and Heterogeneous Integration

Chiplet integration is a technique that involves partitioning a sizable monolithic chip into smaller, specialized functional units referred to as chiplets. This approach can bring several advantages, including enhanced yield rates, shortened development timelines, and the flexibility to mix and match chiplets for customized applications. Examples of successful implementation of chiplet technology can be seen in AMD's Ryzen processors and Intel's Foveros technology.

In parallel, heterogeneous integration entails blending diverse components, often with varied materials, technologies, and functions, onto a single chip or package. A prime example is in data centers, where CPUs, GPUs, and specialized accelerators like FPGAs or artificial intelligence (AI) processors are integrated within a single server or rack.

D. Neuromorphic Computing and Quantum Computing

The future of computer architectures and computing approaches is poised to revolutionize various computational functions. Two potential breakthroughs in this realm are neuromorphic computing and quantum computing.

Neuromorphic computing strives to mimic the intricate architecture of the human nervous system by utilizing electronic analog circuits and specialized components like oxide-based memristors. Unlike traditional sequential Von Neumann computer setups, neuromorphic architectures embrace parallel computing to effectively replicate the brain's computational prowess.

In contrast, quantum computing marks a monumental departure from classical computing norms. It envisions a foundation built on quantum gates and logic, harnessing qubits – units capable of occupying multiple states at once due to superposition. This remarkable property empowers quantum computers to tackle complex calculations and tap into quantum parallelism, a realm well beyond the capabilities of classical systems.

1.1.4 Preview

With a history spanning over 50 years, the IC design industry has witnessed substantial progress. This evolution encompasses various design platforms, increased automation within the design process through the use of comprehensive electronic design automation (EDA) toolsets, and innovative hardware description solutions.

In this chapter, we will provide an overview of the fundamental knowledge and background necessary for hardware descriptions and IC design. This includes essential information and prerequisites for RTL design using Verilog hardware description language (HDL).

1.2 Introduction to IC Industry

1.2.1 What Are FPGA and ASIC?

Before delving into the introduction to the IC industry, it is important to understand the two main categories in the field of IC design: FPGAs and ASICs development. Most IC design companies we see today, such as Intel, AMD, NVIDIA, Qualcomm, and Broadcom, are primarily known for their ASIC design capabilities, but many of them also have dedicated FPGA development teams.

An ASIC is a specialized chip designed for a specific purpose, tailored to perform dedicated functions. Its digital circuitry is composed of permanently connected cells, flip-flops, and memory blocks fabricated in silicon. In modern applications, ASICs often take the form of system-on-chip designs, integrating multiple components like CPUs, memory blocks, interfaces, and peripherals into a single chip. The ASIC design and fabrication process involves a very complex procedure, starting from the register-transfer level coding to the final chip manufacture steps.

On the contrary, an FPGA is a semiconductor device that boasts configurable logic components, which encompass programmable logic blocks (CLBs), reconfigurable interconnects, and dedicated hard-silicon blocks designed for specific functions. The fundamental building blocks within CLBs typically consist of look-up tables (LUTs), multiplexers, and flip-flops, providing FPGAs with an exceptionally versatile and adaptable nature. FPGA development leverages existing hardware resources, making the FPGA design flow generally simpler compared to the more complex ASIC design flow. However, it's essential to highlight that the initial stages of the design flow for both ASICs and FPGAs share considerable similarities. In practice, an ASIC design flow often includes the utilization of FPGAs during development for functional and timing validation, allowing engineers to test their designs in a real hardware environment.

Both ASICs and FPGAs play critical roles in the semiconductor industry, each catering to different application requirements. ASICs offer the advantage of high performance and power efficiency for specific tasks, while FPGAs provide flexibility and reconfigurability, making them suitable for prototyping, rapid development, and applications that require frequent updates or changes to the design.

1.2.2 Brief introduction to ASIC and FPGA industry

1.2.2.1 ASIC industry

The ASIC design and manufacturing industry can be divided into three primary business models: integrated device manufacturers (IDMs), semiconductor foundries, and fabless design companies.

- IDMs are companies that not only design and offer their own IC products but also operate their semiconductor fabrication facilities. Prominent examples of IDM companies include Texas Instruments, Intel, and Samsung.
- Semiconductor foundries, on the other hand, specialize in manufacturing ICs for other companies and typically do not engage in IC design themselves. Notable foundries include TSMC, Globalfoundries, UMC, and SMIC.
- Fabless companies, also known as design houses, focus primarily on IC research and development. They collaborate with semiconductor foundries to fabricate their designed chips using the **tape-out** process. Some well-known fabless companies include AMD, NVIDIA, and Qualcomm.

1.2.2.2 FPGA industry

While many IC design companies primarily prioritize ASIC design, it's common practice for them to maintain FPGA development groups. This approach stems from the inherent risks associated with fabricating an ASIC, as it involves significant time and financial investments. FPGA testing assumes a pivotal role in the ASIC design flow, serving as a crucial step for functional verification and timing checks on actual hardware before fabricating the ASIC.

In the past, the two main FPGA vendors dominating the industry were Xilinx and Altera. However, there have been significant developments in recent years. In 2015, Intel acquired Altera, bringing the FPGA expertise of Altera under its umbrella. Similarly, in 2020, AMD acquired Xilinx, consolidating its position as a major player in both the ASIC and FPGA markets. These acquisitions have solidified the positions of Intel and AMD as the largest FPGA vendors, strengthening their overall presence in the semiconductor industry.

1.2.3 EDA tools and vendors

In the realm of IC design, EDA vendors offer extensive toolsets that address a wide range of tasks crucial to IC development. These tools encompass diverse functions, including simulation, synthesis, layout, and timing analysis, among others. They play an integral role in facilitating various aspects of IC design and testing.

This book primarily concentrates on introducing the ASIC/FPGA design flow at the front-end, with a specific focus on the RTL coding and simulation procedures. Consequently, this section aims to provide a concise overview of the simulator and synthesis tools that find application in both ASIC and FPGA development. These tools are fundamental to the design and verification processes, ensuring the creation of robust and functional ICs.

1.2.3.1 ASIC simulation and synthesis

The major EDA tool providers for ASIC design are Synopsys, Cadence, and Siemens EDA (formerly known as Mentor Graphics before being acquired by Siemens in 2017). All three companies not only offer EDA tools but also serve as IP vendors. As of 2021, Siemens EDA operates as a division within Siemens following the acquisition of Mentor Graphics.

A. ASIC Simulation

Among the popular simulation solutions widely used by most semiconductor companies are Synopsys VCS®, Cadence NC-Verilog®, and Siemens EDA ModelSim®. An advanced version of ModelSim is QuestaSim®. These tools provide essential functionalities for verifying and validating the design before moving on to the synthesis and layout stages.

B. ASIC Synthesis

The synthesis process is a critical step in the ASIC design flow, enabling conversion from RTL code into gate-level netlists with specified constraints and optimization settings. Synopsys provides its Design Compiler® as a core synthesis solution, while Cadence offers the Genus Synthesis Solution®, and Siemens delivers the Precision RTL Synthesis® as their respective synthesis tools.

1.2.3.2 FPGA simulation and synthesis

As mentioned earlier, two of the largest FPGA vendors are Intel and AMD. Now, let's delve into the key EDA tools offered by these vendors.

A. FPGA Simulation

Functional verification in both ASIC and FPGA designs involves the critical step of design simulation, a task that can be efficiently carried out using commonly shared simulators. Although FPGA development tools typically include built-in simulators, it is often recommended to utilize industry-standard options like Synopsys VCS, Cadence NC-Verilog, and Siemens EDA ModelSim due to their exceptional efficiency and effectiveness in ensuring functional verification.

B. FPGA Synthesis and Implementation

After simulation, RTL synthesis and implementation are typically performed using FPGA tools like AMD Vivado® or Intel Quartus®, which are specifically tailored for FPGA designs rather than ASIC synthesis and layout. The FPGA synthesis process converts the RTL design into a netlist, while implementation involves various steps to place and route the netlist onto the available FPGA device resources. Once the placement and routing processes are complete, the final bitstream is generated. This bitstream is the configuration file that contains all the necessary information to program the specific FPGA board, effectively implementing the designed circuit in a real-world scenario.

It's important to note that the names and offerings of EDA tools may have evolved since the last update of this section. Therefore, it is advisable to consult the latest information and documentation provided by the respective EDA vendors to access the most current tool offerings and capabilities.

1.2.4 Hardware design solutions

In both ASIC and FPGA development workflows, the hardware description phase stands out as a pivotal milestone. This stage bears the primary responsibility for shaping the final hardware design's functionalities and performance attributes, influencing key parameters such as chip size (gate/slice count), computational latency (in clock cycles), speed and throughput, maximum operational frequency (MOF), and power consumption within the IC. In this section, we will explore four prevalent design approaches: RTL design using HDL, high-level synthesis, various code generators, and SystemVerilog.

1.2.4.1 Hardware description language

Over the years, both the industry and academia have presented various hardware design approaches. Yet, the predominant and dependable route for describing hardware lies within the realm of register-transfer level, or RTL design. Referred to as RTL design in the industry, HDL coding stands tall as the prevalent hardware design method used across both ASIC and FPGA development.

In the realm of IC design, HDL coding stands as the dominant solution. It empowers designers to delineate the flow of data among registers and its processing through combinational logic. HDL-based description enjoys widespread adoption owing to its precision, flexibility, and suitability for hardware synthesis. This approach offers a robust and meticulous means of capturing hardware behavior and effectively translating it into physical implementation, solidifying its status as a fundamental and indispensable element of the IC design process.

A. HDL

Since the invention of the first IC in 1958, various methods and techniques have been proposed to miniaturize electronic circuits and improve the manufacturing process, resulting in the rapid growth of the IC design industry following Moore's Law. A significant milestone in IC design was the introduction of HDLs which enable designers to describe detailed design specifications using computer programs at a high level of abstraction, without delving into the implementation details at the transistor level. The use of HDLs has revolutionized the way complicated ICs are designed, as synthesis tools can generate low-level circuits and schematics based on the HDL programs, significantly improving efficiency and productivity in IC production.

The first HDL, called Verilog, was introduced by Gateway Design Automation in 1985. Four years later, Cadence Design Systems acquired Gateway Design Automation. Cadence is a major player in electronic design today, offering software, hardware, and IP solutions in the field of IC design. In 1990, Cadence made Verilog HDL accessible to the public. Then, in 1995, Verilog was established as the IEEE Standard 1364. The most recent widely used version in the industry, IEEE 1364-2001, was launched in 2003, with a minor update in 2005.

Alongside Verilog HDL, another hardware description language, Very High-Speed IC HDL (VHDL), was standardized in 1987 as IEEE 1076-1987. An updated version, IEEE 1076-1993, was released in 1994.

B. Verilog vs. VHDL

Both Verilog and VHDL languages are capable of describing hardware operations, including timing and concurrency, constructing testbenches for simulation, and synthesizing into gate-level circuits. Once proficient in one of the languages, it becomes relatively easy to adapt to the other.

One primary difference between the two is their syntax and form. Verilog bears similarities to the C programming language, while VHDL is more akin to Ada. In typical designs, Verilog is often regarded as simpler to program, while VHDL enforces stricter syntax rules. Additionally, VHDL finds significant use in certain industries such as aerospace and government applications, while Verilog is more prevalent among leading ASIC design companies. Given Verilog's widespread adoption in the industry today, this book will focus on introducing Verilog as the primary hardware description language.

1.2.4.2 High-level synthesis (HLS)

Efforts to promote accessibility and popularity of IC design in academia have resulted in transforming complex hardware problems into simpler computer science problems. One approach involves embracing higher-level languages like C or C++ to describe digital circuits, which can then be automatically converted into HDL code using HLS tools. This method allows designers to work at a higher abstraction level, making it more intuitive and familiar for software developers.

However, a significant concern with these design solutions is the potential lack of certain hardware-related descriptions, such as timing and performance considerations, which are critical from a hardware design perspective. HLS tools, developed by prominent EDA companies like Synopsys, Mentor Graphics, and NEC Corporation in Japan, offer valuable capabilities, but they are not universally adopted by major IC design companies yet.

As a result, achieving successful and efficient IC designs using HLS requires striking a balance between the advantages of higher-level languages and the necessity for hardware-specific optimizations. Continuous advancements in HLS tools and methodologies may play a significant role in further improving the adoption and effectiveness of this approach in modern IC design.

1.2.4.3 HDL code generators

An alternative option for RTL design is the development of code generators utilizing domain-specific languages, which provide parametrizable solutions for describing complex circuits and generating HDL code compatible with standard ASIC and/or FPGA tools. Several existing hardware design tools, such as Chisel [3], PyMTL [13], PyRTL [4], and Spatial [10], enable hardware generation by elevating the level of abstraction and allowing users to work with Python or Scala as their main programming language. While Chisel, PyMTL, and PyRTL focus on circuit-level abstractions, Spatial raises the level of abstraction to a reduced set of algorithmic patterns specifically targeted at matrix-matrix multiplication, a key component in machine learning algorithms.

Chisel, in particular, is one of the open-source embedded domain-specific language developed at the University of California, Berkeley. It extends the Scala programming language with hardware construction primitives and offers parameterizable solutions for describing intricate circuits. Although Chisel may not include certain hardware-oriented information commonly found in RTL descriptions, its use significantly reduces the entry barrier for hardware design by enabling the description of circuits at higher-level abstractions. This approach opens up opportunities for students, researchers, and engineers from diverse backgrounds to engage in hardware development.

Notably, Chisel-based programming offers an efficient and effective solution for Verilog code generation. It's noteworthy that all floating-point (FP) operators presented in this book stem from Chisel-based designs [17]. This highlights the prowess of code generators in streamlining hardware development across diverse applications, establishing them as valuable tools for hardware designers and researchers in the field.

1.2.4.4 SystemVerilog

SystemVerilog is an extension of Verilog-2005, designed specifically for hardware description and verification tasks. With support for structured and object-oriented programming paradigms, SystemVerilog combines the strengths of Verilog and C++. In 2005, SystemVerilog was standardized by IEEE as IEEE 1800. It underwent further updates and improvements in 2012, resulting in IEEE 1800-2012.

One of the significant advantages of SystemVerilog is its class-based testbench framework, which surpasses traditional Verilog testbenches in terms of dynamic behavior and reusability. Additionally, SystemVerilog facilitates advanced verification methods, including coverage-driven verification, constrained random verification, and assertion-based verification.

Built on SystemVerilog, two well-known verification methodologies are created: the Universal Verification Methodology (UVM) and the Verification Methodology Manual (VMM). These methodologies offer class libraries that streamline various aspects of the verification process, automating sequences and data management. Among these methodologies, UVM has gained widespread adoption in the industry and is now the more commonly used verification approach.

In summary, Verilog HDL continues to maintain its position as the dominant hardware description solution in the IC design industry, with SystemVerilog widely adopted for IC verification purposes. In the academic sphere, High-Level Synthesis serves as a prominent choice for research projects requiring higher-level hardware descriptions. Furthermore, domain-specific languages like Chisel have emerged as valuable additions to RTL design generation, offering increased flexibility and productivity for creating standardized and consistent designs.

As a result, this book primarily focuses on introducing Verilog HDL as the principal hardware description language. Additionally, Chisel-generated designs are employed to construct the FP datapath and numerical applications, demonstrating their practical utility.

1.3 FPGA and ASIC Design

1.3.1 FPGA and ASIC implementations

A. FPGA Slice

As previously mentioned, ASIC designs are tailored and manufactured for specific silicon functions. In contrast, FPGA development involves utilizing fundamental CLBs along with other resources such as digital signal processors (DSPs) and memory blocks through placement and routing processes. Each CLB element in an FPGA may comprise a pair of slices, with each slice containing LUTs and storage elements capable of performing logic, arithmetic, and read-only memory (ROM) functions. For example, a typical slice found in Xilinx's 7 Series FPGA includes four LUTs for logic operations, eight storage elements, and various multiplexers [2].

B. Combinational Circuit on ASIC/FPGA

Although Verilog code can be shared among different hardware platforms, the actual implementations on ASIC and FPGA are completely different due to the code conversion schemes employed for each platform. For instance, a single-bit half adder can be expressed using the following Verilog code. In Verilog, the symbol "∧" denotes an XOR gate, and "&" represents an AND gate. For an ASIC design, the synthesis result would be a specific combinational circuit consisting of an XOR gate and an AND gate, as shown in Figure 1.1(a).

```
1  //Verilog Description of Single-Bit Half Adder
2  module Comb_Circuit (input  A, B,
3                       output SUM, CARRY);
4  assign SUM   = A ^ B ;
5  assign CARRY = A & B ;
6  endmodule
```

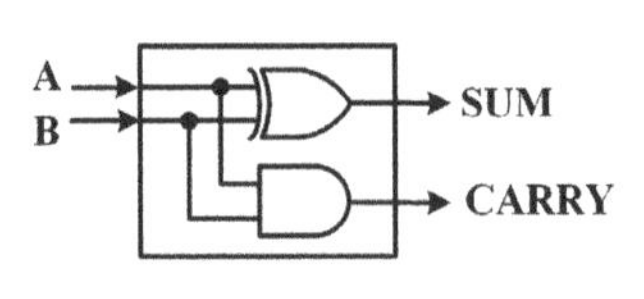

(a) ASIC Implementation

LUT1: XOR Function

A	B	SUM
0	0	0
0	1	1
1	0	1
1	1	0

LUT2: AND Function

A	B	CARRY
0	0	0
0	1	0
1	0	0
1	1	1

(b) FPGA Implementation with LUT

FIGURE 1.1
Half adder implementation with ASIC and FPGA.

Unlike the ASIC implementation using specific logic gates, the FPGA implementation uses LUTs to perform logic functions. Specifically, the two inputs of the first LUT are mapped with "A" and "B" inputs in the Verilog design, and the output is mapped with the "SUM" output, as illustrated in Figure 1.1(b). In the first LUT, whenever inputs "A" and "B" differ, the output "SUM" is filled with binary ones; otherwise, it is filled with binary zeros, behaving as an XOR logic with a LUT. Similarly, the second LUT functions as an AND gate, such that only when both inputs "A" and "B" are binary ones, the output "CARRY" is filled with binary one. For the remaining three combinations of inputs "A" and "B", the output "CARRY" is filled with binary zeros.

C. Sequential Circuit on ASIC/FPGA

Another example demonstrates ASIC and FPGA implementations for a sequential circuit. The circuit comprises a combinational part that includes XOR and AND gates (lines 6–7) and a sequential register (lines 9–12). The combinational part is described using two *assign* blocks, while the sequential register is designed with an *always* block. The details of using *assign* and *always* blocks will be introduced in the following chapters. Here, we focus on the difference between the implementations of ASIC and FPGA.

```
1  //Verilog Description of Sequential Circuit
2  //with a Single-Bit Half Adder and a Register
3  module Seq_Circuit (input                A, B,
4                      input                CLK ,
5                      output reg [1:0]     Q    );
6  wire SUM   = A ^ B ;
7  wire CARRY = A & B ;
8
9  always @(posedge CLK) begin
10   Q[1] <= CARRY  ;
11   Q[0] <= SUM    ;
12 end
13 endmodule
```

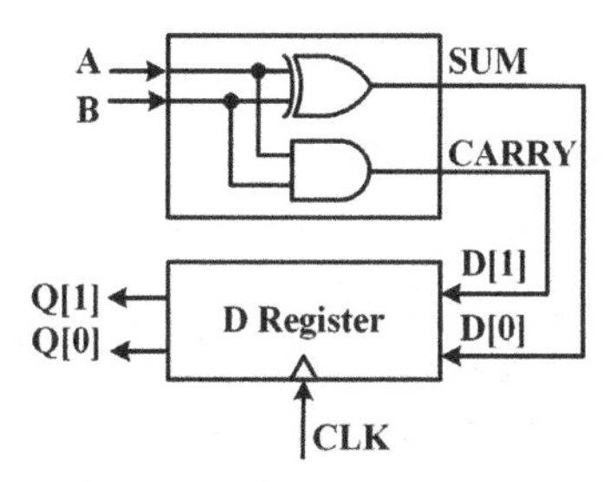

FIGURE 1.2
Half adder and register implementation with ASIC.

In an ASIC design, a specific combinational circuit and a 2-bit register would be generated. As shown in Figure 1.2, the circuit's inputs are single-bit "A" and "B", and its output is the 2-bit "Q". The internal signals are used to connect the combinational and sequential circuits, with the "SUM" signal connected to the most-significant bit (MSB) "D[1]" and the "CARRY" signal connected to the least-significant bit (LSB) "D[0]". The sequential circuit involves a 2-bit register to store the outputs "SUM" and "CARRY".

For the FPGA implementation, two LUTs and a 2-bit register would be used to perform the same functions. The FPGA result depends on the selected part or board of the project, and different FPGA parts/boards may have varying resource utilization.

1.3.2 Comparison of FPGA and ASIC

A. FPGA Advantages

FPGA and ASIC development each come with their distinct pros and cons. The advantages unique to FPGA designs are summarized in Table 1.2. Notably, FPGA designs excel in achieving a quicker time-to-market compared to ASIC designs, owing to their streamlined design process. By bypassing many of the manufacturing stages inherent in ASIC development, FPGA designs employ EDA tools to expedite tasks like placement, routing, and timing analysis. This expedites the design cycle, enabling faster implementation and showcasing on FPGAs. A pivotal attribute of FPGA design is its field reprogrammability, making FPGAs a prime fit for scenarios where frequent updates or modifications are essential.

Overall, FPGA designs offer a compelling combination of flexibility, reconfigurability, and faster time-to-market, making them well-suited for various applications, especially in the early stages of project development and/or for low-volume products that require frequent updates or design changes.

TABLE 1.2
Advantages of FPGA implementation.

Advantages	**Benefits**
Faster time-to-market	No layout, masks, and other manufacturing steps
Simpler design cycle	EDA tools can handle the placement, routing, and timing analysis
Reprogrammability	A new bitstream can be uploaded frequently

B. ASIC Advantages

Table 1.3 provides an overview of the advantages associated with ASIC designs. ASICs are tailored to specific design specifications, offering full custom capability. However, due to the expensive fabrication process, ASICs are generally preferred only for very high-volume IC products, allowing for achieving reduced unit costs.

Furthermore, ASICs typically demonstrate superior resource utilization, such as gate count and chip size, along with more efficient power consumption compared to FPGAs. These benefits arise from the application-specific design and specialized manufacturing process employed in an ASIC development. Additionally, ASICs tend to outperform FPGA implementations in terms of speed, thanks to their ability to leverage the latest chip fabrication technology and customize the design to precisely match the specific application requirements. This optimization maximizes overall performance for the targeted use case.

Overall, ASIC designs offer significant benefits in terms of cost-effectiveness, resource utilization, and performance, making them the preferred choice for high-volume and performance-critical applications. The selection between ASIC and FPGA depends on the specific needs and priorities of the given project.

TABLE 1.3
Advantages of ASIC implementation.

Advantages	**Benefits**
Full custom capability	ASIC is manufactured to specialized design
Lower unit costs	For very high-volume designs
Smaller size, higher speed, and low power	Latest technology and custom design

1.3.3 ASIC development: MPW vs. Fullset

A. MPW vs. Fullset

The ASIC design process is renowned for its extensive development cycle and the considerable expenses tied to chip manufacturing. Hence, it's commonly advised to initiate with an **MPW** (multi-project wafer) iteration of ASIC fabrication before transitioning to a Fullset. All semiconductor foundries, including notable ones like TSMC and Globalfoundries, extend the MPW service. This approach, marked by its cost efficiency, permits multiple IC designs from diverse clients to coexist on a single wafer for production. Particularly advantageous for smaller design teams and companies, this method allows for design testing and low-volume prototype creation prior to embarking on full-scale production commitments.

Conversely, a **Fullset** constitutes the comprehensive collection of design files and specifications indispensable for the production of a single IC. Once a design has undergone testing and validation via an MPW fabrication, a Fullset becomes the go-to choice for extensive commercial manufacturing. It encompasses all essential design files and intricate particulars needed to accurately and individually fabricate the designated IC.

In summary, MPW fabrication serves as an efficient and cost-effective means for initial testing and prototype production, while a Fullset is crucial for large-scale commercial production, ensuring that the finalized IC design can be fabricated accurately and efficiently. By leveraging both MPW and Fullset approaches, fabless design companies can pre-verify their ASIC designs and achieve successful outcomes for future large-volume fabrication.

B. ASIC Fabrication Cost

Providing a glimpse into the substantial expenses linked to ASIC fabrication, we present a concise approximation of the ASIC design cost in 2023. For instance, a basic chipset spanning one square millimeter and utilizing 16–24 nanometer technology from Global Foundries could span from tens of thousands of dollars (for MPW) to millions of dollars (for Fullset). It's crucial to acknowledge that these figures may fluctuate based on factors like technology preference, chip dimensions, mask set costs, project-specific demands, and various marketing-related factors.

Due to the high-design complexity and intricate timing processing involved, there is a considerable risk of failure when fabricating an ASIC. In the front-end of the design flow, RTL verification requires achieving 100% functional coverage and more than 95% code line coverage. On the back-end, a timing violation could result in disparities between the RTL design and the synthesized gate-level netlist, causing design functionalities to deviate from expectations. Consequently, each step of the ASIC design process must be meticulously approached and fully verified before proceeding with the chip fabrication.

1.3.4 FPGA development and applications

FPGA development serves two primary applications in the industry. First, FPGA verification and prototyping play a crucial role within the ASIC design flow, facilitating risk mitigation associated with ASIC fabrication and contributing to time and cost reduction. Secondly, FPGA designs find popularity in various specific applications where production volume is low. These applications span across aerospace and defense, automotive, data centers, medical devices, video and image processing, telecommunication, data communication, high-performance computing, industrial sectors, and more.

Recent advancements in FPGA development have witnessed the rise of software-hardware co-design platforms integrating FPGAs and CPUs, such as the Xilinx Zynq FPGA and Intel HARPv2. These platforms serve as demonstrations and accelerators for specific algorithm processing tasks, catering to evolving computing demands.

The integration of ASIC and FPGA technologies continues to drive innovation in the IC industry, meeting the diverse needs of modern applications and industries. This book provides a comprehensive understanding of the hardware descriptions with Verilog HDL and its implementations in both ASIC and FPGA platforms, empowering readers to harness the full potential of this powerful hardware description language in their projects.

1.4 Fundamentals of Verilog HDL Design

1.4.1 Multi-level description with Verilog HDL

A. Multi-Level Designs

In general, hardware can be described at four levels of abstraction: specification level, RTL level, gate level, and physical level. The initial step in the ASIC design flow is to create a design specification, which outlines various aspects of the design, such as input/output (IO) configurations, timing diagrams, block diagrams, functional registers, and more. As a simple example, Figure 1.3(a) provides a design specification for a single-bit inverter. It basically includes a block diagram, a truth table, and a timing diagram. The function of the inverter is to invert the value of the input.

Figure 1.3(b) depicts the Verilog description of the inverter using the operator "~". While the Verilog standard does support gate-level and transistor-level descriptions, its primary purpose is to describe hardware at the register-transfer level. Industrial EDA tools can automate the design translation from RTL code into the gate-level and transistor-level netlist, which includes the detailed circuit and timing information. This automation streamlines the design

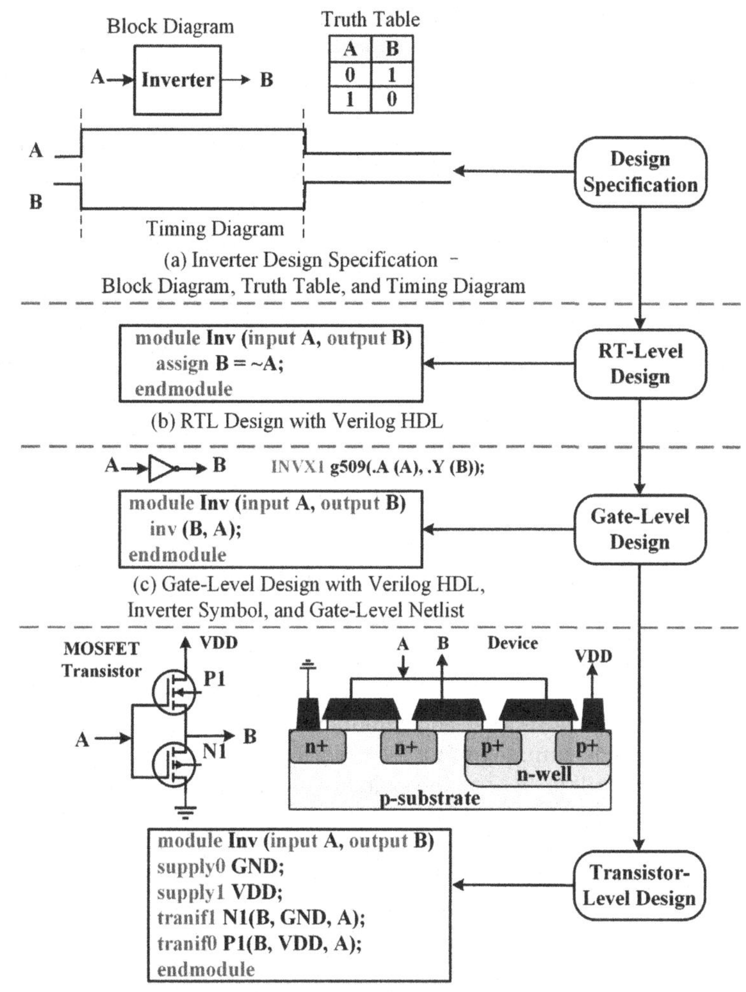

FIGURE 1.3
Multi-level Verilog HDL descriptions.

process and allows for a more efficient translation from RTL to the lower-level descriptions required for FPGA implementation and ASIC fabrication.

As a case study, Figure 1.3(c) presents the synthesis result of the gate-level netlist. The netlist features a logic cell labeled $INVX1$, designed to implement the unique design $g509$. The logic cell has input and output ports named "A" and "Y", respectively, which are connected to the corresponding Verilog design

IOs "A" and "B". Additionally, the Verilog description in gate-level is provided in this figure.

Finally, Figure 1.3(d) illustrates the transistor-level schematic, consisting of one PMOS transistor (denoted as P1) on top and one NMOS transistor (denoted as N1) on the bottom. The source terminal of the PMOS is connected to VDD, while the source terminal of the NMOS is connected to ground. Both transistors share a common gate terminal, which serves as the input "A", and their drain terminals are interconnected to form the output "B". When the input "A" is at logic low (binary zero), the NMOS transistor is OFF, and the PMOS transistor is ON. Consequently, the output "B" is pulled up to logic high (binary one) through the VDD. Conversely, when the input "A" is at logic high (binary one), the NMOS transistor is ON, and the PMOS transistor is OFF. Consequently, the output "B" is pulled down to logic low (binary zero). This behavior allows the circuit to function as an inverter. In the same figure, the physical implementation on a p-substrate, as well as the Verilog description in transistor level is provided as well.

CMOS, PMOS, and NMOS Transistors

As previously mentioned, CMOS serves as the foundational building block for digital circuits and plays a pivotal role in modern semiconductor technology. CMOS technology incorporates both P-type (PMOS) and N-type (NMOS) metal-oxide-semiconductor field-effect transistors (MOSFETs) on a single chip, enabling the creation of complementary logic circuits. Here, we will delve into the switching behavior of PMOS and NMOS transistors.

The PMOS transistor, as illustrated by the P1 transistor in Figure 1.3(d), features three terminals: the gate (positioned in the middle), the source (located at the top), and the drain (situated at the bottom). Essentially, when the gate terminal is binary zero, it creates a conducting channel between the source and the drain (channel ON status); conversely, when the gate terminal is binary one, the channel remains open (channel OFF status).

In contrast, the NMOS transistor, exemplified by the N1 transistor in Figure 1.3(d), comprises three terminals: the gate (positioned in the middle), the drain (located at the top), and the source (situated at the bottom). In essence, when the gate terminal is binary one, it establishes a conducting channel between the drain and the source (channel ON status); conversely, when the gate terminal is binary zero, the channel remains open (channel OFF status).

B. Verilog Description in Gate Level and Transistor Level

This section delves deeper into the Verilog designs on gate-level and transistor-level circuits. For instance, the Verilog code below demonstrates how to call an inverter with the output "B" and the input "A" at the gate level.

```
1  //Verilog Description in Gate-Level Design
2  module Inv (input  A ,
3               output B);
4  inv (B, A);
5  endmodule
```

Further, the corresponding Verilog code at the transistor level is shown below. Similar to the gate-level description, the design code "tranif1 N1(B, GND, A)" (line 7) indicates that an NMOS named "N1" is mapped to a transistor "tranif1" and lists the drain, source, and gate inputs in that order. Thus, in line 7, it shows that the drain terminal is connected to the "B" output, the source terminal is connected to ground, and the gate terminal is connected to the "A" input. This indicates that an NMOS transistor turns on when the gate "A" is logic high and turns off when the gate "A" is logic low.

```
1  //Verilog Description in Transistor-Level Design
2  module Inv (input  A,
3               output B);
4  supply0 GND;
5  supply1 VDD;
6
7  tranif1 N1(B, GND, A)
8  tranif0 P1(B, VDD, A)
9  endmodule
```

Likewise, the design code "tranif0 P1(B, VDD, A)" (line 8) denotes that a PMOS named "P1" is mapped to a transistor "tranif0" and lists the drain, source, and gate inputs in that order. Therefore, line 8 shows that the PMOS's drain terminal is connected to the "B" output, the source terminal is connected to VDD, and the gate terminal is connected to the "A" input. This indicates that a PMOS transistor turns on when the gate "A" is logic low and turns off when the gate "A" is logic high. This code implements the same function as the CMOS schematic in Figure 1.3(d).

Low-level representations, while informative, are not an efficient approach for complex hardware description. On the other hand, register-transfer level design has gained widespread adoption in both industry and academia due to its ability to strike a balance between design simplicity and efficient synthesis for ASICs and FPGAs. As the dominant hardware description solution, Verilog HDL has proven to be robust and effective, making it suitable for a wide range of design projects, from small-scale circuits to large and intricate design systems. Hence, this book exclusively focuses on RTL design using Verilog HDL, providing readers with a powerful and versatile toolset to describe hardware functionality and behavior.

1.4.2 Fundamentals of combinational circuits

This section provides an introduction to fundamental combinational circuits, encompassing NAND, NOR, tri-state inverter, and multiplexer. Each logic gate is presented with its corresponding symbol, truth table, and the CMOS schematic to aid in understanding its behavior and implementation.

A. Construction of NAND Gate

Figure 1.4 depicts a 2-input NAND gate and its corresponding CMOS schematic. The schematic consists of two PMOS transistors (P1 and P2) in parallel on the top and two NMOS transistors (N1 and N2) in series on the bottom. When either input "A" or "B" is binary zero, the corresponding PMOS(es) will be ON, and the two NMOS transistors in series will be OFF. As a result, the output "C" will be pulled up to VDD or binary one. Only when both "A" and "B" inputs are binary ones, the two NMOS transistors in series will be ON, and both PMOS transistors in parallel will be OFF. The output "C" will be pulled down to GND or binary zero.

Using the same schematic, a N-input NAND gate can be constructed by connecting $N\times$ PMOS transistors in parallel on the top and $N\times$ NMOS transistors in series on the bottom. If any of the inputs is binary zero, the output

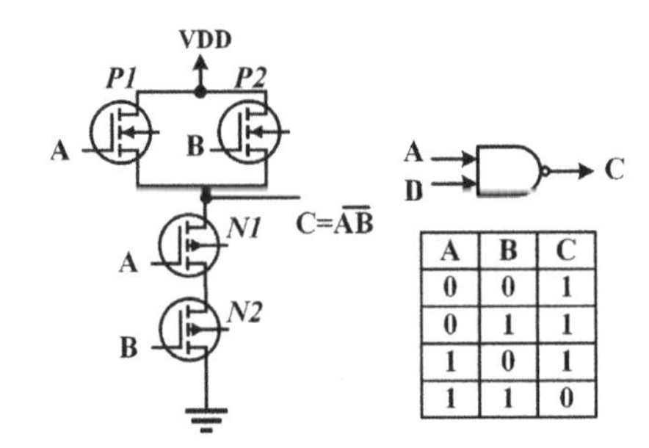

A	B	C
0	0	1
0	1	1
1	0	1
1	1	0

FIGURE 1.4
NAND gate implementation at the transistor level.

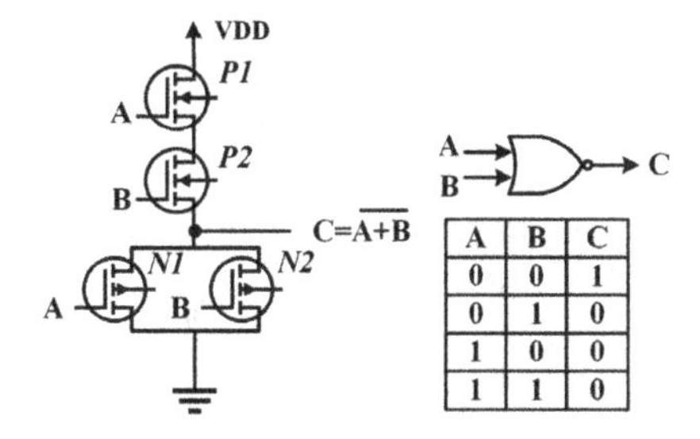

A	B	C
0	0	1
0	1	0
1	0	0
1	1	0

FIGURE 1.5
NOR gate implementation at the transistor level.

will be pulled up to VDD through the PMOS(es) in parallel. Only when all N inputs are binary ones, the output will be pulled down to GND via the NMOS transistors in series.

B. Construction of NOR Gate

Figure 1.5 shows the physical implementation of a 2-input NOR gate, which consists of two PMOS transistors (P1 and P2) in series on the top and two NMOS transistors (N1 and N2) in parallel on the bottom. Only when both inputs "A" and "B" are binary zeros, the two PMOS transistors in series will be ON, and both NMOS transistors in parallel will be OFF. The output "C" will be pulled up to VDD. If either "A" or "B" is binary one, the corresponding NMOS(es) will be ON, and the two PMOS transistors in series will be OFF. The output will be pulled down to binary zero.

Using the same schematic, an N-input NOR gate can be constructed by connecting $N\times$ PMOS transistors in series on the top and $N\times$ NMOS transistors in parallel on the bottom. If any of the inputs is binary one, the output will be pulled down to GND through the NMOS transistors in parallel. Only when all N inputs are binary zeros, the output will be pulled up to VDD via the PMOS transistors in series.

It's worth noting that by adding an inverter to the output of a NAND/NOR gate, an AND/OR gate can be constructed, respectively.

C. Construction of Tri-State Inverter

Moving on to the construction of a tri-state inverter, its CMOS schematic, shown in Figure 1.6, includes two PMOS transistors in series on the top (P1 and P2) and two NMOS transistors in series on the bottom (N1 and N2). When the enable signal "EN" is binary zero, both the P2 and N1 transistors are OFF, thus the two series paths, VDD-P1-P2-B, and GND-N2-N1-B, are both OFF. Consequently, the output "B" is left floating or in a high-Z state.

When the enable signal "EN" is binary one, both the P2 and N1 transistors are ON, and the output "B" is determined by the input "A". If "A" is binary

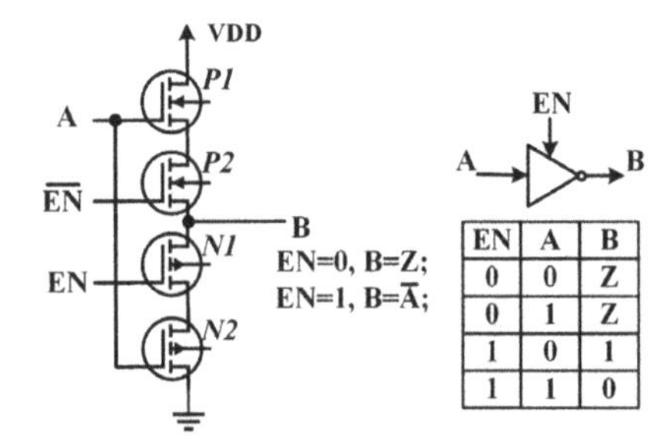

EN	A	B
0	0	Z
0	1	Z
1	0	1
1	1	0

FIGURE 1.6
Tri-state inverter implementation at the transistor level.

zero, the P1 transistor will be ON and the N2 transistor will be OFF, causing the output "B" to be pulled up to VDD. Conversely, if "A" is binary one, the N2 transistor will be ON and the P1 transistor will be OFF, resulting in the output "B" being pulled down to binary zero.

It's worth noting that by adding an inverter to the output of a tri-state inverter, a tri-state buffer can be constructed.

D. Construction of 2-to-1 Multiplexer

Figure 1.7(a) shows the physical implementation of a 2-to-1 multiplexer, which is composed of two stages. The first stage is an inverting multiplexer that includes four PMOS transistors (P1, P2, P3, and P4) on the top and four NMOS transistors (N1, N2, N3, and N4) on the bottom. Note that the drain terminals of transistors P1 and P3 are connected. The second stage is an inverter that inverts the output of the first stage, labeled as "D", to produce the final output labeled as "C".

When the selection signal "S" is binary zero, P1 and N1 transistors will be ON, and P2 and N3 transistors will be OFF, as shown in the equivalent circuit in Figure 1.7(b). In this case, the first-stage output "D" is decided by the input "A". When the input "A" is binary zero, P4 will be ON, and N2 will be OFF. The output "D" will be pulled up to VDD, and the final output "C" will be inverted into binary zero. When the input "A" is binary one, P4 will be OFF, and N2 will be ON. The output "D" will be pulled down to GND, and the final output "C" will be inverted into binary one. Therefore, when the selection signal "S" is binary zero, the output "C" will be the same as the input "A", and the path from "A" to "C" conducts.

In contrast, when the selection signal "S" is binary one, N3 and P2 will be ON, and N1 and P1 will be OFF, as shown in the equivalent circuit in Figure 1.7(c). The first-stage output "D" will be decided by the input "B". When the input "B" is binary zero, P3 will be ON and N4 will be OFF. The output "D" will be pulled up to VDD, and the final output "C" will be inverted into binary zero. When the input "B" is binary one, N4 will be ON and P3 will

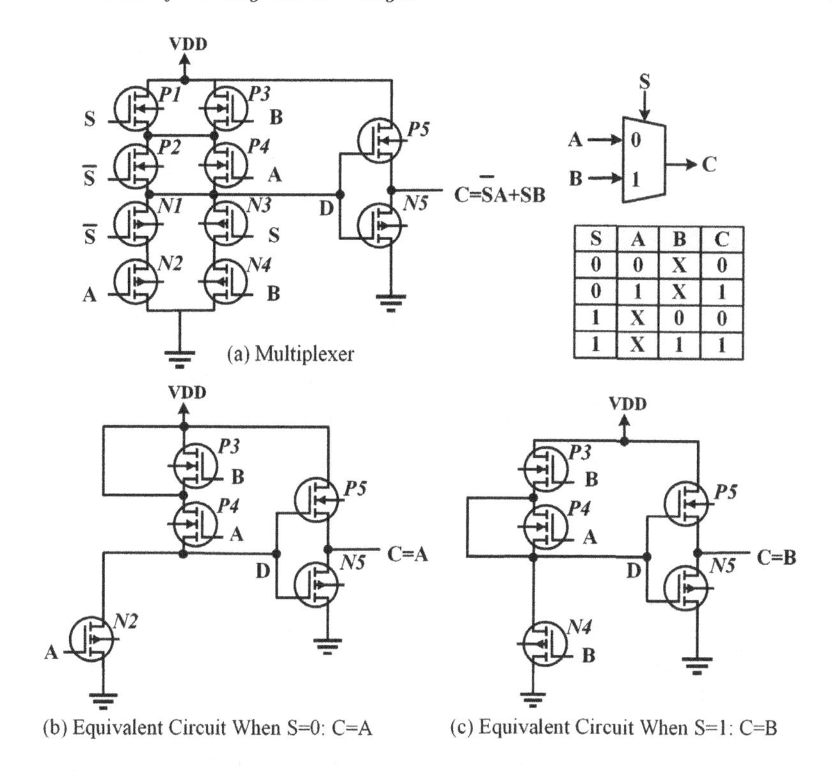

S	A	B	C
0	0	X	0
0	1	X	1
1	X	0	0
1	X	1	1

(a) Multiplexer

(b) Equivalent Circuit When S=0: C=A

(c) Equivalent Circuit When S=1: C=B

FIGURE 1.7
2-to-1 multiplexer implementation at the transistor level.

be OFF. The output "D" will be pulled down to GND, and the final output "C" will be binary one. Therefore, when the selection signal "S" is binary one, the output "C" will be the same as the input "B", and the path from "B" to "C" conducts. Overall, the implementation behaves a 2-to-1 multiplexer.

1.4.3 Fundamentals of sequential circuits

Sequential circuits are pivotal in register-transfer level designs, acting as storage units to preserve and hold data. There are two fundamental types of storage elements utilized in these circuits: latches and registers. These storage components are foundational for crafting intricate digital circuits and facilitate the persistent storage of information, thereby contributing to the register-transfer behavior of hardware circuits.

A. Construction of D-Latch

A D-latch is a basic sequential circuit that can be transparent or opaque based on the enable input "EN". When the latch is transparent ("EN=1"), the output

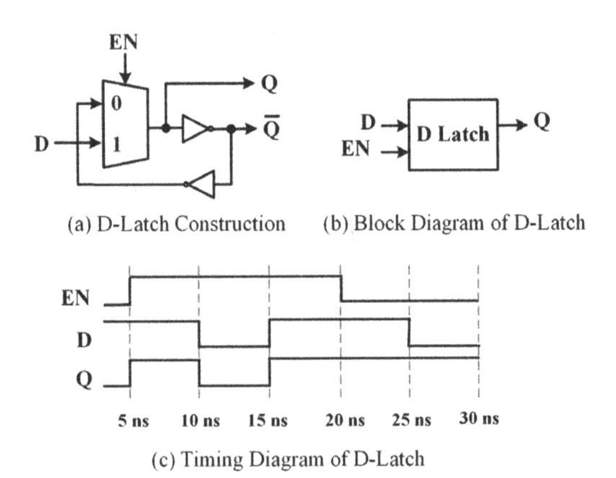

FIGURE 1.8
D-latch implementation.

"Q" follows the input data "D". When the latch is opaque ("EN=0"), the output "Q" retains its previous value, memorized from past values.

A typical latch can be built using a 2-input multiplexer and two inverters, as shown in Figure 1.8. When "EN" is binary one, the "Q" output is assigned the value of the input "D". When "EN" is binary zero, the feedback path with two inverters in series is used to hold the current state of "Q".

As illustrated in the timing diagram in Figure 1.8, the input "EN" switches from binary zero to one at 5 ns and remains until 20 ns. Subsequently, the output "Q" follows the input "D", transitioning from binary zero to one at 5 ns, one to zero at 10 ns, and then back to one at 15 ns. This behavior is due to "EN" being asserted during this time period. After 20 ns, when the "EN" signal falls to binary zero, the "Q" output ignores the change of the input "D" (which occurs at 25 ns) and retains its value at binary one until the end of the simulation. The D-latch thus demonstrates transparent and opaque behaviors based on the "EN" signal, making it a fundamental building block for sequential circuits.

As a consequence, a D-latch stores and transfers its input to its output exclusively when the enable input is asserted. Additionally, it is worth noting that a D-latch may include a global reset signal to initialize the data stored within it. This reset signal is used to clear the latch and set its output to a known state when required.

B. Construction of DFF/Register

A D-flip-flop (DFF) is a crucial building block of sequential circuits. It features a single data input denoted as "D", a clock input labeled as "CLK", and an output named "Q". Unlike a D-latch, which is controlled by a level-triggered enable signal, a DFF operates differently. The DFF is designed to store and

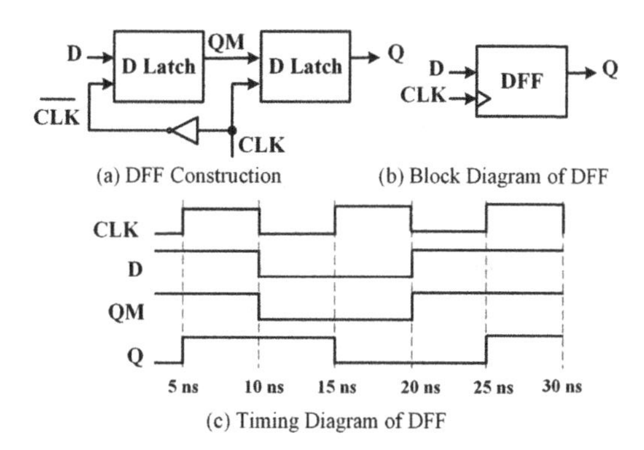

FIGURE 1.9
DFF implementation.

synchronize the value of its input "D" at the rising edge of the clock "CLK". This means that when the clock signal transitions from low (binary zero) to high (binary one), the input "D" is captured and stored in the DFF, and the output "Q" reflects the stored value.

A typical DFF can be constructed using two level-sensitive latches and one inverter, as shown in Figure 1.9. The first latch is called the master latch, and the second is called the slave latch. When the clock signal "CLK" is low (binary zero), the master latch follows the input "D" due to the inverted clock input, while the slave latch maintains its previous value.

At the rising clock edge (when "CLK" transitions from low to high), the slave latch becomes transparent, and its output "Q" follows the output of the master latch "QM". At the same time, the master latch becomes opaque and retains its previous value. In this state, the master latch blocks the input "D" from affecting the final output "Q". When the clock signal falls from high (binary one) to low (binary zero), the slave latch holds its value, and the master latch starts sampling the input "D" again.

As a consequence, a DFF stores and transfers its input to its output exclusively during the rising clock edge. Although there are flip-flops that capture their input on the falling edge, they are less commonly used in the ASIC design. Additionally, it is worth noting that a DFF may include a global reset signal to initialize the data stored within it.

DFF and Register

In Verilog description, a single-bit flip-flop is typically referred to as a DFF, while a multi-bit flip-flop is often denoted as a register. However, for the sake of clarity and consistency throughout this book, we will use the term **register** for all RTL design and simulation purposes. This will help in maintaining a uniform and unambiguous terminology throughout the text.

In summary, a register is a vital sequential circuit element that captures and stores the value of its input "D" on active clock edges. This behavior makes it an essential component in register-transfer design circuits for synchronization and storing input data.

1.5 Attributes of Verilog HDL Design

1.5.1 Key attributes of hardware description

A. Hardware Construction: Block Diagram

Software programming involves executing specific tasks and operations on hardware but is not directly part of the hardware itself. It interacts with the hardware through the operating system and device drivers. Conversely, Verilog coding is specifically tailored for describing hardware components, such as logic gates, registers, arithmetic units, and memory blocks. The primary objective of Verilog coding is to delineate the behavior and structure of hardware components, which will subsequently be implemented in tangible electronic circuits. Consequently, Verilog programming has a direct impact on hardware performance and resource utilization, as each statement in the design code corresponds to physical circuits. Therefore, **establishing a clear linkage between the code and the specific hardware being described is imperative when coding with Verilog**.

Prior to commencing HDL coding, designers should outline a clear block diagram to visualize the hardware components and their interconnections. This diagram serves as a blueprint for the hardware design, ensuring that the Verilog code accurately reflects the intended hardware behavior, data paths, and interfacing.

B. Hardware Behaviors: Timing Diagram

Another crucial design requirement when writing Verilog is to adhere to key attributes of hardware behaviors, such as concurrency and timing controls. In hardware descriptions, all design statements can be executed in parallel due to the concurrent nature of hardware execution. Therefore, timing controls play a vital role in synchronizing and coordinating the behavior of hardware components within the entire system.

Hardware design incorporates various mechanisms to manage timing, including clock-cycle-related counters and timers, finite state machines to transition operations between different states, ready-valid bus protocols for data transactions, enable-finish handshaking for data processing, request-grant mechanisms for bus arbitration, and many more. Proper timing management is essential in digital systems to prevent issues like address/data collision, race conditions, and timing violations, ensuring smooth and efficient operation of the hardware.

1.5.2 Concurrency

Concurrency is one of the key hardware behaviors which is not present in software to the same extent. To illustrate this, consider the following C code and Verilog design. In the C program, assuming that "A=1" and "B=2" initially, the fourth line adds "A" and "B" and assigns the result (3) to "C". Subsequently, the sixth line adds the updated value of "C" (which is 3) and "B" (which is still 2) and assigns the result (5) to "D". Therefore, after the execution of this program in a sequential manner, "C" holds the value 3 and "D" holds the value 5.

```
1  int A = 1;
2  int B = 2;
3  //Add A and B and assign the result to C
4  int C = A + B;
5  //Add the updated C and B and assign the result to D
6  int D = C + B;
```

In contrast to the sequential nature of software execution, Verilog statements execute concurrently, making it fundamentally different. The following Verilog code describes similar design statements with two Verilog *assign* blocks.

```
1  module Binary_Adder (input  [31:0] A, B,
2                       output [31:0] C, D);
3  // 32-bit binary adder with inputs A and B, output C
4  assign C = A + B;
5  // 32-bit binary adder with inputs C and B, output D
6  assign D = C + B;
7  endmodule
```

Both the inputs, "A" and "B", and the outputs, "C" and "D", are 2-bit IOs as declared in lines 1–2 following the design module's name "Binary_Adder". As

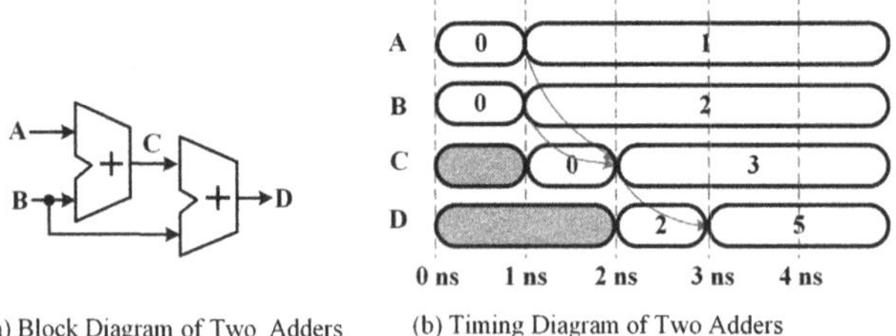

(a) Block Diagram of Two Adders (b) Timing Diagram of Two Adders

FIGURE 1.10
An example of hardware concurrency: two cascading adders.

a result, the Verilog descriptions using the arithmetic operator "+", denoted in lines 4 and 6, can be synthesized into two 32-bit binary adders, as depicted in Figure 1.10(a). Specifically, each *assign* block functions as a 32-bit binary adder. The inputs "A" and "B" are fed into the first adder, generating the output "C". Simultaneously, the value of "C" (which is the previous result of "A+B") along with input "B" is fed into the second adder, resulting in the output "D".

To clarify the concurrency behavior of hardware descriptions, let's delve into the timing diagram shown in Figure 1.10(b). Assume that the inputs "A" and "B" are reset to hexadecimal zeros, and the initial values of "C" and "D" are unknown due to the 1 ns delay of the hardware adders. At t=1 ns, the initial values of "A" and "B" drive the output "C=0". Concurrently, the computation of "D=C+B" begins in parallel with "C=A+B" within the time interval of 0–1 ns. Since "C" is unknown during this period, "D" is also undefined at this stage.

Moving on to t=2 ns, "C" is updated to the value 3 as the sum of "A=1" and "B=2". Simultaneously, "D" is updated to the value 2, resulting from the addition of "B=2" and the previous value of "C=0" (from t=1 ns). At t=3 ns, "C" retains its value 3, as both "A" and "B" maintain their previous values. Meanwhile, "D" is updated to the value 5, reflecting the addition of "B=2" and the updated value of "C=3" (from t=2 ns).

1.5.3 Timing diagram

As mentioned earlier, the concurrent nature of hardware execution makes timing considerations crucial during register-transfer level design. A valuable tool for analyzing signal transitions over time, particularly in sequential circuits experiencing clock cycle delays, is the timing diagram. A timing diagram captures the duration from when the input to a design block becomes stable and valid for a change to when the output of that block stabilizes and becomes valid for a change.

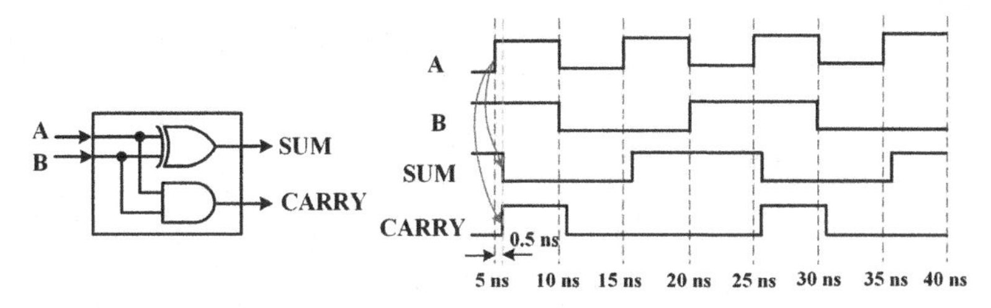

(a) Block Diagram of Single-Bit Half Adder (b) Timing Diagram of Single-Bit Half Adder

FIGURE 1.11
Timing diagram of combinational circuit: single-bit half adder.

A. Data Transitions within Combinational Circuit

To illustrate this concept, let's revisit the single-bit half adder design depicted in Figure 1.11(a). The half adder comprises an XOR gate and an AND gate, responsible for generating the "SUM" and "CARRY" outputs.

Figure 1.11(b) presents a timing diagram that captures signal transitions over a 40 ns period, accounting for a 0.5 ns delay in both the XOR and AND gates. Initially, the inputs are configured as follows: "A=0" and "B=1". At 5 ns, input "A" undergoes a transition from binary zero to one, while input "B" remains unchanged at its initial value. Due to the 0.5 ns latency associated with the XOR and AND gates, both the "SUM" and "CARRY" outputs maintain their previous values at 5 ns. Consequently, the "SUM" output transitions from binary one to zero after a 0.5 ns delay, occurring at 5.5 ns, as determined by the computation "SUM=A XOR B". Simultaneously, the "CARRY" output switches from binary zero to one, following the calculation "CARRY = A AND B". Likewise, at 10 ns, both "A" and "B" inputs transition from binary ones to zeros, causing the "CARRY" output to shift from binary one to zero while the "SUM" output remains at binary zero with a 0.5 ns delay.

In summary, our observation underscores that the outputs of combinational logic respond exclusively to alterations in their inputs. The timing diagram serves as a clear depiction of how output transitions are influenced by variations in the "A" and "B" inputs, factoring in the delay inherent to combinational logic. Profound comprehension of the combinational circuit delay holds paramount importance in the realms of register-transfer designs, meticulous timing analysis, and the precise tracking of signals during simulation waveforms.

B. Data Transitions within Sequential Circuit

Timing diagram is particularly important in the design of sequential circuits, as these circuits are closely tied to state transitions and exhibit behavior that depends on the timing and sequencing of clock cycles. Sequential circuits have memory elements (such as registers and RAMs) that store information and

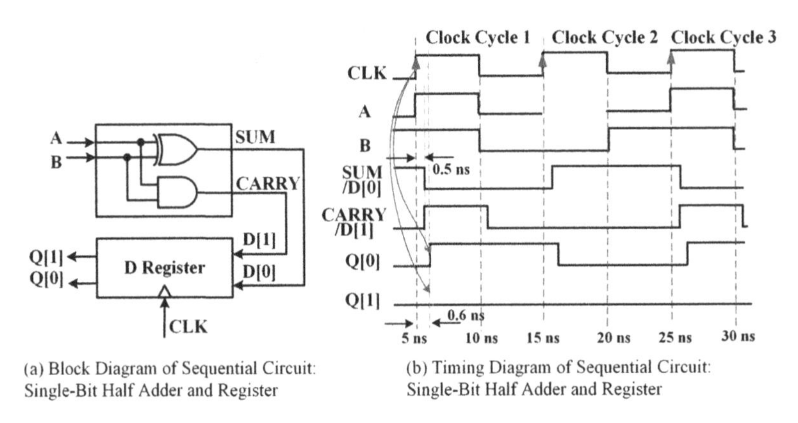

(a) Block Diagram of Sequential Circuit: Single-Bit Half Adder and Register

(b) Timing Diagram of Sequential Circuit: Single-Bit Half Adder and Register

FIGURE 1.12
Timing diagram of sequential circuit: single-bit half adder and register.

update their states over clock signals, leading to a sequential flow of data processing and operations.

As an example, we recall the sequential circuit that interfaces a half-adder with a 2-bit register, which can be depicted in Figure 1.12(a). Specifically, the register inputs, MSB "D[1]" and the LSB "D[0]", are connected with the "CARRY" and "SUM" signals, respectively. The register outputs "Q[1]" and "Q[0]" serve as the sequential circuit outcomes.

Figure 1.12(b) illustrates the operation of the sequential circuit. The XOR gate result, or data on "SUM/D[0]", will be binary one only when the inputs "A" and "B" are different. The AND gate result, or data on "CARRY/D[1]", will be binary one only when the inputs "A" and "B" are both binary ones. Assume that both XOR and AND gates perform with a 0.5 ns delay.

Assume that the register has a 0.6 ns delay from the rising clock edge to the output "Q[1]" and "Q[0]", also known as the clock-to-Q delay. At the initial rising clock edge at 5 ns, "Q[1]" captures and mirrors the data from "D[1]", while "Q[0]" captures and mirrors the data from "D[0]". At this point, due to the combinational delay, "D[1]=0" and "D[0]=1". As a result, at 5.6 ns, "Q[1]=0" and "Q[0]=1" are updated and maintain their values throughout the entire first clock cycle until the next rising clock edge at 15 ns.

At the second rising clock edge at 15 ns, "D[1]=1" and "D[0]=0", so after 0.6 ns, "Q[1]=1" and "Q[0]=0" are updated within the second clock cycle. Until the third rising clock edge at 25 ns, both "D[1]=0" and "D[0]=0", so at 25.6 ns, "Q[1]=0" and "Q[0]=0" are updated.

C. Register-Transfer Level Design

As mentioned earlier, HDL coding primarily revolves around register-transfer level descriptions. In RTL design, the primary emphasis lies in describing the hardware's behavior and data operations at the register-transfer level, which

TABLE 1.4
Verilog signal values.

Value	Interpretation
X	Unknown
0	0
1	1
Z	High Impedance

involves representing data flows and computational logic between registers. This design approach breaks down lengthy functional data chains into smaller datapaths, each composed of combinational logic and/or arithmetic operators. The behavior of these datapaths is defined by specifying how data is transferred between registers and how the combinational circuits process data streams.

A general RTL design block is shown in Figure 1.13, illustrating the "register-in and register-out" design pattern. All inputs to the combinational circuits originate from the first register output, and all outputs from the combinational circuits are stored back into another register. Combinational logic is responsible for data processing and computation, while registers serve as memory elements storing data within the sequential circuit and synchronizing the entire design system over clock cycles.

This RTL design approach allows for a clear representation of data flows and logical operations between registers, as well as ensuring proper functioning and timing requirements for digital circuits. In Chapter 11, we will explore the timing constraints in more detail, gaining a deeper understanding of their role in achieving correct and reliable circuit behavior.

1.5.4 Signal values

In this section, we will introduce the four signal values defined in the Verilog standard. Alongside the traditional binary zero and one, Verilog HDL supports two additional signal values – high impedance (high Z) and unknown (X), as depicted in Table 1.4.

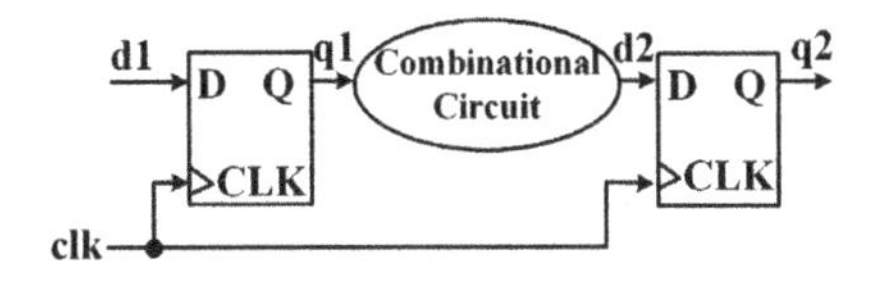

FIGURE 1.13
Register-transfer level design with verilog HDL.

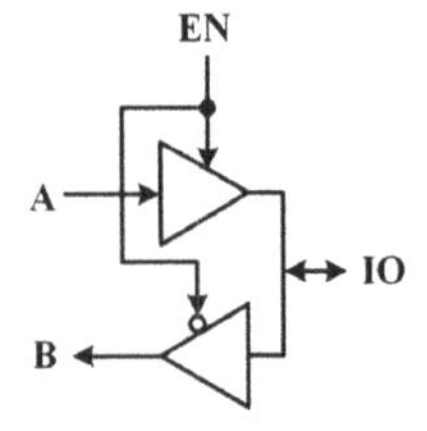

FIGURE 1.14
Bi-directional IO.

1.5.4.1 High-Z value

In subsection 1.4.2, we introduced the high-Z value, which is used to describe a high-impedance circuit or an open circuit. This value is commonly employed to model a tri-state buffer, which acts like an open circuit when the tri-state buffer is disabled.

In practical ICs, tri-state buffers are typically used in the design of bi-directional IOs. For example, in Figure 1.14, the "IO" signal on the right side can be driven by the input "A" when "EN=1". In this case, the signal "IO" functions as an output with a conducted path from the input "A" to the output "IO", while the path from "IO" to "B" is in a high-Z status (an open circuit).

The opposite occurs when "EN=0". The "IO" signal can drive the "B" output, which functions as an input when "EN=0". In this case, the path from "IO" to "B" is conducted, and the "A" to "IO" path is in a high-Z status (an open circuit).

Tri-state buffers enable multiple devices to share a common bus, and they play a critical role in digital circuits when multiple outputs need to be connected to the same bus without causing conflicts. This bi-directional behavior allows for efficient communication between different parts of a digital circuit.

1.5.4.2 X value

The value of X indicates the status of an unknown value and can be either zero or one. For instance, Figure 1.15 illustrates a multi-drive bus comprising $n\times$ bi-directional circuits. When all enable signals, "EN1", "EN2", $\cdots$, "ENn", are binary ones, the paths "I1 to IO", "I2 to IO", $\cdots$, "In to IO" are all connected, resulting in a multi-drive situation in the IO bus where multiple sources can drive the "IO" output simultaneously. If all sources are binary zeros/ones, the output can be a specific binary zero/one without any conflict. However, when the sources are different, some binary zeros and others binary ones, the output "IO" behaviors an unknown X status, which can be either binary zero or one.

In contrast, when all enable signals "EN1", "EN2", $\cdots$, "ENn", are binary zeros, the paths "IO to O1", "IO to O2", $\cdots$, "IO to On" are all connected. Consequently, the input "IO" can drive the outputs "O0", "O1", $\cdots$, "On".

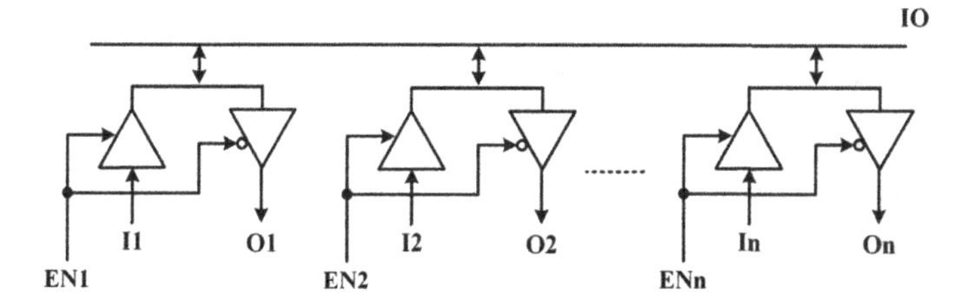

FIGURE 1.15
Multi-drive bus and bi-directional IO.

While the Verilog Standard supports the use of unknown (X) values, designing synthesizable circuits with unknown values is not advisable. Because unknown values can lead to unpredictable behavior during synthesis and simulation. In ASIC designs, all signals should have specific and well-defined values, either binary zeros or ones, or high-Zs when representing tri-state buffers. It is essential to avoid using unknown values in synthesizable designs to ensure predictable and reliable behavior of the hardware. Similarly, when designing a testbench, it is highly recommended to initialize all inputs with known and specific values, either binary zeros or ones. Simulations should not include any unknown signals in the waveform.

1.5.4.3 Muti-drive results

Although multiple drivers are not allowed in RTL design, Verilog supports modeling their effects. Table 1.5 summarizes the resulting behavior when two or more sources drive the same signal. If one source drives binary zero and the other drives binary one, the output will be driven to an unknown X value. If one source is binary X, the output will always be unknown X, regardless of the other source. If one source is a high-Z value, the output will be driven only by the other source: if the other source is binary one, the output will be binary one; if the other source is binary zero, the output will be binary zero. If the two sources do not conflict, the output will be a specific binary zero or one.

TABLE 1.5
Verilog multi-drive results.

Input/Input	0	1	X	Z
0	0	X	X	0
1	X	1	X	1
X	X	X	X	X
Z	0	1	X	Z

2

IC Design Flow

Chapter 2 presents the integrated circuit (IC) design flow, illustrated in Figure 2.1. Its aim is to provide an overview of the key steps involved in IC design, which are facilitated by various electronic design automation (EDA) tools catering to different aspects of the process. Broadly, the design flow can be divided into two main phases: the front-end and the back-end.

In the **front-end** phase, designers focus on the design and verification process in the behavioral description stages, encompassing tasks like creating a design specification, register-transfer level (RTL) coding, system-on-chip (SoC) integration, functional verification at both the intellectual property (IP) level and SoC level, and the verification on field-programmable gate arrays (FPGAs). This phase involves describing and verifying the functionality and structure of RTL design. On the other hand, the **back-end** phase is focused on the physical implementation of the IC. It encompasses critical tasks such as synthesis, which converts the RTL design into a gate-level netlist, and layout, which involves creating the final layout of the IC. This phase also includes crucial stages such as timing analysis, design rule check, design for test, and more. Additionally, some procedures during and after chip tapeout, such as packaging and testing, are briefly introduced.

Throughout the IC design flow, various essential concepts and industry-specific terms/acronyms are commonly used to describe specific design steps. This chapter serves as an introduction to these design details and procedures, offering readers a foundational understanding of the overall IC design flow. Subsequent chapters will delve into the front-end design steps in greater detail, presenting various projects that provide comprehensive insights and practical knowledge.

2.1 Design Specification

The initial step in the IC design flow involves establishing the SoC architecture and defining design specifications for all design components. System design considerations are derived from electronic product requirements, which encompass various factors. These include addressing low-power design needs for applications like battery-driven and portable devices, meet-

DOI: 10.1201/9781003187080-2

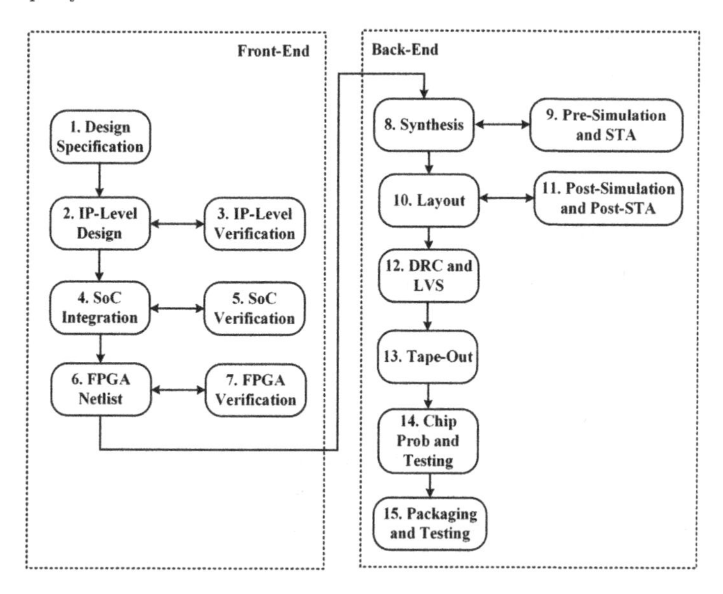

FIGURE 2.1
IC design flow.

ing high-speed/throughput requirements such as hardware accelerators for high-performance servers and clusters, adhering to chip size constraints, integrating third-party IPs like CPUs, DDR, interfaces, and peripherals, ensuring the quality of data processing results for tasks like audio and video processing units and neural network engines, among others. From the perspective of IC design, designers must adeptly navigate challenges to achieve an optimal design that strikes the right balance between chip size, speed, and power efficiency. This often necessitates making trade-offs among these dimensions.

In the SoC design specification, the SoC functionality is divided into multiple design components known as IP blocks. The SoC architecture focuses on the system topology and bus protocol to enable seamless IP block integration. Design teams can develop IPs in-house or obtain them from third-party vendors like Synopsys, Cadence, ARM, MIPS, etc. Third-party IPs may be available in different forms, such as hardware description language (HDL) code, gate-level netlists, or hard macros. Ensuring that the hard macros' technology aligns with the SoC design is crucial for maintaining functional integrity and reliability.

Each IP, whether it's self-designed or obtained from a third-party source, necessitates its own set of design specifications. For self-designed IPs, these specifications should encompass comprehensive details such as block diagrams, specific implementation guidelines, timing diagrams, bus interfaces, connectivity requirements, and functional register configurations. In the case of third-party IPs, the design specifications are primarily concentrated on IP configuration, the integration and interconnection of modules, and the inclusion of suitable simulation models for accurate representation within the overall system.

2.2 RTL Design in IP Level

2.2.1 IP-level design

A. IP Design Categories

The RTL coding begins with the IP-level designs, which can be categorized into two main types: SoC-related designs and algorithm-based designs.

SoC-related designs are primarily concerned with constructing the SoC architectures, managing various modules located on the control bus, moving data between memories and design modules through the data bus, and handling internal/external interfacing. Examples of such designs include bus bridges/wrappers, master/slave bus interfaces, CPUs, DMA (direct memory access), memory controller, interrupt controller, and industry-standardized interfaces and peripherals like UART, I2C, SPI, SDIO, GPIO, WatchDog, and Timer. These designs generally rely on industry standards, so software coding with identical functions is not typically required.

On the other hand, algorithm-based designs involve the implementation of mathematical models using hardware. This requires the design of digital circuits that can perform numerical operations in hardware, rather than relying on software algorithms executed on a general-purpose processor. However, the software codes, such as using C/C++, Matlab, or Scala, are still required to provide the golden models for hardware design and verification. The golden models serve two primary purposes: first, to provide hardware design references and demonstrate the quality of results and functions (such as structural evaluation, precision, error percentage, etc.) for future hardware, and second, to prepare golden results for RTL verification.

B. Algorithm Design Needs for Hardware Description

Achieving the intended design goals for algorithms demands a close correspondence between the algorithmic design (software code) and the final hardware arrangement. It guarantees that the software code furnishes accurate direction for hardware application and design configurations. To illustrate, consider the MATLAB code: $grayscale\_image = rgb2grayscale(rgb\_image)$. This function simplifies color image conversion, transforming an color image (represented as the rgb_image matrix) into a grayscale image (represented as the $grayscale\_image$ matrix). However, this abstraction obscures the intricate floating-point (FP) multiplications of each color pixel with corresponding weights and the FP additions to accumulate the results. Consequently, such software code lacks the capacity to guide hardware design decisions.

Another concern of the example is that the MATLAB function employs default double-precision calculations, which might be a divergence from the precision demanded by hardware design. Such precision mismatch could yield substantial differences when contrasting outcomes between the RTL design and the established golden model, thereby introducing considerable uncertainty during the verification phase.

Rather, hardware implementations often necessitate design optimizations involving low-precision and approximate implementations of FP adders and multipliers to balance result quality against hardware cost. Thus, establishing a close alignment between algorithmic design and hardware implementation is vital for precise translation and optimal performance.

2.2.2 Hardware design considerations

In this section, we use an example to illustrate a hardware design considerations in the algorithm and the design structure.

A. Algorithmic Design: Accuracy vs. Hardware Cost

First, in the process of designing a golden model, it becomes crucial to seek an optimal design that not only meets the accuracy requirements outlined in the design specifications but also minimizes hardware costs. To attain this goal, various design optimizations can be employed during the algorithm coding phase. These optimizations encompass techniques like floating-point to fixed-point conversion, utilizing approximate and/or imprecise design methods to conserve hardware resources, and striking a balance between quality and hardware to achieve the desired algorithmic accuracy.

For instance, a FP matrix-vector multiplication can be mathematically written as

$$\begin{pmatrix} z_0 \\ z_1 \end{pmatrix} = \begin{pmatrix} x_0 & x_1 \\ x_2 & x_3 \end{pmatrix} \cdot \begin{pmatrix} y_0 \\ y_1 \end{pmatrix} \tag{2.1}$$

In the initial hardware design, the numerical operations involves providing software codes for identical functions that accept six FP data inputs (x_0, x_1, x_2, x_3, y_0, and y_1) and produce two FP outputs (z_0 and z_1). However, unlike traditional software programming, the algorithms provided to the IC design and verification team must select the appropriate precision for the computation (e.g., integer, FP8, FP16, FP32, FP64, FP128, etc.) and determine the desired levels of approximation for the hardware designs. Higher design accuracy typically leads to increased hardware costs.

B. Hardware Structural Design: Speed vs. Area-Power

The next step is to translate the software design functions into RTL design. This step is crucial as it involves implementing the design functions in Verilog

HDL to create an equivalent hardware representation of the software program. Simulation and verification are essential during this phase to ensure functional equivalence between the software code and the RTL design. Furthermore, hardware performance is primarily determined by the RTL descriptions, so designers must consider various design techniques, such as parallel computing, pipeline structures, design reuse, clock-gating, and many more, to optimize design efficiency and improve the hardware performance. It is also critical to adhere to IC design rules and guidelines when writing RTL code.

In the provided example, Figure 2.2 showcases two design structures. Assume that each FP operator, either the FP adder or FP multiplier, takes one clock cycle from its input to its output. The first design structure, depicted in Figure 2.2(a), adopts a straightforward parallel implementation utilizing four FP multipliers and two FP adders. With a streaming width of four, this design can process four sets of FP data inputs ("x0-x3" and "y0-y3") in each clock cycle. Consequently, it achieves a total of six FP operations per clock cycle (or six FLOPs per cycle), comprising four FP multiplications and two FP additions.

Alternatively, a sequential structure, depicted in Figure 2.2(b), can be used to save chip area and power dissipation at the expense of reduced speed. In this design, the multiplication-addition circuit (MAC) can be reused twice for each four sets of FP data inputs. Specifically, the inputs are divided into two groups due to a reduced streaming width of two. The first group ("x0-x1" and "y0-y1") is fed into the engine in one clock cycle, and the second group ("x2-x3" and "y0-y1") is fed into the engine in the following clock cycle in a pipelined manner. As a result, the first output ("z0") of the MAC circuit can be obtained over two clock cycles, while the second output ("z1") is a pipelined output with an additional clock cycle. This sequential design achieves three FP operations per clock cycle (or three FLOPs per cycle), including two FP multiplications and one FP addition.

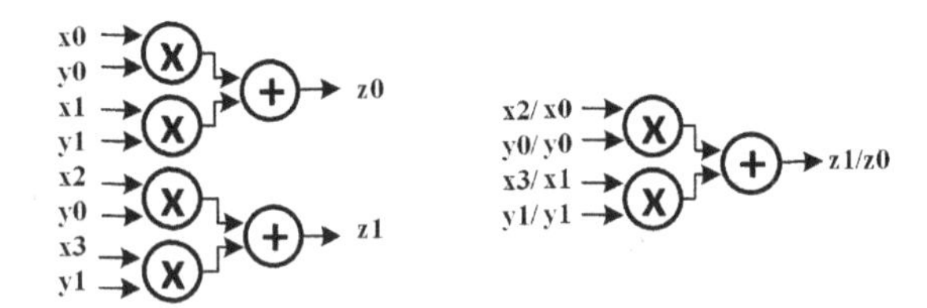

FIGURE 2.2
Parallel design and sequential design structures for matrix-vector multiplier.

2.2.3 Code editors

Before delving into design simulation, let's explore some popular code editors commonly used in the IC design industry, including Vim, Emacs, and WinEdit. These text editors play a crucial role in designing ICs and building simulation environments, as they are used to edit various files including scripts, Verilog codes, SystemVerilog testbenches, golden models/results, and more.

Among the three code editors, **Vim** is recommended in this book for several reasons. Firstly, Vim is an open-source text editor available since 1976. Secondly, Vim is known for its fast and lightweight performance, making it suitable for editing large files or source code. It is primarily used in the Unix/Linux environment but can also be used on Mac OS and Windows OS. Thirdly, Vim supports a wide range of programming languages and file formats, including ".v" (for Verilog code) and ".sv" (for SystemVerilog code) files, and can automatically detect file extensions and apply the appropriate syntax highlighting. Fourthly, Vim is highly customizable using the ".vimrc" source file, allowing users to tailor their editor experience. For example, users can configure the font and font size to their preference, as shown in the following example:

```
1 set guifont=Consolas:h11
```

More importantly, Vim is used in this book due to its extensive collection of powerful commands that can greatly enhance programming efficiency. By mastering these commands, users can streamline their code editing process in Vim, leading to improved efficiency and productivity. While there are numerous Vim commands available, this book focuses on introducing the most commonly used ones, ensuring that readers gain a solid foundation in utilizing Vim effectively. For a more comprehensive reference, **Appendix A** of this book provides a detailed overview of these commands, allowing readers to delve deeper into the full potential of Vim.

2.3 RTL Verification in IP Level

In the front-end of the IC design flow, the design verification process is crucial to ensure that all the RTL design features and functions meet the requirements specified in the design specifications.

2.3.1 Simulation testbench with Verilog HDL

This book offers a comprehensive introduction to utilizing Verilog HDL for the creation of simulation testbenches and automated verification environments. Functional verification is an important part of the IC design process, involving iterative rounds of debugging, code modifications, and the execution of numerous test cases to enhance functional and code coverage. Therefore, automation plays a vital role in simplifying this iterative simulation process.

To illustrate the concept, a basic testbench for a FP adder design is presented in Figure 2.3(a). This testbench consists of a **bus functional model (BFM)** that generates simulation stimuli, a **golden model** and/or some **golden files** for result comparison, and a **monitor** that automatically checks the results from the RTL design against those from the golden model. The simulation log/report displayed in Figure 2.3(b) shows that "1000000 test cases pass" and "0 test cases fail!" within all the 1,000,000 test cases. If any failures occur, the simulation waveform, depicted in Figure 2.3(c), can be used to trace the signals and designs, facilitating the identification and resolution of any design bugs.

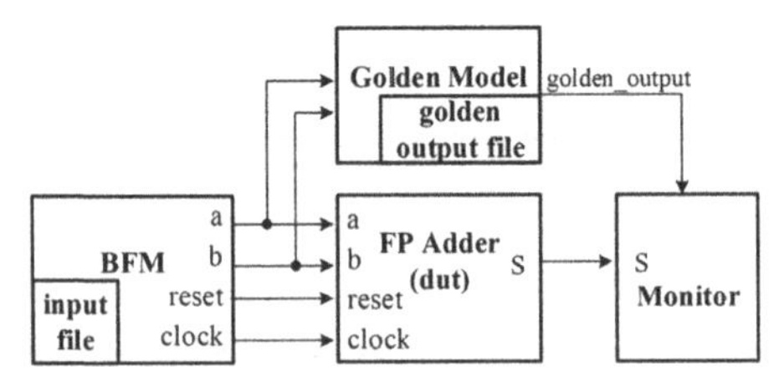

(a) General Simulation Testbench

```
# For all the 1,000,000 test cases, there are      1000000 test cases pass,            0 test cases fail!
# ------------------------------------ FP adder design simulation passed cases ------------------------
# For all the 1,000,000 passed test cases, there are      659016 test cases exactly the same!
# For all the 1,000,000 passed test cases, there are      297848 test cases with different 3-0 mantissa bits!
# For all the 1,000,000 passed test cases, there are       37530 test cases with different 7-0 mantissa bits!
# For all the 1,000,000 passed test cases, there are        5606 test cases with different 15-0 mantissa bits!
# ------------------------------------ FP adder design simulation fail cases --------------------------
# For all the 1,000,000 failed test cases, there are           0 test cases with different sign bit!
# For all the 1,000,000 failed test cases, there are           0 test cases with different exponent bits!
# For all the 1,000,000 failed test cases, there are           0 test cases with different 22-16 mantissa bits!
# ------------------------------------ FP adder design simulation summary ------------------------------
```

(b) Simulation Report/Log

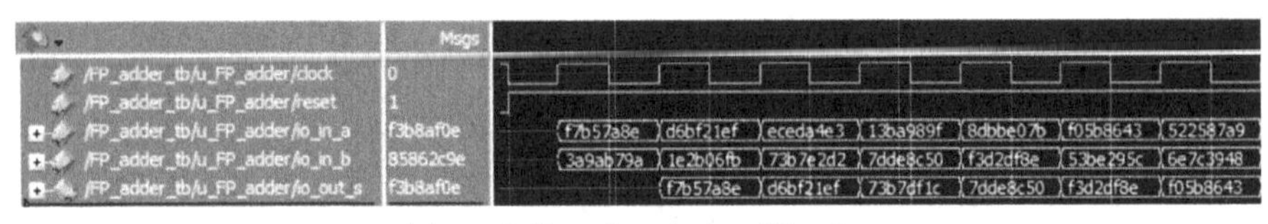

(c) ModelSim Simulation Waveform

FIGURE 2.3
Simulation example of FP adder.

Manually feeding in all one million single-precision inputs and comparing their corresponding outputs is impractical. Therefore, the input data is randomized, and output data verification is automated. The simulation process itself is also automated through the use of a TCL script.

2.3.2 Verification methods and methodologies

To construct a robust verification environment, a combination of software languages such as C/C++, Scala, Vera, and Matlab may be utilized alongside the Verilog testbench. These languages enable the creation of sophisticated verification components, including monitors, bus functional models, and scoreboards, to simulate and validate the behavior of RTL **design-under-test (DUT)**.

In addition to the Verilog testbench, the industry employs various verification methods and methodologies to ensure the quality and reliability of RTL designs. For example, the **SystemVerilog** verification language and **Universal Verification Methodology (UVM)** have been widely adopted in the industry to establish standardized and reusable verification environments.

SystemVerilog extends the capabilities of Verilog HDL and includes features specifically developed for verification, such as assertions, coverage models, and constrained-random stimulus generation. UVM, on the other hand, provides a set of methodologies, libraries, and guidelines that facilitate the creation of scalable and reusable verification components.

A. Coverage-Driven Verification

The **Coverage-Driven Verification (CDV)** method, supported by SystemVerilog, plays a vital role in ensuring the quality and assessing the completeness of verification process. CDV incorporates both functional coverage and code coverage metrics to evaluate the effectiveness of the verification efforts.

- **Functional coverage** aims to verify that all design functionalities described in the verification plan have been exercised by the RTL design. It provides qualitative measurements of how thoroughly the design has been tested, focusing on verifying specific groups and combinations of all the design features.

- **Code coverage** metrics, on the other hand, provide quantitative measurements of how much of the RTL code has been executed during simulations. Common code coverage metrics include line coverage, statement coverage, branch coverage, and condition coverage. These metrics offer insights into the comprehensiveness of the verification process and assist in identifying sections of the design that might need further testing or redundant code.

While there is no universally defined standard for passing functional verification, achieving **100% functional coverage** and a **line coverage of over**

95% are widely recognized as good practice. However, the specific criteria for passing verification may vary depending on the design team or project requirements.

B. Constrained-Random Test

In the verification process, test cases serve as simulation stimuli for the RTL DUT. These test cases can be classified into two main types: direct tests and random tests.

- **Direct tests** are designed to verify specific features and functionalities outlined in the design specifications. Executing all direct tests ensures that the functional coverage reaches 100%, effectively covering all desired design functionalities.
- **Random tests**, on the other hand, focus on exploring different combinations of design features, particularly targeting abnormal or corner cases that cannot be directly tested.
- Furthermore, SystemVerilog provides support for **Constrained-Random Verification (CRV)**, allowing stimuli to be constrained and guided to effectively target the verification plan, functional coverage goals, and code coverage increases.

During the verification process, RTL designers typically focus on conducting simulations for direct tests, while verification engineers take charge of running random tests that target abnormal scenarios and meet line coverage objectives. Direct test cases often account for over 70% of line coverage, and constrained-random tests can further elevate line coverage to approximately 90%. Achieving the final 5% increment beyond 90% may require manual analysis of the coverage report and the creation of specific corner cases to address the remaining coverage gaps. Here is an example to provide insight into the coverage contributions from various test cases; however, it's crucial to acknowledge that the specific simulation coverage can vary based on many factors such as project complexity, verification plans, and other relevant considerations.

When a design bug is identified and fixed during the verification process, it's essential to conduct a **regression test**. A regression test typically comprises a set of direct and random tests aimed at verifying that the code changes made to address the bug do not adversely affect other design features and functions. Additionally, the regression test can serve as a means to collect function and code coverage metrics, offering a comprehensive evaluation of the verification process and highlighting any potential coverage gaps or areas that require further improvement.

C. Assertion-Based Verification

Assertion-Based Verification (ABV) is a technique primarily used for validating circuit design against specifications during simulation. Assertions, articulated in SystemVerilog, express specific design requirements, embodying a part of the SystemVerilog standard as defined by IEEE 1800. These assertions can be incorporated directly into the RTL code for validation, or integrated into the testbench as part of the broader verification process.

Below is an example of assertion design that delineates a bus handshaking that necessitates the grant (termed as "gnt" in the code) signal must be asserted within 1 to 100 clock cycles after the request (termed as "req" in the code) is asserted. Running this assertion during simulation evaluates the bus handshaking mechanism. In the event of a violation of this requirement, the simulator will pinpoint an error, detailing the specific simulation time where the discrepancy occurred.

```
1  //Request-grant handshaking within 1-100 cycles
2  mbus_req_gnt: assert property (@posedge clk) req |->
3  ##[1:100] gnt;)
4
5  else $error ("Assertion failed at %0t ns",$time);
```

D. UVM

UVM has been an extensively adopted de facto standard in the field of IC design verification. Its purpose is to establish a standardized approach for developing and validating testbenches that effectively test and verify the functionality of the RTL design.

By incorporating principles from object-oriented programming (OOP), UVM enables the creation of modular and reusable testbench components, which can be combined to form a comprehensive verification environment. UVM is based on the SystemVerilog library and facilitates constrained random stimulus generation, as well as function and code coverage techniques, thereby facilitating the development of reusable verification components.

2.3.3 EDA tools for simulation

In the industry, several RTL simulators are commonly utilized, including Synopsys VCS, Cadence NC, and **Siemens ModelSim**. For the design simulation examples in this book, Siemens ModelSim serves as the basis. Executing a simulation with ModelSim involves eight steps, which can be performed using either the GUI window or a TCL script. A comprehensive explanation of both simulation approaches is provided in **Appendix B** as a software tutorial.

Debussy is a widely employed tool for signal tracing in simulation waveforms. Debussy's waveform dump is compact in size and easily shared between verification and design teams for debugging purposes. The waveform dump generated by Debussy employs the ".fsdb" file format, while ModelSim uses the ".vcd" (value change dump) format. The dumped waveform encompasses transition activities of all inputs, signals, and logic elements, making it valuable not only for functional simulation but also for evaluating dynamic power consumption.

2.4 SoC Integration

Upon completion of IP-level verification, all the design components can be integrated into a SoC utilizing prevalent industry-standard bus architectures. Only IP components with compatible bus interfaces can be successfully integrated. Typical industrial bus protocols encompass the advanced high-performance bus (AHB) [11] and advanced eXensible interface (AXI) [12] from ARM Holdings, Wishbone from Silicore Corporation [15], open core protocol (OCP) from OCP international partnership [14], etc.

For instance, as depicted in Figure 2.4, an SoC is assembled using the Advanced Microcontroller Bus Architecture (AMBA) AXI as the high-speed bus and AMBA AHB as the low-speed bus for peripherals with lower bandwidth requirements. High-speed design modules, such as the SoC-related designs (the USB host/device and the Wi-Fi component) and algorithm-based designs (the neural engine, numerical accelerator, and the image/video processing unit), are positioned on the AXI bus.

Conversely, peripherals with lower-speed data movement, including serial interfaces (GPIO/SPI/I2C/UART), timers, watchdogs, and NOR/Serial/-NAND flash controllers, are placed on the AHB bus. A bus bridge is employed to connect the high-speed and low-speed buses, minimizing system power consumption. The CPU governs all IP modules located on the SoC buses, while

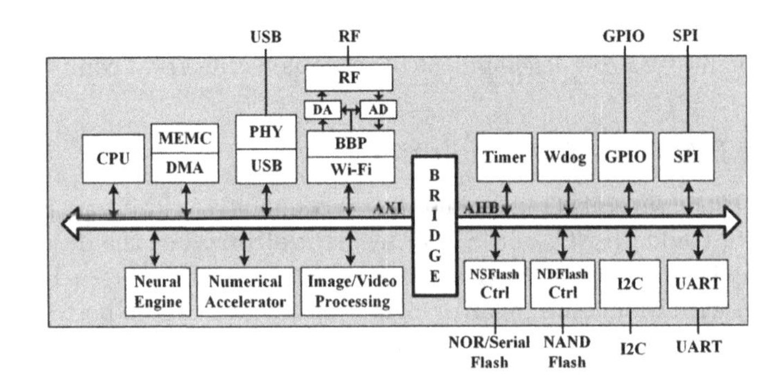

FIGURE 2.4
An example of SoC integration.

data exchange between IP modules and memory is primarily facilitated by DMA in conjunction with the memory controller (MEMC).

Alternatively, users have the flexibility to design their own SoC bus architecture for IP integration. However, it's essential to implement bus wrappers to facilitate the conversion of bus protocols between different interfaces, including the self-designed bus and the interfaces of third-party IP components. In Chapter 8, a master-slave bus is introduced as a self-developed control bus for interfacing slave devices with the microprocessor through the bus structure.

2.5 SoC Verification

SoC verification is similar to functional verification at the IP level but with a focus on testing the interconnection and integration of the entire system. When utilizing the UVM, the verification components developed at the **IP tree** can be reused in the **SoC tree**. Once the verification environment is established, the fundamental procedure for system-level verification is outlined below:

- **Run simulations** by assigning a unique ID seed, observe the log file, and dump the corresponding waveform if any bugs are detected. The unique ID seed allows for the reproduction of random stimuli, enabling the reexecution of simulation with the same test case after resolving design bugs.
- **Collaborate with designers** to debug the waveform dump. If the bug is originating from the testbench, the verification engineer should address and rectify it. If the bug is stemming from the design code, the verification engineer should report it to the designers. Design bugs can be present in both the IP tree and the SoC tree.
- Designers should **resolve the design bugs** and rerun the simulation with the unique ID seed. After passing the specific test case, regression testing is necessary in both IP design and SoC integration to ensure that the bug fix does not impact other design components and design features.
- After resolving the bug, **update the code with a new version** using **version control software** such as **Git** and SVN (Subversion). Version control is especially significant for SoC verification, as it involves many designers and verification engineers contributing to the SoC tree concurrently.
- **Collect functional coverage and code coverage** during system-level verification, although the coverage metrics will be lower than those achieved during IP-level verification. As mentioned earlier, IP design and verification primarily focus on design functions and features, whereas

SoC verification is primarily concerned with system connectivity and integration.

2.6 FPGA Netlist

2.6.1 FPGA development and assessment

After the functional verification phase is complete, the RTL code can be utilized to demonstrate the design's functionality on a specific FPGA board in a real-world context. This involves additional steps such as synthesis and implementation to generate an FPGA bitstream.

The synthesis process converts the RTL design into a netlist that can be mapped using available FPGA resources. Subsequently, the implementation phase involves placing and routing the netlist into a bitstream understood by the physical FPGA. This process ensures that the design is executed onto the FPGA device in an optimal manner, considering factors such as resource utilization and timing constraints. It's crucial to emphasize that FPGA synthesis and implementation are configurable to accommodate various design specifications. These specifications may encompass high-performance optimization, low-power design, area optimization, constraints on DSP utilization, and other relevant factors.

The FPGA synthesis and implementation report provides detailed information about the resource utilization of the design, including the number of slices, Block RAMs, DSPs, and other FPGA components utilized. As illustrated in Figure 2.5(a), the synthesis report indicates that the design utilizes 401,948 slices of look-up tables (LUTs) and 90,864 slices of registers. Additionally, in Figure 2.5(b), it is shown that the design uses 192 Block RAMs. Figure 2.5(c) demonstrates that 1,048 DSPs are employed in the design.

These numbers provide insights into the utilization of critical resources in the FPGA, allowing designers to evaluate the efficiency of their design in terms of LUTs, registers, Block RAMs, and DSPs. By analyzing the resource utilization, designers can optimize their designs and ensure efficient utilization of available FPGA resources.

2.6.2 MOF and timing constraints

In addition to facilitating application demonstrations, FPGA implementation allows for the testing of timing constraints and estimation of the **maximum operational frequency (MOF)**. This can be achieved by analyzing the FPGA timing reports, which provide insights into various aspects of the timing information and analysis. It includes examining timing constraints, identifying

1. CLB Logic

Site Type	Used	Fixed	Prohibited	Available	Util%
CLB LUTs*	401948	0	0	1303680	30.83
LUT as Logic	393756	0	0	1303680	30.20
LUT as Memory	8192	0	0	600960	1.36
LUT as Distributed RAM	0	0			
LUT as Shift Register	8192	0			
CLB Registers	90864	0	0	2607360	3.48
Register as Flip Flop	90864	0	0	2607360	3.48
Register as Latch	0	0	0	2607360	0.00
CARRY8	10914	0	0	162960	6.70
F7 Muxes	544	0	0	651840	0.08
F8 Muxes	256	0	0	325920	0.08
F9 Muxes	0	0	0	162960	0.00

(a) Resource Utilization on LUT/Register Slices

2. BLOCKRAM

Site Type	Used	Fixed	Prohibited	Available	Util%
Block RAM Tile	192	0	0	2016	9.52
RAMB36/FIFO*	192	0	0	2016	9.52
RAMB36E2 only	192				
RAMB18	0	0	0	4032	0.00
URAM	0	0	0	960	0.00

(b) Resource Utilization on RAMs

3. ARITHMETIC

Site Type	Used	Fixed	Prohibited	Available	Util%
DSPs	1048	0	0	9024	11.61
DSP48E2 only	1048				

(c) Resource Utilization on DSPs

FIGURE 2.5
FPGA synthesis and implementation report.

critical paths, and determining the worst negative slack (WNS) using timing reports generated by tools like AMD Vivado.

Using the WNS timing, the achievable MOF can be approximately estimated using the formula:

$$MOF = \frac{1}{T_R - T_{WNS}} \tag{2.2}$$

where T_R denotes the reference clock period and T_{WNS} represents the amount of slack in each clock cycle. For instance, we assume that a design specification calls for a reference clock frequency of 250 MHz, which corresponds to a minimum clock period of $\frac{1}{250 \text{ MHz}} = 4$ ns. Hence, T_R can be set to 4 ns. In order to estimate corresponding T_{WNS} through the Vivado synthesis, the design constraint file can be used to create a clock object with the desired period and waveform characteristics. The following command demonstrates the creation of a clock named "clk" with a period of 4.0 ns, associated with the "clk" signal in the design's IO:

```
create_clock -add -name clk -period 4.0 -waveform {0 2}
[get_ports {clk}];
```

After creating the clock object, the timing report, including the value of T_{WNS}, can be generated by running the FPGA synthesis process with the desired clock period. Assuming the reported T_{WNS} is 0.683 ns, the design circuit can achieve an MOF $= \frac{1}{T_{\mathrm{R}}-T_{\mathrm{WNS}}} = \frac{1}{4-0.683} = 301.5$ MHz, meeting the design goals.

2.6.3 EDA tools for FPGA demonstration

The FPGA tools commonly used in the industry include AMD Vivado and Intel Quartus. While this book primarily focuses on the RTL design and simulation aspects within the IC design flow, it recognizes the importance of establishing a connection between RTL design and real hardware to develop effective Verilog code. To illustrate this connection, therefore, the book extensively employs **AMD Vivado** in its design examples in Chapter 6.

FPGA tools serve multiple purposes, including verifying the equivalence between the design block diagram and RTL analysis results, estimating the resource utilization and MOF of FPGA designs, and assisting users in establishing the link between Verilog descriptions and real hardware components.

To help readers effectively utilize AMD Vivado, **Appendix C** provides a comprehensive tutorial consisting of seven steps. These steps cover various aspects, from project creation to final netlist programming on specific FPGA boards. By following this tutorial and leveraging the capabilities of AMD Vivado, readers can enhance their understanding of the design flow and gain practical experience in implementing FPGA designs.

2.7 FPGA Verification

2.7.1 Importance of FPGA verification

FPGA verification is crucial for two main reasons following RTL simulation. Firstly, it serves to confirm that the design not only functions as intended but also complies with the physical constraints imposed by the chosen FPGA device for implementation. While RTL verification primarily concentrates on verifying the logical behavior of the design within an idealized environment, FPGA verification extends its scope to evaluate the design's performance and physical realization on real hardware. This encompassing verification considers timing considerations and resource utilization, ensuring that the design operates correctly within the context of real hardware platforms.

Secondly, FPGA verification serves as a practical method for establishing and validating the functionality of the hardware design within its intended execution environment. For instance, RTL verification in complex scenarios such as emulating wireless communication channels using SystemVerilog or other languages can be challenging. It involves complex channel modeling, incorporating realistic noise and interference models, simulating the propagation of wireless signals traversing obstacles such as walls and floors, implementing various modulation and demodulation schemes, and more. However, leveraging FPGA technology allows the design bitstream to be run in an actual wireless setting. This can be simply achieved by transmitting wireless data from an FPGA board functioning as a transmitter to an additional FPGA board serving as a receiver. By utilizing the real wireless channel, this approach provides a highly effective method for verifying hardware designs in real-world scenarios.

2.7.2 Challenges of FPGA verification

FPGA debugging can present greater challenges compared to RTL verification due to several factors. One significant challenge arises from the limited availability of comprehensive verification methods and strategies specifically tailored for FPGA verification, which are more abundant in the realm of RTL verification. Additionally, analyzing simulation waveforms in the context of FPGAs can be more intricate, further complicating the debugging process. To enable effective waveform analysis, the use of additional equipment such as logic analyzers and oscilloscopes becomes necessary. Last but not the least, FPGA verification may involve dealing with timing issues that may not consistently replicate. Timing glitches and intermittent failures can occur, where a design may pass testing correctly in some instances but fail in others due to corner cases and timing violations specific to certain data paths.

Given these difficulties, it becomes crucial to ensure thorough verification of RTL code before progressing to FPGA implementation. This entails extensive utilization of random testing and code coverage to detect functional bugs at the RTL itself, prior to the physical design and implementation on the FPGA. By thoroughly verifying the design functions and achieving high levels of functional and code coverage at the RTL, designers can mitigate the debugging challenges encountered during FPGA development and ensure a smoother transition to the physical implementation stage.

2.8 ASIC Synthesis

2.8.1 ASIC synthesis and script file

Application-specific integrated circuit (ASIC) synthesis refers to the process in which the RTL description is translated into a gate-level netlist, forming the basis for chip layout. To illustrate the ASIC synthesis concept, we provide the RTL design of a 4-bit counter below. The details of the design syntax will be explored in later chapters. In this section, our focus lies on the synthesis process, which involves transforming the RTL code into a gate-level netlist.

```
1  //A 4-bit Counter Design with Verilog HDL
2  module counter_4b (input              rst,
3                     input              clk,
4                     input        [3:0] ini,
5                     output reg   [3:0] cnt);
6  always @ (negedge rst, posedge clk) begin
7    if (~rst) begin
8      cnt <= ini        ;
9    end else begin
10     cnt <= cnt + 4'h1;
11   end
12 end
13 endmodule
```

ASIC synthesis is typically automated using script files to minimize operational errors. Below is an example script file to facilitate this process. It is a basic illustration and can be customized according to the specific needs of a design. Before executing the synthesis tool, the script should specify the directory path (line 1) and the series libraries utilized for synthesis (lines 2–4).

```
1  search_path = {"." "./synthesis" }
2  link_library= {"*" "./synthesis/tcb773stc.db" }
3  target_library= {tcb773stc.db}
4  symbol_library= {tcb773s.sdb}
5
6  read -format verilog {"./counter_4b.v"}
7  current_design = "counter_4b"
8  create_clock -name "clk" -period 100 -waveform {"0" "50"}
9  set_dont_touch_network find( clock, "clk")
```

```
10 set_dont_touch_network find( port, "rst")
11 write -format db -hierarchy -output "./counter_4b.db"
12 write -format verilog -hierarchy -output "./counter_4b.v"
13 ......
```

In the following lines, the Verilog design file ("./counter_4b.v") and the current project ("counter_4b") are specified (lines 6–7). A clock signal named "clk" is created in line 8, with a period of 100 ns, equivalent to a frequency of 10 MHz. It is crucial to designate critical timing signals, such as clock and reset, as "don't touch" (lines 9–10) to prevent any optimization or alteration during synthesis. This precaution is necessary because the clock tree and reset network will not be established until the layout process. By preserving these signals during synthesis, we ensure their integrity and proper functionality in the final implementation. In lines 11 and 12, finally, the commands write the synthesized design into a database file "counter_4b.db", and write out the gate-level description of the design in Verilog format to the "counter_4b.v" file.

2.8.2 Gate-level netlist and synthesis report

By executing the script, the synthesis tool, such as Synopsys Design Compiler, can autonomously translate the RTL code into a gate-level netlist. For example, the gate-level description provided below outlines the cells, registers, and wires encompassing the entire design. It provides more detailed insights into the utilization and interconnections of logic cells, including inverters, XOR gates, D-flip-flops, and many more.

```
1  module counter_4b (input rst,clk,ini_3,ini_2,ini_1,ini_0,
2                     output cnt_3,cnt_2,cnt_1,cnt_0);
3
4  wire n96, n97, n98, n99, n100, n101, n102,
5  n103, n104, n105, n106, n107, n108, n109,
6  n110, n111, n112, n113, n114, n115, n116,
7  n117, n118, n119, net5, n120, n121, n122;
8
9  INV0 U21 ( .I(ini_3), .ZN(n108) );
10 INV0 U22 ( .I(ini_2), .ZN(n109) );
11 INV0 U23 ( .I(ini_1), .ZN(n110) );
12 INV0 U24 ( .I(ini_0), .ZN(n111) );
13 XNR2D1 U25 ( .A1(n106), .A2(net5), .ZN(n120) );
14 NR2D0 U26 ( .A1(n107), .A2(n97), .ZN(n106) );
15 ND2D0 U27 ( .A1(cnt_1), .A2(cnt_0), .ZN(n107) );
16 INV1 U28 ( .I(n98), .ZN(n119) );
```

```
17 INV1 U29 ( .I(n99), .ZN(n118) );
18 INV1 U30 ( .I(n100), .ZN(n117) );
19 INV1 U31 ( .I(n101), .ZN(n116) );
20 INV1 U32 ( .I(n102), .ZN(n115) );
21 INV1 U33 ( .I(n103), .ZN(n114) );
22 INV1 U34 ( .I(n104), .ZN(n113) );
23 INV1 U35 ( .I(n105), .ZN(n112) );
24 NR2D1H U36 ( .A1(rst), .A2(n108), .ZN(n105) );
25 NR2D1H U37 ( .A1(ini_3), .A2(rst), .ZN(n104) );
26 NR2D1H U38 ( .A1(rst), .A2(n109), .ZN(n103) );
27 NR2D1H U39 ( .A1(ini_2), .A2(rst), .ZN(n102) );
28 NR2D1H U40 ( .A1(rst), .A2(n110), .ZN(n101) );
29 NR2D1H U41 ( .A1(ini_1), .A2(rst), .ZN(n100) );
30 NR2D1H U42 ( .A1(rst), .A2(n111), .ZN(n99) );
31 NR2D1H U43 ( .A1(ini_0), .A2(rst), .ZN(n98) );
32 XOR2D1 U44 ( .A1(n107), .A2(n97), .Z(n121) );
33 XOR2D1 U45 ( .A1(cnt_0), .A2(cnt_1), .Z(n122) );
34 DFCSN1 cnt_reg_3 ( .D(n120), .CP(clk), .SDN(n112), .
35 CDN(n113), .Q(cnt_3), .QN(net5) );
36 DFCSN1 cnt_reg_2 ( .D(n121), .CP(clk), .SDN(n114),
37 .CDN(n115),     .Q(cnt_2), .QN(n97) );
38 DFCSN1 cnt_reg_1 ( .D(n122), .CP(clk), .SDN(n116),
39 .CDN(n117),     .Q(cnt_1) );
40 DFCSN1 cnt_reg_0 ( .D(n96), .CP(clk), .SDN(n118),
41 .CDN(n119), .Q(cnt_0), .QN(n96) );
42 endmodule
```

The schematic view of the 4-bit counter design is depicted in Figure 2.6. This illustrates the numerous logic gates deployed for combinational addition functions and the four flip-flops serving as sequential memory elements.

Similar to the output from FPGA synthesis, the ASIC synthesis report summarizes the resource usage in terms of the number of ports and cells, along with the total cell area. For instance, the synthesis report displayed below indicates that a design requires 10 IO ports, 36 nets which represent the connections between the different components, and 25 cells which represent the basic building blocks (e.g., logic gates, flip-flops).

```
1 //Synthesis Report
2 Library(s) Used:
3 tcb773stc (File: ./synthesis/tcb773stc.db)
4 Number of ports: 10
5 Number of nets: 36
6 Number of cells: 25
7 ......
```

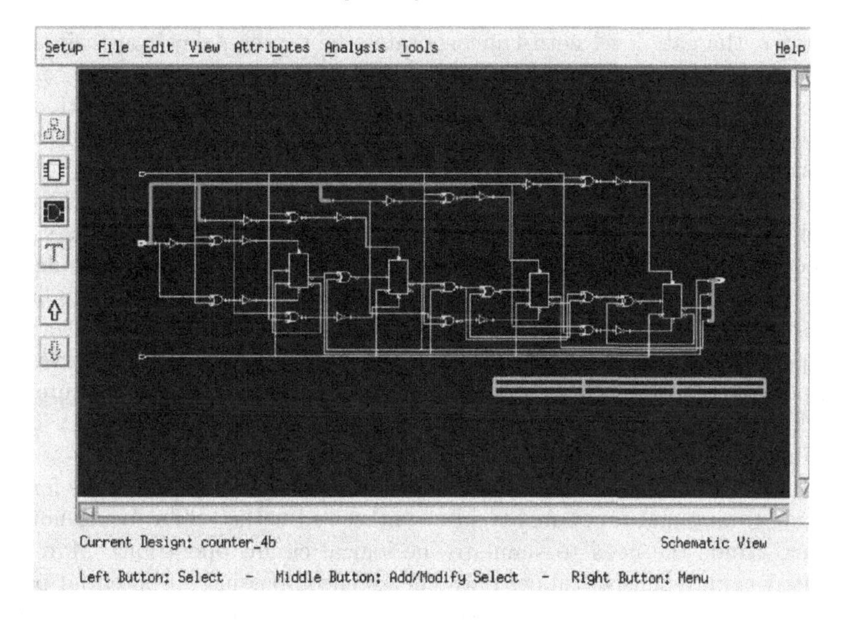

FIGURE 2.6
The schematic view of ASIC synthesis.

2.8.3 EDA tools for ASIC synthesis

ASIC synthesis is carried out using synthesis tools that employ standard cell libraries, constraints, and primarily RTL code to generate a gate-level netlist. Synopsys Design Compiler stands as a prominent ASIC synthesis tool widely embraced in the industry. Moreover, open-source toolchains such as OpenROAD are integral components of the broader initiative to advance open-source IC design [1]. OpenROAD specifically concentrates on developing a fully autonomous toolchain for digital SoC synthesis and layout generation, contributing to the open-source ecosystem.

FPGA verification plays a crucial role in ensuring design functionality; however, it does not necessitate putting the ASIC synthesis process on hold until FPGA verification is completed. In practice, ASIC synthesis and FPGA verification are often carried out simultaneously to overlap the design cycle.

2.9 Pre-simulation and static timing analysis

2.9.1 Pre-simulation vs. static timing analysis

RTL design primarily focuses on behavioral model level descriptions, without including timing information associated with logic cells and registers. In contrast, the gate-level netlist incorporates the timing delay/constraints and resource utilization, including IO ports, logic cells, registers, and wire connections. The objective of timing analysis is to preserve the same functionality between the netlist and the RTL design, with the added inclusion of timing information provided by the cell library.

To verify the functional equivalence between the RTL design and the gate-level netlist, therefore, a **pre-layout simulation** is performed using simple direct tests due to its slower nature compared to RTL design simulation. This simulation, known as **pre-simulation**, takes place before the chip layout phase. Additionally, **static timing analysis (STA)** is an essential process for performing critical timing checks of the synthesized netlist. It serves the purpose of confirming that the design functions accurately within the intended clock frequency, and that data propagation throughout the IC occurs without infringing upon setup and hold time limitations.

Pre-simulation and STA stand as two widely employed approaches for validating timing and functionality in a synthesized netlist. STA, being notably faster, avoids the need to simulate the logical circuit operations. Moreover, it offers comprehensive timing path checks encompassing all potential paths, not solely the logical conditions triggered by specific test vectors. Nevertheless, while STA excels at timing verification, it cannot assess the functionality of a circuit design, a task better suited for pre-simulation.

2.9.2 Pre-simulation example

A. Creating and Annotating SDF File for Pre-Simulation

To facilitate the pre-simulation incorporating delay information of the logic cells and registers, it is necessary to create a file containing these timings when executing the ASIC synthesis. This file format is known as **standard delay format (SDF)** and can be generated using the command *write_sdf* in the synthesis scripts. As an example below, it can generate an SDF file named "counter_4b.sdf" by executing the command in line 2: "write_sdf counter_4b.sdf" during the operation using Synopsys Design Compiler.

```
1  //Create SDF file during Design Compiler
2  write_sdf counter_4b.sdf
3
4  //SDF Annotation in testbench
5  $sdf_annotate("./counter_4b.sdf",tb)
```

To enable the pre-simulation in the verification environment, furthermore, the SDF file needs to be included in the testbench using the command shown in line 5: "$sdf_annotate("./counter_4b.sdf",tb)". This step is commonly referred to as **annotation**. Once the annotation is complete, the netlist can be subjected to pre-simulation on simulators.

During pre-simulation, several considerations come into play to build the verification environment. For instance, if the functionalities of third-party IPs are required, simulation models can be utilized to replace the black boxes, allowing their inclusion in the simulation. Moreover, it is important to note that the timing of the clock and reset signals can be disregarded during pre-simulation since the clock tree will be utilized in the future layout phase, handling the timing aspects. In cases where the fanouts of the clock and reset paths are excessively high, it may be necessary to insert an empty module to replace these paths. This is because a single path may be unable to drive a large number of fanouts, necessitating the use of additional buffering or dividing the fanout to ensure proper functionality.

B. Netlist Pre-Simulation Waveform vs. RTL Simulation Waveform

An example of pre-simulation waveform is depicted in Figure 2.7(a), generated using Siemens ModelSim. It is evident from the waveform that the output signal "cnt" from the register undergoes a bit delay following the rising clock edge. In contrast, the RTL simulation waveform presented in Figure 2.7(b) does not exhibit such a delay. This distinction arises because the RTL simulation stage lacks the delay information of clock-to-register output, unlike the pre-simulation stage.

2.9.3 Static timing analysis and timing failures

A. Timing Check Failures and ECO

Timing check failures can arise from various factors. A frequent cause is the existence of a lengthy data path through which signals traverse multiple logic elements, eventually falling within the setup window of registers. This situation can lead to setup time violations, potentially causing metastable signals to enter registers and consequently affecting the design's functionality. To

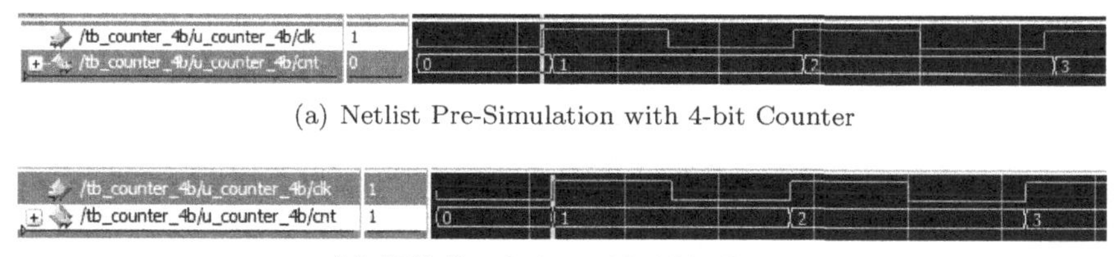

(a) Netlist Pre-Simulation with 4-bit Counter

(b) RTL Simulation with 4-bit Counter

FIGURE 2.7
Netlist pre-simulation vs. RTL simulation.

address this problem, one effective approach is to partition the extended data path into multiple shorter segments, thereby reducing the length of each individual path. In addition to setup time violations, timing issues can also arise in the form of hold time violations. These occur when the timing delay is insufficient for the signal to remain stable after the active clock edge. Detailed discussions on timing constraints and asynchronous design rules are presented in Chapter 11.

Timing check failures can necessitate a return to RTL coding to rectify timing violations. Given the time-to-market pressures inherent in IC design projects, it's essential to streamline the debugging process. This often leads to the utilization of the **engineering change order (ECO)** process during this phase of the IC design cycle. In general, ECO means that designers can sometimes make modifications directly to the gate-level netlist to rectify design and timing issues, instead of making corrections back to the RTL code. Utilizing ECO can significantly reduce the time-to-market, making it an appealing solution for RTL designers. However, it is important to note that the netlist can be complex and challenging to comprehend, increasing the fabrication risk associated with ECO.

B. Static Timing Analysis and EDA Tools

Pre-simulation involves simulating the design at the gate-level netlist before the physical layout is generated. Like the RTL simulation, it uses testbenches and input stimuli to stimulate the netlist and check its logical correctness. STA, on the other hand, is a method of validating the timing performance of the synthesized netlist by checking all possible paths for timing violations. It breaks a design down into timing paths, calculates the signal propagation delay along each path, and checks for violations of timing constraints inside the design and at the IO interface.

Synopsys Primetime is one of the most popular tools used for STA in IC industry. It operates by performing several steps: reading synthesized netlist along with timing constraint files; building timing model with delay information of each gate, register, and net in the design; analyzing all paths from each and every start point to each and every end point and compares it against the constraint that exists for that path; and providing detailed reports indicating timing violations, critical paths, slack values (how much time a path can be delayed without causing a violation), and overall design performance.

2.10 Layout

2.10.1 What is layout?

After pre-simulation and timing checks on the synthesized netlist, layout engineers move on to the chip layout phase. This step entails the transformation of the gate-level netlist into a detailed physical representation through the use of **automated placement and routing (APR)** techniques. The layout process encompasses the following key responsibilities:

- **Placement:** All VCC and GND paths, design blocks, IPs, IO pads around the chip layout, and clock and reset trees are strategically positioned within the chip area, considering the design constraints. The placement ensures optimal utilization of the available space and minimizes signal delays.

  Figure 2.8 illustrates the clock and reset trees both before and after the layout process. During the synthesis process, third-party IPs within the netlist are treated as black boxes, representing their functional behavior without specifying their physical implementation. Similarly, the clock and reset trees are not explicitly represented in the netlist.

  In contrast, in the layout phase it becomes necessary to insert all the IPs into suitable locations on the chip, while also incorporating the clock and reset trees into the chip layout. It is evident that the tree after layout typically exhibits increased fanouts and longer signal paths compared to the tree used in the pre-simulation stage.

- **Routing:** After the placement phase is completed, engineers proceed to establish connections between the placed components, forming a comprehensive chip-level network. The routing process entails identifying the most efficient signal paths while adhering to timing, power, and area constraints.

- **GDSII File Generation:** The output of the placement and routing process is a file in the **graphic data system II (GDSII)** format. This file contains a detailed description of the chip layout, including the positions and interconnections of all the components. The GDSII file serves as the final representation of the chip's physical design.

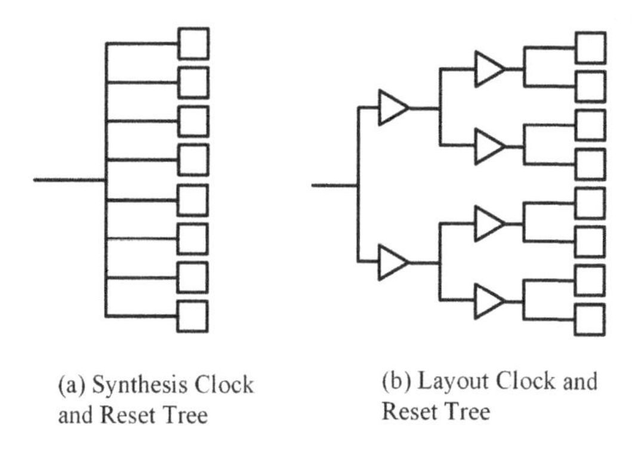

FIGURE 2.8
Clock and reset tree.

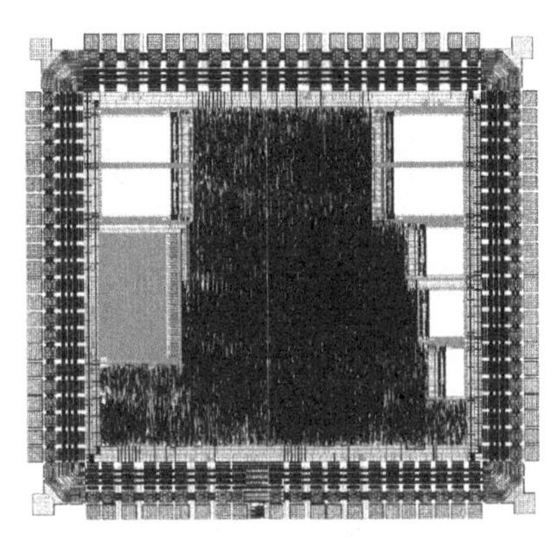

FIGURE 2.9
Layout schematic.

2.10.2 Pad-limited and core-limited chip size

An example of the layout schematic is depicted in Figure 2.9. The final size of the chip can be determined in two distinct ways, referred to as **pad limited** or **core limited** scenarios.

In the case of core-limited sizing, the chip size is primarily determined by the gate count or the overall complexity of the core circuitry. On the other hand, in the pad-limited scenario, the chip size is primarily influenced by the number of IO connections or the amount of peripheral circuitry. By identifying the dominant factor, the layout engineer can ensure that the chip's dimensions are optimized accordingly to accommodate either the core circuitry or the IO connections effectively.

Overall, the layout process plays a crucial role in translating the gate-level netlist into a physically implementable design, ensuring that the chip meets performance, power, and area requirements while considering the constraints imposed by the design specifications.

2.11 Post-Simulation and Post-STA

Following the layout phase, a simulation is conducted using the updated netlist to accommodate any netlist changes. This phase is referred to as **post-layout simulation** or **post-simulation**. The post-simulation waveform encompasses wire delays and the delays introduced by placed and routed cells and registers. Concurrently, **post-layout STA**, or **post-STA**, is carried out. During this stage, post-STA is performed using the layout information. This entails considering not only the logical gate delays but also the physical attributes of the layout, including wire lengths, parasitic capacitances, and resistances. Such an approach offers a more precise evaluation of the design's timing behavior.

2.11.1 Post-simulation

A. Creating and Annotating SPF File for Post-Simulation

To calculate the wire delay accurately, the parasitic capacitance and resistance resulting from the layout must be taken into consideration. This information is typically provided in a file called the **standard parasitic format (SPF)**. Similar to the pre-simulation, the SPF file needs to be annotated into the test-bench in order to perform post-simulation in the verification environment. By incorporating the SPF file, the post-simulation waveform takes into account the influence of parasitics on the overall design performance.

B. Post-Simulation Timing

During the post-simulation stage, clock skew can be detected by analyzing the waveform. Clock skew refers to the variation in arrival times of the clock signal at different components within the design circuit. Identifying and addressing clock skew is crucial for maintaining proper synchronization and timing accuracy.

It is possible to encounter post-simulation failure due to timing violations caused by placement and routing. Some violations may be insignificant and do not affect the overall functionality of the design. In such cases, they can be safely ignored. However, significant violations need to be addressed to ensure proper operation and adherence to the design specifications. Detailed discussions on clock skew and timing constraints will be presented in Chapter 11.

Overall, post-simulation provides valuable insights into the timing characteristics and design performance, taking into account the effects of layout-related parasitics. It helps identify potential issues such as clock skew and timing violations, allowing for necessary adjustments and optimizations to achieve a robust and reliable physical design.

2.11.2 Post-STA

Post-STA is conducted at the layout netlist level, where the physical arrangement of gates, wires, and other components on the chip is taken into account. This analysis considers factors like wire capacitance, resistance, and parasitic effects that can influence the design's timing. By incorporating these physical characteristics, post-STA provides a more precise assessment of the design's timing attributes based on its actual physical realization. This step is crucial to validate that the design's performance aligns with initial expectations, accounting for wiring delays and other physical phenomena.

In essence, STA is typically executed on the synthesized netlist for the initial timing assessment, while post-STA is carried out on the layout netlist to validate the timing once the design is physically implemented. Both forms of analysis are indispensable to ensure that the design fulfills its timing specifications and operates correctly.

2.12 DRC and LVS

Before proceeding to fabricate the chip, it is essential to perform **design rule check (DRC)** to ensure that the layout adheres to the specified design rules, which are necessary for flawless fabrication. These rules include criteria related to manufacturing-related constraints such as spacing and metal layer usage. The DRC process checks for violations of these rules and flags any design elements that do not meet the specified requirements. EDA tools such as Synopsys IC Validator and Siemens Calibre are widely used to provide comprehensive capabilities for DRC, including advanced checks and rule-based analysis.

Meantime, **layout versus schematic (LVS)** check is conducted to verify the accuracy of the layout by comparing it against the corresponding schematic representation of the design. The goal of LVS is to ensure that the physical implementation of the design matches the intended functionality captured in the schematic. It verifies that the same circuit topology, devices, connections, and electrical properties are present in both representations.

EDA tools like Synopsys IC Validator and Siemens Calibre offer comprehensive suites that encompass LVS verification capabilities. Designers can utilize these tools to verify the accuracy of layouts against schematics. This verification process necessitates input files, namely the graphic data system (GDS) layout file and the schematic netlist.

- GDS files have a distinct purpose in manufacturing: they furnish essential data required for fabricating actual silicon wafers. These files play a guiding role in the lithography process, ensuring that intricate patterns are precisely transferred onto the wafer, thereby crafting the tangible components of the circuit.

- In contrast, the schematic netlist typically emerges from a higher-level schematic design. This netlist functions as a benchmark against which the layout netlist is compared. While schematics have an established connection with analog circuit design, it's noteworthy that they can also be pertinent to specific facets of digital design and/or mixed-signal circuits.

Figure 2.10 illustrates the LVS verification flow. Initially, a netlist is extracted from the GDS layout file using a layout extraction step. Subsequently, the IC Validator NetTran utility is employed to generate an ICV netlist, which serves as the basis for comparison between the layout and schematic. This

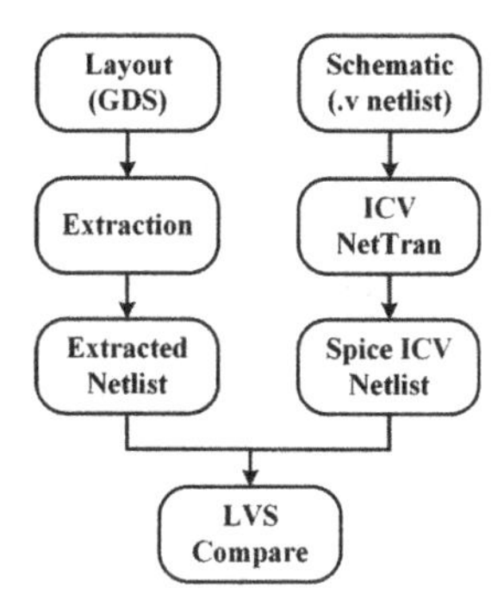

FIGURE 2.10
LVS flow.

utility also generates an equivalence point file, which aids in the comparison process. The extracted netlist is then compared with the schematic netlist, verifying various aspects such as the number of devices in the schematic versus the layout, the number of nets in the schematic versus the layout, and the types of devices in both the schematic and layout. Finally, result reports are generated, providing information on any discrepancies or inconsistencies identified during the LVS check.

2.13 Tape-Out

Once DRC and LVS verification have been successfully completed, the next and final step in the IC design flow is known as **Tape-Out** or **Tapeout**. This term originated in the 1970s when the design information was recorded onto a magnetic tape and sent to the foundries for manufacturing.

Tape-out represents the point at which the final design files, including the layout and associated data, are prepared and sent to the semiconductor fabrication facility, commonly known as **Fab**, for the chip fabrication process. Some of the well-known Fabs or foundries in the semiconductor industry include TSMC (Taiwan Semiconductor Manufacturing Company), Samsung Foundry, GlobalFoundries, UMC (United Microelectronics Corporation), SMIC (Semiconductor Manufacturing International Corporation), etc.

2.13.1 Chip tape-out procedure

Chip tape-out involves complex and sequential steps to produce complete electrical or photonic circuits on semiconductor wafers. While the detailed chip fabrication process is highly intricate and involves numerous steps, it generally includes processes such as photolithography, etching, doping, and many other specialized techniques. These processes are performed in cleanroom environments and require precision and expertise to ensure the successful creation of the final ICs.

It's important to note that the chip fabrication process itself falls beyond the scope of IC design flow, as it involves specialized manufacturing techniques and equipment. The focus of the IC design flow is primarily on designing and verifying the functionality, performance, and layout of the circuit. Therefore, the fabrication process is briefly introduced below:

- **Mask Creation:** Masks are created based on the layout design. Each layer of the design has its own corresponding mask.
- **Photolithography:** As shown in Figure 2.11, the wafer is aligned with a mask using a photo aligner. Ultraviolet light is then passed through the mask, exposing the wafer to create patterns. This step is known as photolithography. The term "photolithography" combines "photo" (light), "litho" (stone), and "graphy" (print), representing the process of "printing with light". In short, photolithography transfers geometric patterns from the mask to a photosensitive chemical photoresist on the substrate.
- **Etching:** The etching process selectively removes material from the wafer's surface that is not covered by the photoresist, creating the desired patterns.
- **Doping:** Doping involves altering the electrical characteristics of the silicon by introducing specific materials. For example, doping can involve introducing atoms with one less electron than silicon (creating P-type regions) or atoms with one more electron than silicon (creating N-type regions). This process allows for the creation of P-type and N-type regions in the silicon.
- **Repetition:** The above steps are repeated for each layer, sequentially building up the complete circuit structure.

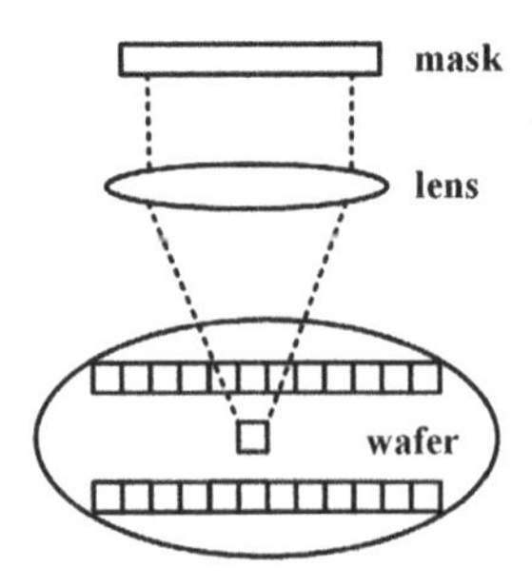

FIGURE 2.11
Photolithography.

2.13.2 MPW/shuttle and full-set

A. Chip Tape-Out Cycle and Cost

The period between tape-out and chip readiness is commonly referred to as the **non-recurring engineering (NRE)** phase. This pivotal stage typically spans several months. It involves substantial costs, influenced by factors such as the chosen technology and the complexity of the design. Reflecting on the fabrication costs outlined in Chapter 1, adopting 16–24 nanometer Global Foundry technology for a square millimeter of chip real estate can result in expenses ranging from tens of thousands to millions of US dollars.

Consequently, a meticulous design flow is imperative to ensure a successful outcome. This entails rigorous design verification, timing analysis, design rule checks, and manufacturability assessment. Any flaws or deficiencies in the design can result in costly setbacks, necessitating revisions or potentially leading to the complete failure of the chip fabrication process. Such failures can have repercussions beyond financial implications, including delays in bringing the final product to market, which could result in missed opportunities and vital deadlines.

B. MPW/Shuttle

It is not possible for IC design teams to provide an absolute guarantee of complete chip verification with 100% certainty throughout the design flow. Therefore, to address the risks associated with tape-out, many foundries provide the **multi-project wafer (MPW)** service as a means of mitigation. This service allows different chips or projects from various design teams or companies to share the same wafers, reducing individual tape-out risks. The MPW arrangement facilitates cost reduction by distributing the mask set price among multiple projects, making it more affordable for each participant. The MPW process is also known as a **Shuttle** in the industry.

C. Full-Set

Given various steps involved throughout the entire design flow, which can potentially introduce failures, the first tape-out is often done as an MPW/Shuttle run due to its lower cost compared to a formal NRE process. This preliminary step allows for early testing and identification of any issues, mitigating risks and enhancing the overall reliability of the final chip.

On the other hand, the subsequent tape-out following the MPW/Shuttle run is referred to as a **Full-Set**, representing a complete mask set for the specific design. The full-set tape-out typically occurs after functional testing and validation of the MPW/shuttle chips.

2.14 Chip Prob and Design for Test

2.14.1 Chip prob and tester

After chips have undergone the tape-out process, the next step in the semiconductor manufacturing flow is testing, which is carried out using specialized testing equipment known as testers. The primary purpose of a tester is to assess the functionality of the semiconductor chip before it is packaged and mounted on a printed circuit board (PCB). Specifically, a tester performs various testing steps, including:

- **Powering on the Chip:** The tester applies power to the chip, allowing it to operate under normal operating conditions.
- **Resetting the Chip and Sending Input:** The tester resets the chip to a known state and sends prepared input data to the chip. These input patterns are typically stored in memory cards.
- **Observing and Comparing Output:** The tester observes the output of the chip and compares it with the expected result. This comparison ensures that the chip is functioning correctly and producing the desired output.
- **Identifying Mismatches:** If any mismatches occur between the observed and expected results, it indicates potential functional errors or discrepancies in the chip's operation.

More details related to chip testing are shown in Figure 2.12. The tester consists of several insertion cards, each of which contains multiple channels. Each channel can serve as either an input or output probe, allowing the tester to send data patterns into the chip or receive data patterns from the chip during the testing process. The input and output patterns used for testing are stored in the memory cards associated with the tester. These patterns serve as test vectors, providing specific data sequences to evaluate the chip's performance and functionality.

On the right hand side of the figure, the probe card is depicted for connecting the channels of the tester with the individual chips on a wafer. The probe card is designed to be movable, allowing it to be positioned and connected to different chips for testing purposes. The pins on the probe card make contact with the pads on the chip, enabling the transmission and reception of input and output patterns.

2.14.2 DFT and ATPG

A. Color-Coding for Tested ICs

Multiple sets of test patterns are provided for testing purposes, with each set specifically targeting different functions. It is possible for chips to pass certain sets of tests while failing others. Although they may fail in some function tests, these chips can still be valuable for specific applications.

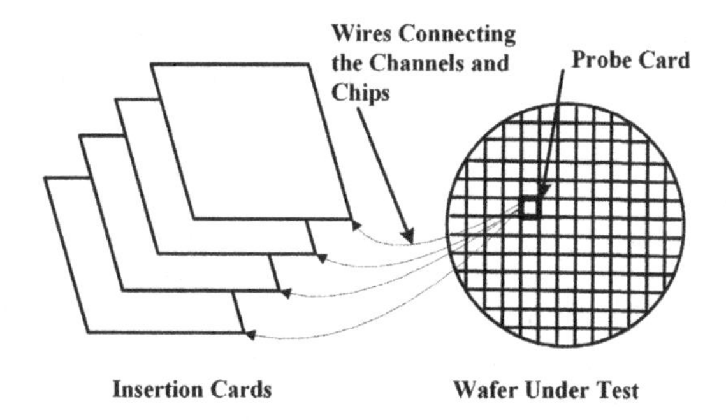

FIGURE 2.12
Chip probe and tester.

Testing results are documented by the tester, associating them with a unique serial number on the wafer. To distinguish chips with different test results, they are marked with different colors. The recording of testing results and the use of color-coding assist in identifying the performance of individual chips and their suitability for specific purposes. This information aids in making informed decisions about chip utilization based on their testing outcomes.

B. Testing Coverage and DFT

The challenge of achieving functional coverage in testing cells and nets with test patterns is significant. Typically, to ensure thorough testing, the toggle rate of nets, transitioning from binary zeros to ones and from ones to zeros, should reach around 70–90%.

To aid in generating test patterns, additional circuits can be incorporated into the chip. These circuits often include scan chains and scan flip-flops. The process of generating test patterns is referred to as **auto test pattern generation (ATPG)**, and the knowledge and techniques involved in test pattern generation are collectively known as **design for test (DFT)**.

2.15 Packaging and Testing

After the chip probing process, the wafer is ready to be cut along the scribe lines. Defective chips are identified by marked ink and will be discarded, while the functional ones will proceed to the **packaging** stage. To optimize costs, only a small number of pads around the chip are led out as pins, with some pads reserved specifically for testing purposes.

Once the chips are packaged, they undergo another round of testing. Many of the test patterns from the chip probing stage can be reused. However, since not all the testing pads are accessible, some test patterns may not be utilized during this stage. This testing process is known as the **final test**. The same tester is used, but the probe card is replaced with a socket on the handler. The IC is placed onto the socket and tested. After the IC is mounted in the actual system, it may undergo further testing in its real-world environment.

By following this approach, designers can minimize the likelihood of unexpected failures in the deployed ICs and ensure the highest possible quality and performance in real-world applications. The thorough testing at different stages of the design and manufacturing process contributes to the overall success and reliability of the IC.

In conclusion, this section outlines the key front-end design steps and offers a high-level overview into the subsequent back-end processes, encompassing chip tape-out, testing, and packaging. However, it's crucial to acknowledge that IC fabrication can entail significantly more intricate and detailed procedures. The IC industry is in a perpetual state of evolution, with emerging design methodologies, tools, and manufacturing technologies constantly shaping the design flow. This book aims to furnish readers with a foundational framework for understanding IC design steps, with a central emphasis on introducing the front-end aspects of IC design.

2.16 An Example of IC Design Flow

Figure 2.13 presents an outline of a standard ASIC design cycle, covering approximately one year, using a RISC-V design project as a specific example. It offers insights into the essential design team and resources required for a project of moderate size. It's crucial to acknowledge that different teams or companies may have diverse requirements and considerations that could affect the timeline and team sizes. The example provided aims to depict a typical scenario and may not align precisely with every company's specific needs.

2.16.1 Front-end phase

A. Front-End Phase and Timeline

In the front-end, the design cycle commences by creating design specifications for all the design modules, encompassing system architectures and various submodules for the system integration. This phase spans around 40 days. During this stage, designers provide essential hardware design details, including design block diagrams, timing diagrams, design IOs, and other necessary information, in preparation for RTL coding.

The subsequent phase entails approximately 35 days dedicated to RTL design and simulation for all submodules, followed by about 15 days for in-

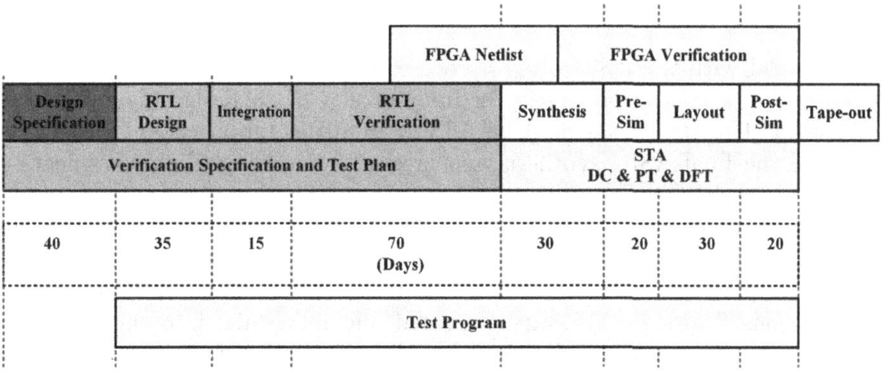

FIGURE 2.13
An example of ASIC design cycle.

tegrating all submodules together as the RISC-V top module. During the RTL coding, testbenches and test plans are prepared concurrently to facilitate functional verification of each submodule as well as the final integrated RISC-V.

The RTL design for submodules and the system integration together span around 50 days, while the comprehensive functional verification phase extends over 120 days, overlapping with the RTL design and integration cycle. As the design code and testbenches are being developed, the design and verification specifications must be regularly updated to align with the progress and changes in the RTL programs.

B. Front-End Design Team

Throughout the RTL design, integration, and verification stages, designers and verification engineers work in parallel to create the RTL code and corresponding verification environments. The functional verification phase, which encompasses comprehensive test cases and testbenches, is a critical component within the ASIC design cycle. It is conducted throughout the entire design phase and requires a team size of approximately 3–5 times the number of RTL designers. For example, a typical team composition for a design project of this scale may consist of around two designers and six verification engineers. The increased number of verification engineers compared to designers highlights the importance and complexity of verification activities.

By conducting thorough functional verification, designers and verification engineers can significantly reduce the risk of design functional flaws, ensuring that the RTL design meets the specified requirements and functions correctly. This extensive verification process plays a pivotal role in enhancing the overall quality and reliability of the ASIC design.

C. FPGA Verification

In parallel with the ASIC design process, FPGA implementation and verification can be conducted to test the functionality of the design in a real-world scenario. The RTL code used for FPGA implementation should closely resemble the final ASIC synthesis and layout to ensure a reliable verification between the FPGA prototype and the intended ASIC implementation.

However, due to resource limitations on the FPGA board, certain design modules may need to be adjusted or replaced by IP blocks to accommodate the system design within the FPGA platform. For example, in FPGA implementations, Block RAMs can be created and integrated into the design to replace the register files that were originally described in RTL code. This substitution can help reduce the utilization of register slices in the FPGA and can lead to more efficient utilization of available resources. In some cases, large-scale projects may require the use of multiple FPGA platforms to establish the complete system.

It's important to note that FPGA implementation and verification are not always mandatory and their inclusion in the design flow depends on project-specific requirements and constraints. The decision to incorporate FPGA implementation and verification is typically made based on specific projects and the level of confidence in the RTL verification results such as functional and code coverage.

2.16.2 Back-end phase

A. Back-End Phase and Timeline

Once functional verification is complete, the ASIC design flow progresses to the back-end phase, which encompasses gate-level netlist generation, chip layout, and timing-related verification. Specifically, the ASIC synthesis (with Synopsys Design Compiler, or DC in the figure), which involves converting the RTL code into gate-level netlist and timing analysis, typically takes about 30 days. Simultaneously, pre-simulation and static timing analysis (with Synopsys PrimeTime, or PT in the figure) are conducted to ensure the equivalence between the synthesized netlist and the RTL design.

Following the synthesis stage, the layout process takes place for an additional 30 days, which includes the placement and routing of cells, registers, IPs, and the consideration of wire delay. This step is crucial for optimizing the physical layout of the chip and ensuring proper signal propagation. Once the layout is complete, post-simulation and post static timing analysis are performed to verify the netlist after layout, ensuring that the functional behavior of the design is maintained.

B. Back-End Design Team

The back-end phase encompasses a span of approximately 100 days and concludes with the tape-out of the chip. To effectively execute the back-end phase, a team of ten engineers with specialized skills in synthesis, layout, and static timing analysis is required for this project. These engineers possess the expertise to optimize the design, ensure proper placement and routing of cells, registers, and IPs, and address any timing-related concerns. Their efforts contribute to the successful completion of the back-end phase and pave the way for the subsequent manufacturing process.

During the back-end phase, verification engineers carry out pre-simulation and post-simulation activities using the same verification environment established during the functional verification in the front-end. These activities involve employing direct testing methods to validate the design. The focus of the test patterns is to traverse the longest data path within the system design to ensure the timing checking over critical paths. Pre-simulation is performed to verify the equivalence between the synthesized netlist and the RTL design, ensuring that the functionality of the design is maintained. Post-simulation, on the other hand, is conducted to verify the netlist after layout, taking into account the impact of wire delay and assessing the overall design performance. Once the back-end phase concludes, the IC design flow is considered complete, and the focus shifts to tape-out and subsequent chip packaging and testing processes.

2.17 Basic Qualifications Needed For IC Design and Simulation

This section provides a summary of the key qualifications needed by entry-level positions in the front-end IC design. These qualifications are outlined in Table 2.1. It is important to note that the information presented may vary depending on specific company/team requirements and job roles.

2.17.1 RTL design and simulation

As chip design engineers working in the front-end, it is essential to have a strong understanding of the ASIC/FPGA design flow, as depicted in the second row of Table 2.1. Proficiency in either Verilog or VHDL is a prerequisite for designing and simulating digital circuits using HDLs. The selection between Verilog and VHDL is contingent on project-specific requirements and the preferences of the company.

In addition to HDL proficiency, it's essential to be skilled in using code editors such as Vim, Emacs, and/or WinEdit for writing design code and testbenches. Given that ASIC design and simulation frequently take place on Linux/Unix operating systems, a good grasp of basic Linux/Unix commands is also a prerequisite. Furthermore, scripting languages are pivotal in constructing the verification/simulation environment, so possessing basic coding skills in scripting languages like TCL, Perl, and/or Makefile is advantageous.

TABLE 2.1
Key qualifications of industry needs for entry-level roles in the front-end IC design.

Industry Needs	Details
Design Flow	ASIC and/or FPGA Design Flow
Design and Simulation	Verilog/VHDL
	Editor – Vim/Emacs/WinEdt
	Basic Linux Commands
	Scripting languages such as TCL, Perl, makefile, etc.
Verification (Simualtion)	Development of test plans, testbenches, verification (simulation) environments, bus functional models, monitors, test cases, etc.
	Simulator – Synopsys VCS/Cadence NC/ Siemens ModelSim
FPGA Development	FPGA tool – AMD Vivado or Intel Quartus
	FPGA prototype and debug
Project Related	Bus architecture – AMBA AXI, etc.
	SoC peripherals such as SPI, I2C, UART, GPIO, SDIO, Timer/Watchdog, etc.
	RTL IP, Vivado/Quartus IP, and design reuse and integration
	Basic designs components and design rules, such as clock domain crossing (CDC), FIFO, DMA, bus bridges/wrappers, etc.
	Timing constrains and MOF

In this book, tutorials on the Vim editor, TCL scripting for RTL simulation, and fundamental Linux/Unix commands are provided in **Appendix A**.

2.17.2 Functional verification (simulation) and FPGA verification

Simulation experience is a key requirement for IC design and verification positions, encompassing the development of test plans, simulation environments, bus function models, scoreboards/monitors, test cases, etc. Proficiency in using simulators is essential for running simulations at the RTL and netlist levels.

The tutorial of Siemens ModelSim is provided in **Appendix B**. Furthermore, experience in FPGA prototyping is valuable, allowing for the rapid validation and exercise of design netlists on real hardware. The AMD Vivado tutorial is provided in **Appendix C**.

2.17.3 Industry design rules and design standards

In order to equip readers with practical and industry-relevant skills, it is vital to incorporate projects that align with industry standards and protocols. For an entry-level position, project experiences should cover fundamental design blocks and design rules, including topics like clock domain crossing (CDC), finite state machines (FSMs), synchronous/asynchronous FIFOs, DMA, bus bridges/wrappers, design constraints pertaining to achievable clock frequency, timing considerations, and more.

Moreover, a crucial aspect involves the in-depth exploration of widely adopted industrial protocols, such as the AMBA AXI architecture, along with pivotal bus peripherals like I2C, SPI, UART, SDIO, GPIO, timer, and watchdog. By incorporating these components, readers can achieve a holistic grasp of fundamental IP-level design concepts utilizing commonly used FSM-Datapath design structures, while also comprehending the intricacies of data transfer across SoC buses.

Furthermore, emphasizing SoC integration in both ASIC and FPGA implementations is crucial for entry-level chip design positions. This encompasses areas such as IP creation, reuse, and integration on ASIC/FPGA platforms. Possessing a solid understanding of the SoC integration process and the ability to incorporate IPs into a larger system design with the AMBA AXI bus are valuable skills for IC designers and verification engineers.

2.17.4 Book introduction

This book primarily focuses on the RTL design, verification, and FPGA evaluation with Verilog HDL. It covers the design flow including seven key steps: design specification, IP-level design using Verilog HDL, IP verification using Verilog HDL, SoC integration, SoC verification, FPGA netlist generation, and FPGA verification and evaluation. Throughout the book, the Vim editor, Siemens ModelSim, and AMD Vivado will be utilized as the development tools.

In addition to Verilog HDL, readers will also receive an introduction to fundamental Linux/Unix commands and TCL scripting. These skills are instrumental in setting up testbenches and building automated simulation environments. This foundation is essential for comprehending the automated verification process, empowering readers to develop and expand their own test plans, test cases, testbenches, and FPGA prototypes for debugging and other purposes.

The book further includes a wide range of industry-centric projects, providing detailed insights into the complexities of designing and simulating SoC interfaces and peripherals, algorithmically driven components, implementations of numerical accelerators, and more. By thoroughly exploring these projects, readers will gain practical, hands-on experience in constructing and validating intricate IC designs. The primary goal of this book is to thoroughly prepare readers for careers as entry-level IC designers and verification engineers, equipping them with the skills needed to meet the evolving demands of the IC design industry.

3

Introduction to Verilog HDL

Chapter 3 introduces the Verilog hardware description language (HDL), covering the fundamentals of synthesizable hardware descriptions using Verilog. project-based learning (PBL) 1–3 demonstrate the design and simulation of basic combinational circuits.

3.1 Background of Verilog HDL

Designing integrated circuits (ICs) with a significant number of transistors posed a daunting challenge in the past, as it necessitated a bottom-up design approach, starting from individual transistors and gradually progressing to complex logic functions. This process was intricate, time-consuming, and prone to errors. However, the advent of HDLs in the 1980s transformed the field of IC design. HDLs facilitate the description of intricate digital circuits at a higher level of abstraction known as the register-transfer level (RTL). Leveraging powerful electronic design automation (EDA) tools, these RTL descriptions can be further translated into gate-level netlists and physical layouts.

Two widely used HDLs, Verilog and VHDL, play a significant role in this transformation. Verilog HDL, in particular, resembling the C programming language, gained widespread adoption among IC companies, making it the focus of this book. In 1995, Verilog HDL was standardized as IEEE 1364, which facilitated IC design and simulation at the register-transfer level of abstraction. A revised version was published in 2001 and has become the most widely used version among Verilog users. Although a further revision was released in 2005, it introduced only a few additional features. Another significant HDL, SystemVerilog, is a substantial extension of Verilog, extensively used for verification and as an extension to IC design. It was first standardized by IEEE in 2005.

While the complete and authoritative definition of Verilog HDL can be found in the Language Reference Manual published by IEEE, this book takes a more focused approach. Instead of covering the entire Verilog syntax, it introduces a small subset of the Verilog standard. The emphasis lies on synthesis constructs using Verilog HDL and some simulation constructs to build a

DOI: 10.1201/9781003187080-3

typical testbench. The primary objective is to prepare entry-level IC designers and/or verification engineers, providing them with the essential knowledge to embark on their roles confidently.

3.2 An overview of Verilog HDL

Starting this section, it's crucial to emphasize that Verilog doesn't function like a software programming language; instead, it serves as a means to describe hardware design structures, interfaces, and hardware behavior. Each synthesizable statement in Verilog carries substantial implications for hardware performance, chip dimensions, speed, and power usage. Beyond its role in hardware design, Verilog also extends its utility to encompass simulation and synthesis domains. For simulation purposes in this book, Siemens ModelSim is employed, while AMD Vivado is utilized to showcase the RTL analysts circuit and the associated resource costs and design performance in FPGA implementations.

3.2.1 Starting with a Verilog design example

To initiate your journey with Verilog, let's take a look at a fundamental design – an example of a single-bit full adder – outlined below. In Verilog, the pivotal components for delineating a digital circuit are encapsulated within the *module – endmodule* construct. On the second line, we define the module's name as "full_adder" and enumerate the design's interfaces or ports within the subsequent parentheses. These ports encompass three single-bit inputs and two single-bit outputs, as depicted in lines 2–6. Each declared port is separated by a comma, except for the last one. For visual reference, Figure 3.1 illustrates the operations, encompassing the creation of a black box labeled "full_adder" and the specification of all design IO ports.

```
1  //Logic Design with Verilog-2005
2  module full_adder (input  a    ,
3                     input  b    ,
4                     input  cin  ,
5                     output sum  ,
6                     output cout);
7  assign sum  = a ^ b ^ cin                      ;
8  assign cout = (a & b) | (a & cin) | (b & cin);
9  endmodule
```

Step 1: Declare IO ports and design name within *module-endmodule*

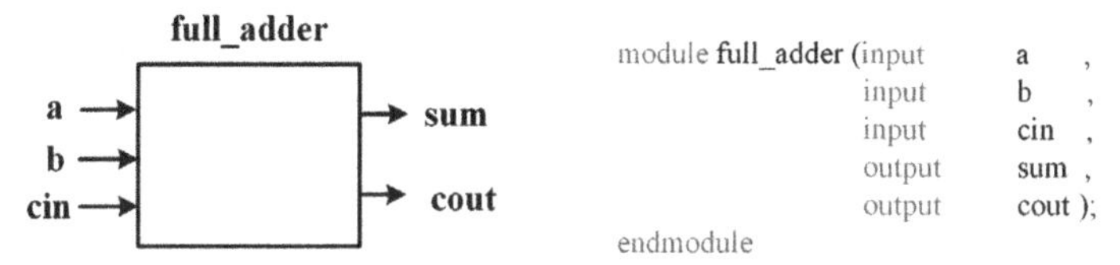

Step 2: Describe functions enclosed by the *module-endmodule*

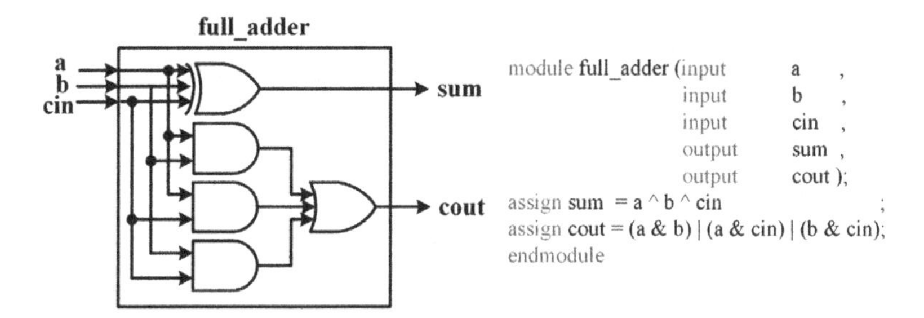

FIGURE 3.1
Hardware description procedure using verilog.

The next step is to describe specific functions including XOR, AND, and OR gates. In line 7, the "sum" output is defined as the result of the Boolean expression "a XOR b XOR cin", while in line 8, the "cout" output is defined as the result of the Boolean expression "(a AND b) OR (a AND cin) OR (b AND cin)". The *assign* keyword is one of the Verilog design blocks used to construct combinational circuits. It's important to note that each line of Verilog code must terminate with a semicolon, except for the last *endmodule*. For visual reference in Figure 3.1, the design code in line 7 is translated into a 3-input XOR gate, while line 8 is converted into a combined circuit that includes three 2-input AND gates and one 3-input OR gate.

This example provides a broad overview of a basic Verilog design. To summarize, a Verilog design begins with *module – endmodule* to name the design and declare its ports. Between *module* and *endmodule*, three types of behavioral blocks can be used to construct (and simulate) the hardware: *assign*, *always*, and *initial*. In this example, the *assign* block is used twice. Table 3.1 lists all three design blocks and the corresponding statements and basic operators that can be used within each block. Subsequent sections will delve deeply into the intricacies of each of these blocks, as expounded in Section 3.5. In these sections, the diverse array of basic operators and statements will be introduced, elaborated upon in Sections 3.4 and 4.1, respectively. For a comprehensive grasp of data types in Verilog, please refer to Section 3.3.

TABLE 3.1
Summary of Verilog HDL description.

<table>
<tr><th colspan="2">Blocks</th><th>Statements</th><th>Basic Operators</th></tr>
<tr><td rowspan="2">assign</td><td>concurrent assign</td><td>Basic Operators</td><td rowspan="5">~ (NOT),
& (AND),
| (OR),
∧ (XOR),
+ (Adder)</td></tr>
<tr><td>conditional assign</td><td>? :</td></tr>
<tr><td rowspan="2">always</td><td>level trigger always</td><td>if-else
case-endcase</td></tr>
<tr><td>edge trigger always</td><td>if-else</td></tr>
<tr><td colspan="2">initial</td><td>if-else
case-endcase
for/while/repeat/forever</td></tr>
</table>

3.2.2 Basic Verilog design syntax

Before commencing the first Verilog program, let's provide an overview of the fundamental syntax, covering aspects such as naming design signals and modules, coding styles, and the practice of adding comments.

- Module and signal names can be composed of characters, digits, and underscores, but they are case sensitive for characters. For example, "reset" and "Reset" are considered different signals. Additionally, signal names cannot start with a number. For example, "2add" is not a valid module or signal name.
- Avoid using keywords as module and signal names, such as *module*, *input*, *output*, *endmodule*, *begin*, *end*, *initial*, *always*, *assign*, *if*, *else*, *case*, and so on.
- Verilog ignores white space such as spaces and tabs, so it is helpful to vertically align the design code to make your program readable, concise, and easy to edit.
- Verilog comments can be single-line or multi-line. Single-line comments are denoted by "//" and multi-line comments are enclosed between "/*" and "*/".

3.2.3 Recommended coding format

Just like in other programming languages, implementing a consistent coding format in Verilog programs offers a range of benefits. These advantages include, but are not limited to, improving code comprehensibility, simplifying the management of large projects, aiding in the identification and correction of bugs, enhancing teamwork efficiency, and more. Verilog code is crafted and managed by diverse teams engaged in the IC design process, including design, verification, synthesis, layout groups, and engineers. Through adherence to standardized coding practices, these teams can cultivate enhanced collaboration, streamline workflows, and attain heightened consistency and reliability in their endeavors.

General Verilog Coding Styles

1. It is highly recommended to use vertical alignment in Verilog code for improved readability and consistency. Align assignment operators = and/or <=, commas to separate IO ports, semicolons at the end of each line, and the first letters of inputs and outputs. Utilizing the Vim editor's visual block editing feature can greatly enhance coding efficiency and overall experience when writing Verilog code in a vertically aligned manner.
2. It is advisable to list all inputs and outputs on separate lines, with each line dedicated to a single port. This practice makes the port declarations more readable and facilitates design instances and connections.
3. To enhance code clarity and understanding, use meaningful names for signals. For instance, consider using "en" for enable, "clk" for clock, and "rst" for reset.
4. To maintain consistency, ensure that the file name, module name, instance name, and testbench name are related. For instance, if the design module is named "full_adder", the file name should be "full_adder.v". When instantiating the design, use the related name such as "u_full_adder" by adding "u_" (short form of unit) in front. If you need to instantiate the design multiple times, use "u1_", "u2_", and so on, to maintain a coherent and systematic naming convention. When creating a testbench, use the related name such as "tb_full_adder" by adding "tb_" (short form of testbench) in front.
5. At the outset of the Verilog text, include essential comments like the design title, description, author, date, version, and revision history. Providing this information helps in identifying and understanding the purpose and context of the design.
6. Within the design code, incorporate as many comments as possible to enhance comprehensibility and maintainability. Meaningful comments clarify the code's functionality, making it easier for both the original developer and others to comprehend and modify the code effectively.

3.3 Verilog Data Types

3.3.1 Verilog constant

Verilog HDL utilizes a value set consisting of four basic values: 0 representing a logic zero; 1 representing a logic one; X representing an unknown logic value; and Z representing a high-impedance state for a tri-state buffer.

To represent a Verilog constant, you can use the format $N'B$, where N denotes the number of bits and B denotes the base. The base can be b for binary, d for decimal, o for octal, and h for hexadecimal. For single-bit constants, the binary format is recommended, while for multi-bit signals, the hexadecimal format is preferable. In designs involving counters and timers, the decimal format is commonly used.

Table 3.2 presents examples to illustrate the Verilog representation of hardware constants. The first example in the table showcases a 3-bit data 3'b101 (size: three, base: binary), equivalent to the decimal 3'd5. The data stored in memory is represented in binary as 3'b101. In the third row, the table illustrates another example of an 8-bit constant 8'b11 (size: eight, base: binary), which is equivalent to the decimal 8'd3. The data stored in memory occupies eight storage bits with a binary value of 3'b00000011.

To enhance readability for lengthy numbers, you can insert the underscore character "_" between every four binary digits. For instance, in the fourth row, the table presents an 8-bit value in binary format 8'b1010_1011. This translates to decimal $2^7 + 2^5 + 2^3 + 2^1 + 2^0 = 128 + 32 + 8 + 2 + 1 = 171$, and the data stored in memory is represented as 8'b10101011.

The last two rows of the table depict examples with non-binary data bases. In line 5, a 3-bit decimal number is represented as 3'd6 (size: three, base: decimal). It is essential to note that the data stored in memory is in binary as 3'b110. In the final example, 8'hac (size: eight, base: hexadecimal), an 8-bit data is represented in hexadecimal base. The data stored in memory is in binary as 8'b1010_1100, yielding an equivalent decimal value of $2^7 + 2^5 + 2^3 + 2^2 = 128 + 32 + 8 + 4 = 172$.

TABLE 3.2
Examples of Verilog constants.

Number	Bits	Base	Decimal	Stored
3'b101	3	binary	5	101
8'b11	8	binary	3	0000_0011
8'b1010_1011	8	binary	171	1010_1011
3'd6	3	decimal	6	110
8'hac	8	hexadecimal	172	1010_1100

3.3.2 Verilog data types: wire and reg

Verilog data types play a crucial role in defining storage and transmission elements for data. While various data types defined by the IEEE 1364 standard can be recognized by simulators and synthesis tools, it is essential to use only the *wire* and *reg* data types for synthesizable Verilog designs. For example, the first line below declares a single-bit signal "a" as a *wire*, while the second line declares a single-bit signal "b" as a *reg*. The third and fourth lines declare byte-sized signals "c" and "d" as the *wire* and *reg* data types, respectively.

```
1 wire        a;
2 reg         b;
3 wire [7:0]  c;
4 reg  [7:0]  d;
```

In Verilog design, the data width of a signal is usually written from the most significant bit (MSB) to the least significant bit (LSB). Additionally, a part of a multi-bit signal can be referenced by a design statement; for instance, "a[7]" represents the MSB of the eight-bit signal "a", "a[0]" represents the LSB, and "a[3:0]" indicates the least significant four bits.

3.3.3 Hardware results for declared data types

It is important to note that the declaration of signals as *wire* or *reg* types in Verilog does not necessarily determine their synthesis as hardware wires or registers. Instead, the specific Verilog descriptions govern the synthesis result. A signal declared as a *wire* type can be synthesized as an internal wire, an output of combinational logic, or even a latch. Similarly, a signal declared as a *reg* type can be synthesized as an output of a latch or a register.

The rule for declaring a data type is straightforward: the left-hand side (LHS) signal in an *assign* block must be declared as a *wire*, and the LHS signal in an *always* or *initial* block must be declared as a *reg*. For instance, in the code below, the signal "e" must be declared as a *wire* (line 6) since it appears on the LHS within an *assign* block (line 9). In Verilog-2005, these two lines (6 and 9) can be combined as "wire e = a&b;". As illustrated in Figure 3.2, the signal "e" is synthesized as an internal wire, connecting the output of the AND1 gate to one of the inputs of both the AND2 gate and the OR2 gate.

```
1 //An Example of Using Wire and Reg Date Types
2 module example_wire_reg (input       a ,
```

```
3                                 input       b ,
4                                 output      c ,
5                                 output reg  d );
6   wire e;
7   reg  f;
8
9   assign e  = a & b ;   //e must be a wire
10  always @(a, b) begin
11    f <= a | b;         //f must be a reg
12  end
13
14  assign c  = e & f ;   //c output is in a default wire
15  always @(e, f) begin
16    d <= e | f;         //d output must be a reg
17  end
18  endmodule
```

On the other hand, the signal "f" must be declared as a *reg* (line 7) since it appears on the LHS within an *always* block (lines 10–12). In Figure 3.2, "f" serves as an internal wire connecting the output of the OR1 gate to one of the inputs of both the AND2 gate and the OR2 gate. Although "f" is declared as a *reg* data type, its hardware synthesis results in an internal wire connection as well. In practice, a simple *assign* block is highly recommended to describe a logic gate, such as "assign f = a | b;" or simply combining and converting line 7 and lines 10–12 as "wire f = a | b;".

In the provided code, the port declaration in line 4 does not specify the data type of the "c" output explicitly. As a result, "c" defaults to the *wire* data type because it is the LHS signal within an *assign* block in line 14. On the other hand, the other output, "d", must be declared as a *reg* (line 5) since it appears on the LHS within an *always* block (lines 15–17). As depicted in Figure 3.2, "c" and "d" are the outputs of the combinational circuit. Likewise, it is important to note that a simple *assign* block is highly recommended to describe logic gates, such as "assign d = e|f;" to define an OR gate. However,

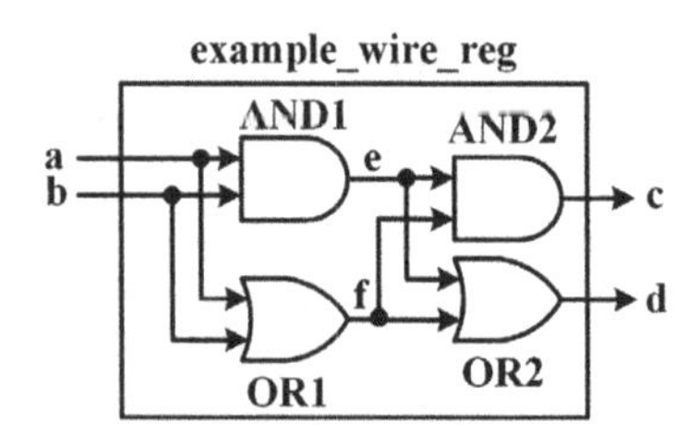

FIGURE 3.2
An example of wire and reg declaration.

if an *assign* block is used for the output "d" assignment, the port declaration in line 5 should be "output d);" to declare the output "d" as a default *wire* data type.

While the Verilog Standard defines various data types, such as *integer*, *real*, *time*, etc., they are recommended for creating testbenches and not for synthesizable designs. In synthesizable Verilog code, only *wire* and *reg* data types are recommended.

3.3.4 Mismatched data width

Matching bit widths on the left-hands-side (LHS) and right-hands-side (RHS) of data assignments is important. If the bit widths are mismatched, simulators follow the rules below:

- If the bit width of the LHS signal is greater than that of the RHS signal, only the lower bits of the LHS signal can be assigned.

  For instance, in the example below, the signal "a" is declared as an 8-bit signal and assigned the value of hexadecimal 8'h12. In the second line, the lower eight bits of the LHS signal "b[7:0]" are assigned the value of signal "a", which is represented as hexadecimal 8'h12. The higher eight bits of the signal "b[15:8]" are left unknown, indicated by the hexadecimal 8'hxx, as they are not explicitly assigned in the RHS.

- If the bit width of the LHS signal is less than that of the RHS signal, the LHS signal will be assigned the value of the lower bits of the RHS signal.

  For example, in the third line below, the signal "c" is assigned the value of the lower four bits of the signal "a[3:0]", which is hexadecimal 4'h2. The remaining higher bits of "a[7:4]" are not considered in the assignment to "c".

```
1 wire [7 :0]  a = 8'h12 ;
2 wire [15:0]  b = a     ; // b is 16'hxx12
3 wire [3 :0]  c = a     ; // c is 4'h2
```

In case of a mismatched data assignment, the simulator can only report a warning and not an error. While a warning does not terminate the simulation, it may lead to failed design features or functions. Therefore, it is crucial to manually check all the warnings to ensure the proper declaration of data types and data width. Below are some design rules recommended.

Design Rules – Data Types

1. Specify the bit width and base when using Verilog constants. Although the default constant is in decimal, it is highly recommended to provide a complete representation, such as 4'd15 instead of just 15.
2. For synthesizable designs, only use *wire* and *reg* data types, although other data types like integer can be synthesized.
3. Address and rectify all the mismatched warnings that arise during simulation. Although mismatched issues may pass the compilation stage, they can trigger simulation warnings that might ultimately result in design feature failures.
4. When designing synthesizable circuits, it's crucial to avoid using unknown states for signals and IOs. All bits of signals and IOs must be explicitly represented by a specific value, which can be either binary zero or one. For tri-state buffers, you should use the high-impedance state (high-Z) where appropriate.
5. When designing a testbench, initialize all inputs of the design circuit with either binary zeros or ones. Even though some inputs may not be utilized for certain test cases, it is recommended to provide specific initial values to avoid unexpected behavior during simulation.

3.4 Basic Verilog operators

Verilog operators are used to perform logic operations on different data types and produce output. These operators are used within Verilog blocks including *assign*, *always*, and *initial*.

3.4.1 Equality and inequality

Table 3.3 lists the equality and inequality operators that can be used for both single-bit and multi-bit signals.

When dealing with single-bit signals, the equality and inequality operators can be simplified using Boolean expressions. For instance, instead of writing "if(sel==1'b1)", the equality operation can be represented as "if(sel)". As demonstrated in the example shown in lines 4–9 below, this means that if "sel" is TRUE (or binary one), "c" is assigned the value of "a". Otherwise,

TABLE 3.3
Equality and inequality.

Operators	**Verilog**	**Examples**
Equality	==	if(a[3:0]==4'h8)
Inequality	!=	if(a[3:0]!=4'h8)

"c" is assigned the value of "b". This Verilog description emulates a 2-to-1 multiplexer.

```
1  module example_equ_inequ();
2  wire a, b, sel;
3  reg  c, d     ;
4  always @(a, b, sel) begin
5    if (sel) begin
6      c <= a;
7    end else begin
8      c <= b;
9    end
10 end
11
12 always @(a, b, sel) begin
13   if (~sel) begin
14     d <= a;
15   end else begin
16     d <= b;
17   end
18 end
19 endmodule
```

Similarly, in lines 13–17, the expression "if(~ sel)" is used to simplify the description of "if(sel==1'b0)". This means that if "Not sel" is TRUE, "d" is assigned the value of "a". Otherwise, "d" is assigned the value of "b". This Verilog description also functions a 2-to-1 multiplexer.

Recommended Coding Styles – Equality and Inequality

1. It is recommended to use a simple Boolean statement "if(~ a)", to describe a condition of "if(a==1'b0)".
2. Likewise, "if(a)" is recommended to describe a condition of "if(a==1'b1)".

TABLE 3.4
Bit-wise logic operators.

Operators	Verilog	Examples
Bit-wise NOT	~	$\sim a$; $\sim a[3:0]$
Bit-wise AND	&	$a\&b$; $a[3:0]\&b[3:0]$
Bit-wise OR	|	$a \vert b$; $a[3:0] \vert b[3:0]$
Bit-wise XOR	∧	$a \wedge b$; $a[3:0] \wedge b[3:0]$

3.4.2 Bit-wise logic operators

The bit-wise operators, including Inverter/NOT, AND, OR, and XOR, are introduced in Table 3.4. They can be used for both single-bit and multi-bit signals. For multi-bit signals, they take each bit in one operand and perform the operation with the corresponding bit in the other operand.
Design example: Here's an example demonstrating the use of bit-wise operators. In lines 4–8, we utilize the AND, OR, XOR, NAND, and NOR operators to assign values to 4-bit LHS signals, namely, "c", "d", "e", "f", and "g". The 4-bit RHS signals, "a" and "b", are declared as the *reg* data type since they are initialized within the subsequent *initial* block in lines 10–27.

```
1  module example_logic_op();
2  //--- Bit-Wise Logic Design Code---//
3  reg  [3:0] a,b;
4  wire [3:0] c = a & b   ; // AND
5  wire [3:0] d = a | b   ; // OR
6  wire [3:0] e = a ^ b   ; // XOR
7  wire [3:0] f = ~(a & b); // NAND
8  wire [3:0] g = ~(a | b); // NOR
9
10 //--- Testbench ---//
11 initial begin
12   a = 4'b1010;
13   b = 4'b0101;
14   #10;
15   $display ("Outputs c=%b, d=%b, e=%b, f=%b, g=%b
16       with Inputs a=%b, b=%b", c, d, e, f, g, a, b);
17 /* Printed log: # Outputs c=0000, d=1111, e=1111,
18 f=1111, g=0000 with Inputs a=1010, b=0101 */
19
20   a = 4'b1100;
21   b = 4'b1101;
22   #10;
23   $display ("Outputs c=%b, d=%b, e=%b, f=%b, g=%b
```

```
24         with Inputs a=%b, b=%b", c, d, e, f, g, a, b);
25  /* Printed log: # Outputs c=1100, d=1101, e=0001,
26  f=0011, g=0010 with Inputs a=1100, b=1101 */
27  end
28  endmodule
```

Testbench: The testbench code provided in lines 10–27 drive the RHS signals "a" and "b" and monitor the outputs to verify the design functionality. For the first test case, the input signals "a" and "b" are assigned binary values 4'b1010 and 4'b0101, respectively, as shown in lines 12–13. Then, for the second test case, "a" and "b" are assigned binary values 4'b1100 and 4'b1101 in lines 20–21. A delay of 10 ns (assuming the *timescale* unit is 1 ns) is introduced between the two test cases, as indicated in line 14. The *$display* task is used to print the simulation log and the values of all signals. The simulation log for both test cases is attached in lines 17–18 and 25–26, respectively.

The Verilog standard defines both logical operators – logical NOT (!), logical AND (&&), and logical OR (||), and bit-wise operators – bit-wise NOT (~), bit-wise AND (&), and bit-wise OR (|). Since bit-wise operators can perform the same functions as logical operators, such as the equivalent expressions "(a==4'h2 && b==4'h3)" and "(a==4'h2 & b==4'h3)", this book will focus solely on the bit-wise operators. This is because both types of operators have identical functions, and the bit-wise operators are shorter to type.

Recommended Coding Styles – Bit-Wise Operators

1. Bit-wise operators can be used for both bit-wise and logical operations since they have identical functions in many cases.
2. The Verilog standard allows for mixing different data types and logical operations. For example, the expression "if(a & (b==4'h4) & |c)" means that the *if* statement is TRUE only if all three conditions ("a==1'b1", "b==4'h4", and the result of a Reduction OR operation on the multi-bit signal "c" is binary one) are TRUE.

3.4.3 Reduction on multi-bit signals

Table 3.5 enumerates all the reduction operators. A reduction operator typically features a single-bit signal on the LHS and expects a multi-bit signal on the RHS. The reduction operator iteratively processes each bit of the RHS operand and produces a single-bit output. When the NOT notation is applied before the reduction operator, it initially conducts a bitwise operation on the RHS and subsequently inverts the result to yield the LHS output.

TABLE 3.5
Reduction on multi-bit signals.

Operators	Verilog	Examples
Reduction AND	&	$\&a[3:0]$
NOT Reduction AND	~&	$\sim\&a[3:0]$
Reduction OR	|	$\|a[3:0]$
NOT Reduction OR	~|	$\sim\|a[3:0]$
Reduction XOR	∧	$\wedge a[3:0]$
NOT Reduction XOR	~∧	$\sim\wedge a[3:0]$

Design example #1: The following design example showcases the reduction operators. Lines 4–8 demonstrate the application of the Reduction AND, Reduction OR, Reduction XOR, NOT Reduction AND, and NOT Reduction OR operators, one after the other. It is important to note that while the RHS signal "a" has a length of four bits, all LHS signals "b-f" are single-bit signals.

```
1  module example1_reduction ();
2  //--- Reduction on Multi-Bit Signals Design Code---//
3  reg [3:0] a;
4  wire b = &a ;    // Reduction AND
5  wire c = |a ;    // Reduction OR
6  wire d = ^a ;    // Reduction XOR
7  wire e = ~&a;    // NOT Reduction AND
8  wire f = ~|a;    // NOT Reduction OR
9
10 //--- Testbench ---//
11 initial begin
12   a = 4'b1010;
13   #10;
14   $display ("Outputs b=%b, c=%b, d=%b, e=%b, f=%b,
15              with Input a=%b", b, c, d, e, f, a);
16 /*Printed log: # Outputs b=0, c=1, d=0, e=1, f=0,
17 with Input a=1010*/
18
19   a = 4'b1100;
20   #10;
21   $display ("Outputs b=%b, c=%b, d=%b, e=%b, f=%b,
22              with Input a=%b", b, c, d, e, f, a);
23 /*Printed log: # Outputs b=0, c=1, d=0, e=1, f=0,
24 with Input a=1100*/
25
26   a = 4'b1111;
27   #10;
```

```
28    $display ("Outputs b=%b, c=%b, d=%b, e=%b, f=%b,
29               with Input a=%b", b, c, d, e, f, a);
30  /*Printed log: # Outputs b=1, c=1, d=0,. e=0, f=0,
31  with Input a=1111*/
32
33    a = 4'b0000;
34    #10;
35    $display ("Outputs b=%b, c=%b, d=%b, e=%b, f=%b,
36                with Input a=%b", b, c, d, e, f, a);
37  /*Printed log: # Outputs b=0, c=0, d=0, e=1, f=1,
38  with Input a=0000*/
39  end
40  endmodule
```

Testbench #1: In the testbench in lines 10–39, four test cases are shown where the multi-bit signal "a" is assigned different values using 4-bit binary literals on lines 12, 19, 26, and 33. Each test case has a delay of 10 ns (assuming the *timescale* unit is 1 ns). The LHS signals of the combinational circuit, "b" through f", are printed using the $*display* task on lines 14, 21, 28, and 35, respectively. The simulation log is attached for each test case in lines 16-17, 23–24, 30–31, and 37–38.

Design example #2: Reduction operators can express a condition when a multi-bit signal reaches a value of all binary zeros. For example, in line 6 below, the statement "if(~|a)" will be TRUE only when all bits in the signal "a" are binary zeros. This condition can also be expressed using the equality operator as "if(a==32'h0)". The synthesis result is simply a multi-input NOR gate.

Similarly, in line 14, the statement "if(&b)" will be TRUE only when all bits in the multi-bit signal "b" are binary ones. This condition is equivalent to using the equality operator as "if(b==32'hffff_ffff)". The hardware result is a multi-input AND gate.

```
1   module example2_reduction ();
2   //--- Reduction on Multi-Bit Signals Design Code---//
3   reg [31:0] a,b;
4   reg        c,d;
5   always @(a) begin
6     if(~|a) begin // Will be true when all bits are zeros
7       c <= 1'b1;
8     end else begin
9       c <= 1'b0;
10    end
```

```
11 end
12
13 always @(b) begin
14   if(&b) begin   // Will be true when all bits are ones
15     d <= 1'b1;
16   end else begin
17     d <= 1'b0;
18   end
19 end
20
21 //--- Testbench ---//
22 initial begin
23   a = 32'h0;
24   b = 32'hffff_ffff;
25   #10;
26   $display ("Outputs c=%b, d=%b, with Inputs a=%h, b=%h",
27             c, d, a, b);
28 /*Printed log: # Outputs c=1, d=1,
29 with Inputs a=00000000, b=ffffffff*/
30
31   a = 32'h0123_4567;
32   b = 32'h89ab_cdef;
33   #10;
34   $display ("Outputs c=%b, d=%b, with Inputs a=%h, b=%h",
35             c, d, a, b);
36 /*Printed log: # Outputs c=0, d=0,
37 with Inputs a=01234567, b=89abcdef*/
38 end
39 endmodule
```

Testbench #2: In the testbench, the first test case initiates "a=32'h0" and "b=32'hffff_ffff" in lines 23–24, which make the two *if* statements evaluate to be TRUE. The printed log in lines 28–29 demonstrates the resulting outcomes, "c=1" and "d=1", for the first test case. In another test case, "a=32'h0123_4567" and "b=32'h89ab_cdef" are assigned in lines 31–32, resulting in the outcomes "c=0" and "d=0" as shown in lines 36–37. Any signals that do not have all binary zeros for "a" or all binary ones for "b" can lead to FALSE results for "c" and "d".

Similarly, the Reduction OR operation, such as "if(|a)", can be used to find a multi-bit signal in which not all bits are binary zeros. The NOT Reduction AND operation, such as "if(~&a)", can be used to find a multi-bit signal in which not all bits are binary ones.

Recommended Coding Styles – Reduction Operators

1. Use a simple NOT Reduction OR, e.g., "if(~|a)", to simplify the equality operation where all bits are zeros, e.g., "if(a==32'h0)".
2. Use a simple Reduction AND, e.g., "if(&a)", to simplify the equality operation where all bits are ones, e.g., "if(a==32'hffff_ffff)".
3. Use a simple Reduction OR, e.g., "if(|a)", to simplify the inequality operation where not all bits are zeros, e.g., "if(a!=32'h0)".
4. Use a simple NOT Reduction AND, e.g., "if(~&a)", to simplify the inequality operation where not all bits are ones, e.g., "if(a!=32'hffff_ffff)".

3.4.4 Relational operators

Table 3.6 displays the relational operators, including *greater than*, *less than*, *greater than or equal to*, and *less than or equal to*. These operators compare the values of two operands and result in a single-bit output of TRUE or FALSE.

In Verilog descriptions, bit-wise operators can be combined with relational operators to perform many conditions. For example, the second line below shows a combination of signal "a" *greater than* hexadecimal 4'h0 AND *less than* hexadecimal 4'h8. The second example in line 6 represents the operation of signal "a" *greater than* or *equal to* hexadecimal 4'h0, AND *less than* or *equal to* hexadecimal 4'h8. In line 9, it demonstrates a combination of signal "a" *less than* hexadecimal 4'ha OR "b" *less than* hexadecimal 4'hb. Here, bit-wise operators are used instead of logical operators, as they have similar functionality but require less typing.

TABLE 3.6
Relational operators.

Operators	Verilog	Examples
Greater Than	>	if(a > b)
Less Than	<	if(a < b)
Greater Than or Equal To	>=	if(a >= b)
Less Than or Equal To	<=	if(a <= b)

```
//Greater than 4'h0 AND less than 4'h8
if(a>4'h0 & a<4'h8 )

//Greater than or equal to 4'h0 AND
//less than or equal to 4'h8
if(a>=4'h0 & a<=4'h8 )

//Signal "a" less than 4'ha OR signal "b" less than 4'hb
if(a<4'ha | b<4'hb)
```

3.4.5 Concatenation

The concatenation operator serves the purpose of merging multiple signals into a fresh multi-bit signal. As demonstrated in Table 3.7, it facilitates the blending of inputs, outputs, internal signals declared as *wire* or *reg*, as well as Verilog constants. It's vital to emphasize that the bit width of the resulting multi-bit signal needs to match the cumulative bit widths of all the signals encompassed within the concatenation operator.

Design example #1: Below is a design example that demonstrates concatenation using three inputs: "a", "b", and "c". In the first example shown in line 6, all three RHS signals are concatenated using the basic concatenation operator to form the LHS signal "vec1". Since the RHS signal widths are 8, 16, and 8 bits for "a", "b", and "c", respectively, the resulting "vec1" must be declared as a 32-bit signal to avoid a width mismatch issue. In the second example in line 7, the higher byte of the 16-bit signal "b" is switched with its lower byte using a concatenation operation. The bit width of the LHS signal "vec2" must match the combined bit width of the RHS signals.

When an expression needs to be repeated several times, a replication constant is used, which is enclosed in braces along with the original concatenation operator. The constant indicates how many times the expression will be repeated. In line 8, the third example shows that the 8-bit signal "a" is repeated twice to form a new 16-bit signal "vec3". The bit width of "vec3" should be twice the bit width of "a".

In the fourth example shown in line 10, a concatenation is performed using a constant and the signal "a". The constant "1'b0" is repeated twice and

TABLE 3.7
Concatenation.

Operators	Verilog	Examples
Concatenation	{}	{a, b, c, 32'hf }

combined with the signal "a", which is concatenated into the lower eight bits of the signal "vec4". The bit width of the signal "vec4" is ten, which is the sum of two bits for the constant "2{1'b0}" and eight bits for the signal "a". Finally, the last example shows the opposite of the previous example, where the constant "2{1'b0}" is concatenated to the least significant two bits of the signal "vec5", and "a" is concatenated into the higher eight bits.

```
1  module example_concatenation ();
2  //--- Concatenation Design Code---//
3  reg  [7 :0] a ;
4  reg  [15:0] b ;
5  reg  [7 :0] c ;
6  wire [31:0] vec1 = {a, b, c}           ;
7  wire [15:0] vec2 = {b[7:0], b[15:8]};
8  wire [15:0] vec3 = {2{a}}              ;
9
10 wire [9:0]  vec4 = {{2{1'b0}}, a}     ;
11 wire [9:0]  vec5 = {a, {2{1'b0}}}     ;
12
13 //--- Testbench ---//
14 initial begin
15   a = 8'h12   ;
16   b = 16'h3456;
17   c = 8'h78   ;
18   #10;
19   $display ("Outputs vec1=%h, vec2=%h, vec3=%h, vec4=%h,
20             vec5=%h, with Inputs a=%h, b=%h, c=%h",
21             vec1, vec2, vec3, vec4, vec5, a, b, c);
22 /* Printed log: Outputs vec1=12345678, vec2=5634, vec3=1212,
23 vec4=012, vec5=048 with Inputs a=12, b=3456, c=78 */
24 end
25 endmodule
```

Testbench #1: In lines 13–24 it shows a testbench with input stimulus of "a–8'h12", "b=16'h3456", and "c=8'h78". The corresponding LHS signals "vec1-vec5" are printed out as shown in lines 22–23. The signal "vec1" displays the combined result of inputs "a-c", the signal "vec2" shows a switching of the higher byte and lower byte of the input "b", and the signal "vec3" depicts a duplication of input "a".

In what follows, let's consider the concatenation results of "vec4" and "vec5" with the initialized signal "a" (binary 8'b0001_0010). By concatenating two binary zeros in front, the signal "vec4" becomes binary 10'b00_0001_0010, which is equivalent to hexadecimal 10'h012. Similarly, the signal "vec5" becomes binary 10'b00_0100_1000 by concatenating two binary zero to the end, which is equivalent to hexadecimal 10'h048.

Design example #2: Concatenation has a wide range of diverse and practical applications within RTL design. For example, it can be utilized in creating shift register using clock-edge-triggered *always* blocks, building primary buses by concatenating secondary buses, and simplifying the design of binary division and multiplication circuits.

Presented below is an illustrative example of binary multiplier design involving a constant input. In line 4, the operation denoted as "$r \times 32$" or "$r \times 2^5$" is demonstrated by appending five zeros to the right-hand side of the 8-bit signal "r". This approach is grounded in the principle that appending a single zero to the least significant bit (equivalent to a left shift by one position) corresponds to multiplication by two. Similarly, introducing a single zero to the most significant bit (equivalent to a right shift by one position) equates to division by two. Hence, the manipulation of multiplying or dividing a number by powers of two can be readily achieved by incorporating zeros to either the left-hand side or right-hand side (or employing left shifts or right shifts) based on the desired exponent. It's important to observe that the signal "rx32" spans 16 bits, necessitating the concatenation of three zeros ahead of the signal "r" as well.

Similarly, in line 5, it shows the operation of "$g \times 64$" or "$g \times 2^6$" by adding six zeros on the RHS and two zeros on the LHS of the signal "g", in order to output the final 16-bit "gx64". In line 9, it demonstrates the operation of "$b \times 8$" or "$b \times 2^3$" by adding three zeros on the RHS and five zeros on the LHS of the signal "b" to obtain the final 16-bit signal "bx8".

```
1  module example_con_mul ();
2  //--- Concatenation Design Code---//
3  reg  [7 :0] r,g,b ;
4  wire [15:0] rx32  = {3'b0, r, 5'b0};
5  wire [15:0] gx64  = {2'b0, g, 6'b0};
6  wire [15:0] bx8   = {5'b0, b, 3'b0};
7
8  //--- Testbench ---//
9  initial begin
10   r = 8'h12;
11   g = 8'h34;
12   b = 8'h56;
13   #10;
14   $display ("Outputs rx32=%h, gx64=%h, bx8=%h with Inputs
15             r=%h, g=%h, b=%h", rx32, gx64, bx8, r, g, b);
16 /*Printed log: Outputs rx32=0240, gx64=0d00, bx8=02b0,
17 with Inputs r=12, g=34, b=56 */
18 end
19 endmodule
```

TABLE 3.8
Logical shift operators.

Operators	Verilog	Examples
Logical Left shift	<<	a << 2
Logical Right shift	>>	a >> 2

Testbench #2: The following testbench in lines 8–18 is provided to validate the design functionality. The signal "r" is initialized to the hexadecimal value 8'h12 (equivalent to decimal 8'd18) in line 10. In lines 16–17, the outcome of multiplying "r" by the integer 32 is calculated and showcased as the hexadecimal value 16'h240 (or in decimal, 16'd576).

Likewise, the outcomes of the multiplication operations of "gx64" and "bx8" are examined in lines 16–17. These results are as follows: "gx64" is calculated as the product of 8'h34 multiplied by 8'h40, yielding 16'hd00, and "bx8" results from the multiplication of 8'h56 and 8'h8, yielding 16'h2b0. In this context, the hexadecimals 8'h40 and 8'h8 correspond to the integers 64 and 8, respectively.

3.4.6 Logical shift

Verilog introduces logical shift operators as outlined in Table 3.8. The left shift operator shifts its left operand to the left by the number of bit positions specified by the right operand. The vacated bit positions are then populated with zeros. Conversely, the right shift operator moves its left operand to the right by the bit count defined by the right operand. This operation results in the vacated bit positions being filled with zeros.

Design example: The provided example showcases the utilization of logical shift operators. Lines 4–5 offer instances of employing logical left shift by two positions and logical right shift by two positions.

```
1  module example_shift ();
2  //--- Logical Shift Design Code---//
3  reg  [3:0] a;
4  wire [3:0] lls2 = a<<2;
5  wire [3:0] lrs2 = a>>2;
6
7  //--- Testbench ---//
8  initial begin
9    a  = 4'h1;
10   #10;
11   $display ("Outputs lls2=%b, lrs2=%b
12                with Input a=%b", lls2, lrs2, a);
13 /* Printed log: # Outputs lls2=0100, lrs2=0000
```

```
14                          with Input a=0001 */
15  end
16  endmodule
```

Testbench: In the provided testbench, line 9 initializes the stimulus as "a=4'h1". The outcomes of the logical shifts, namely "lls2" and "lrs2", are printed out as binary values 4'b0100 and 4'b0000, correspondingly, as evidenced in lines 13–14. This outcome arises because the initial signal "a=4'b0001" is subjected to left and right shifts by two positions, followed by zero-filling.

3.4.7 Arithmetic operators

Verilog delineates binary arithmetic operators in Table 3.9. Among these, the addition operator finds widespread application in both FPGA and ASIC designs. Although the Verilog standard encompasses subtraction, multiplication, and division operators, it is advisable to leverage arithmetic IPs (intellectual properties) furnished by FPGA tools or ASIC foundries for proficient integration. This approach guarantees enhanced performance and optimal resource utilization specifically tailored to FPGA/ASIC implementations.

3.5 Behavioral Verilog Blocks: initial, assign, and always

Verilog introduces three design and simulation blocks: *assign*, *always*, and *initial*. All logic operators (Section 3.4) and Verilog statements (Section 4.1) must be included within these description blocks. In synthesizable designs, only *assign* and *always* blocks are allowed. However, all three blocks, *assign*, *always*, and *initial*, can be used to construct a testbench.

TABLE 3.9
Arithmetic operators.

Operators	Verilog	Examples
Addition	+	a + b
Subtraction	−	a - b
Multiplication	*	a * b
Division	/	a / b

3.5.1 Synthesizable block – assign

The first synthesizable block is the *assign* block, encompassing both concurrent *assign* and conditional *assign* options. Within an *assign* block, execution takes place whenever the signals on the RHS undergo a modification. This behavior closely emulates the functioning of combinational circuits and sequential latches, where the outputs update whenever there are change in the inputs. Both concurrent *assign* and conditional *assign* have the following characteristics:

- The LHS signals must be declared as *wire*.
- They can be used in parallel with other *assign*, *always*, and *initial* blocks.
- They cannot be "nested".

3.5.1.1 Concurrent assign

In preceding sections, various design illustrations employing the concurrent *assign* block have been presented. This type of *assign* block proves useful for depicting diverse scenarios, including a straightforward wire connection (as depicted in line 3 below), a gate-level operation (as exhibited in line 4), or even multiple gate-level operations (as demonstrated in line 5).

```
1  // Examples of Concurrent Assignments
2  wire   c1, c2, c3;
3  assign c1 = a    ; // Ex1: wire c1=a    ; Wire connection ;
4  assign c2 = a&b  ; // Ex2: wire c2=a&b  ; AND Gate        ;
5  assign c3 = a&b|c; // Ex3: wire c3=a&b|c; AND and OR Gates;
```

All the basic operators introduced in Section 3.4 can be used in concurrent *assign* blocks. Note that the signals on the LHS within an *assign* block, whether it is a concurrent *assign* or a conditional *assign*, must be declared as the *wire* type.

In the Verilog IEEE 2005 standard, it is possible to merge the *wire* declaration with the *assign* block description. This is demonstrated in the comments of the third, fourth, and fifth lines: "wire c1=a;", "wire c2=a&b;", and "wire c3=a&b|c;", all of which represent alternative approaches for describing the same design.

3.5.1.2 Conditional assign

A conditional *assign* block is listed in Table 3.10. In the third column, an example is provided to illustrate the use of a conditional *assign*, where if the

TABLE 3.10
Conditional assign.

Operators	Verilog	Examples
Conditional	? :	assign d = a ? b: c;

signal "a" is TRUE, the LHS signal "d" will be assigned as the value of "b"; otherwise, "d" will be assigned as the value of "c". The hardware implementation using conditional operators is dependent on the Verilog descriptions. It could be implemented as a multiplexer, encoder, tri-state buffer, latch, or other circuit structures.

Below are examples of different circuits described using the conditional *assign* block. The first example, depicted in line 4, describes a 2-to-1 multiplexer where the inputs "a" and "b" are selected by the input "sel". When "sel=1'b0", the output "c" takes the value of "a", and when "sel=1'b1", the output "c" takes the value of "b". The corresponding hardware can be seen in Figure 3.3(a).

```
1  //a) A 2-to1 Multiplexer Using Conditional Assignment
2  module example_conditional_mux (input  sel, a, b,
3                                 output c        );
4  assign c = ~sel ? a: b;
5  endmodule
6
7  //b) A Tri-State Buffer Using Conditional Assignment
8  module example_conditional_tri (input        en ,
9                                 input  [1:0] a  ,
10                                output [1:0] b  );
11 assign b = en ? a : 2'bzz ;
12 endmodule
13
14 //c) A Latch Using Conditional Assignment
15 module example_conditional_latch (input  en, d ,
16                                   output q     );
17 assign q = en ? d : q ;
18 endmodule
19
20 //d) A 2-Stage Multiplexer Using Conditional Assignment
21 module example_conditional_nested (input  sel0, sel1,
22                                    input  a,  b,  c,
23                                    output d         );
24 assign d = ~sel0 ? a :
25                ~sel1 ? b: c ;
26 endmodule
```

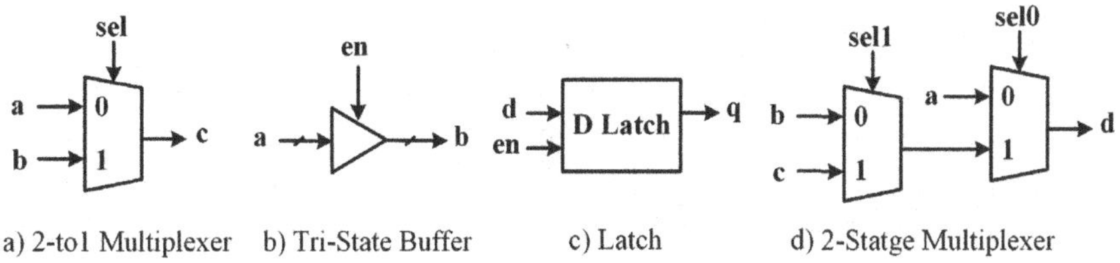

a) 2-to1 Multiplexer b) Tri-State Buffer c) Latch d) 2-Statge Multiplexer

FIGURE 3.3
Hardware results of conditional assignments.

The second example, illustrated in line 11, describes a tri-state buffer where the input "a" can drive the output "b" when "en=1'b1". Otherwise, when "en=1'b0", the data path is in high impedance or an open circuit, represented by high-Z status of "2'bzz". The corresponding hardware can be seen in Figure 3.3(b).

The third example, presented in line 17, describes a function where the current value, denoted as "q=q", is retained if the input "en=1'b0". This specific design necessitates a storage element like a latch or a register. Due to the distinct triggering mechanisms – a register being clock-edge-triggered and a latch being signal-level-triggered – a latch is employed in the conditional *assign* block for the purpose of data storage. Alternatively, when "en=1'b1" the input "d" can be accepted by the latch, represented as "q=d". The corresponding latch hardware, encompassing two inputs and one output, can be visualized in Figure 3.3(c).

Conditional *assign* blocks can be nested, as demonstrated in the final design example within lines 24–25. This configuration comprises two selection inputs, namely "sel0" and "sel1". The initial conditional operation in line 24 takes precedence over the subsequent operation in line 25. This implies that the design outlined in line 24 will be evaluated first, followed by the execution of line 25. When "sel0=1'b0" in line 24, the operation "d=a" is executed. In this scenario, the value of "sel1 is irrelevant due to the higher priority of the first-stage operation. The second-stage conditional operation is only executed when "sel0=1'b1", or equivalently when "∼sel" is FALSE. If "sel1=1'b0", the operation "d=b" is carried out. Conversely, if "sel1=1'b1", the operation "d=c" is executed. The Verilog code succinctly describes a 2-stage multiplexer, and the resultant hardware module is depicted in Figure 3.3(d).

Recommended Coding Styles – Conditional Assignment

1. It is strongly recommended to vertically align the early-stage ":" mark with the further-stage "?" mark when using nested conditional operators. This makes the code more readable, especially for multi-stage conditional *assign* blocks.

2. The conditional *assign* block is useful for describing both combinational and sequential circuits. In combinational circuit design, it is typically used to implement multiplexers or tri-state buffers. In sequential circuit design, it can be used to create the storage element of a latch.

3.5.2 Synthesizable block – always

The second synthesizable Verilog block is the *always* block. An *always* block operates continuously, running when simulation starts and restarting when it reaches the end of the *always* block.

Broadly speaking, an *always* block can be triggered either by level events or edge events. A level-triggered *always* block emulates the behavior of combinational circuits and latches. In this scenario, the outputs (signals on the left-hand side, or LHS, of the *always* block) update whenever the inputs (all signals listed in the *always* trigger list) undergo changes. In contrast, an edge-triggered *always* block primarily mimics the behavior of registers. Here, the output (LHS signals in the *always* block) tracks the input at an active clock edge specified in the trigger list. Alternatively, the output initializes when a reset event occurs.

As depicted in Table 3.11, in the context of level-triggered *always* blocks, all signals capable of triggering the computation and updating the LHS signals in the *always* block must be included in the trigger list, denoted as "@()". Failing to include a signal in the trigger list can lead to simulation errors, as the absent signal won't activate the *always* block. Moreover, an incomplete trigger list within the description might result in unintended latches during synthesis.

In a standard register, the output is capable of updating solely upon detection of an active clock edge and/or an active reset edge (if an asynchronous reset is incorporated). Consequently, the trigger list should exclusively encompass the clock edge and the reset edge. It is important to note that only the edge events corresponding to the clock and reset can be indicated. This is because registers can solely be activated by these two events. While a

TABLE 3.11
Always blocks.

<table>
<tr><th>always Blocks</th><th>Verilog</th><th>Examples</th></tr>
<tr><td>Level Trigger</td><td rowspan="2">always @()</td><td>always @(a, b, c, d)</td></tr>
<tr><td>Edge Trigger</td><td>always @(posedge clk, negedge rst)</td></tr>
</table>

description involving additional signal edges might be appropriate for simulation objectives, it's not practical for practical circuits.

The following characteristics apply to both level-trigger and edge-trigger *always* blocks:

- The LHS signals must be declared as *reg*.
- They can be used in parallel with other *assign*, *always*, and *initial* blocks.
- They cannot be nested.

3.5.2.1 Level-triggered always

Design example #1: As previously mentioned, an *always* block can utilize a level-trigger list to perform operations. For instance, consider a design for a simple AND gate as shown in lines 2–5 below. In line 3, both input signals "a" and "b" for the AND gate must be included in the trigger list. In line 4, whenever the RHS signals "a" and/or "b" change, the LHS signal "c" will be updated through the AND logic.

While this design code is synthesizable and accurately represents an AND gate, the same functionality can be achieved using a simpler *assign* block, such as "assign c1 = a & b;". In the context of combinational logic, it is highly recommended to use concurrent *assign* blocks because they offer better readability and can reduce unnecessary complexity associated with Verilog coding.

```
1  // An Example of Level Trigger List
2  reg c1;
3  always @(a, b) begin
4    c1 <= a & b;
5  end
6
7  // An Example of Missing Signals in Level Trigger List
8  reg c2;
9  always @(b) begin
10   c2 <= a & b;
11 end
```

Let's explore another design scenario involving a trigger list with missing signals. In line 9, the input signal "a" that feeds into the AND gate is missing in the trigger list. This absence can result in a simulation error when changing values for the signal "a", as the event cannot trigger the AND operation.

The synthesis result may vary depending on the specific synthesis tools used. While many modern synthesis tools can infer missing signals and make an effort to resolve such issues, it is still highly advisable to provide a complete and accurate list of triggers when writing Verilog code.

Design example #2: The second example illustrates a prevalent simulation model utilized for clock generation. To simulate the behavior of a clock, time delays are integrated into the code without explicitly defining a trigger list for the *always* blocks. Lines 3–4 underscore the utilization of #5 and #10 delay statements, which mimic a 5 ns delay and a 10 ns delay, respectively (assuming a timescale unit of 1 ns), prior to toggling the clock signals. It is important to note that delay statements are purely for simulation purposes and are not synthesizable. In a real hardware implementation, the actual circuit delay is determined by the design technology or synthesis library, not by the Verilog descriptions.

```
1  // Examples of Clock Generator within Testbench
2  reg clk1, clk2;
3  always #5  clk1 = ~clk1;
4  always #10 clk2 = ~clk2;
5
6  initial begin
7    clk1=1'b0;
8    clk2=1'b0;
9  end
```

Specifically, in line 3, the signal "clk1" is toggled at a frequency of 100 MHz, changing its value to its opposite every 5 ns. Similarly, in line 4, the signal "clk2" is toggled at a frequency of 50 MHz, with its value changing to its opposite every 10 ns. The *always* blocks simulate different clock frequencies, providing a 100 MHz clock for "clk1" and a 50 MHz clock for "clk2" when utilized in a testbench. For proper functionality, both "clk1" and "clk2" must be declared as *reg*, as they are on the LHS signals within the *always* blocks.

Additionally, it's crucial to highlight that initializing both the "clk1" and "clk2" signals (as seen in lines 7–8) is imperative to prevent ambiguous states. The assignments occurring in distinct blocks, both in *always* and *initial* blocks, can result in a situation of multi-driven implementation. While this scenario isn't permissible in synthesizable RTL designs, it can be employed in the creation of testbenches.

Design Rules – Level-Trigger always Block

1. Due to its simplicity and enhanced readability, employing an *assign* block is strongly recommended for detailing combinational circuits. Nevertheless, an exception arises when dealing

with an abundance of nested conditional operators. In such scenarios, utilizing a level-triggered *always* block in conjunction with a *case* statement proves to be a more appropriate approach. The upcoming sections will introduce the *case* statement utilized in *always* blocks.

2. It is crucial to include a complete trigger list when employing level-triggered *always* blocks. The absence of any signal in the trigger list may lead to simulation failures and result in the synthesis generating unwanted latches.

3.5.2.2 Edge-triggered always

Hardware registers can be effectively described using edge-triggered *always* blocks. A design example featuring a simple AND gate and a register is shown in lines 2–5 below. In the trigger list, a rising clock event (represented as "*posedge* clk") is listed to trigger the computation and storage of an AND gate operation. When a rising clock edge occurs, the *always* block is triggered to compute the AND logic from the RHS to the LHS and store the AND gate result into a register. This hardware description combines both a combinational circuit (AND gate) and a sequential circuit (register) since a clock-edge-triggered *always* block is employed.

```
1  // A Design Circuit including AND Gate and Register
2  reg c1;
3  always @(posedge clk) begin
4    c1 <= a & b;
5  end
6
7  // A Recommended Coding Style for the Same Design
8  wire nxt_c2 = a & b;
9  reg c2;
10 always @(posedge clk) begin
11   c2 <= nxt_c2;
12 end
```

To enhance code readability and maintainability, it is advisable to separate the combinational logic and register descriptions when writing Verilog. As an example, in lines 8–12, the circuit design is divided into a combinational AND gate (in line 8) and a register descriptions (in lines 10–12). To facilitate this separation, an internal signal "nxt_c2" is declared to connect the output of the combinational AND gate with the input of the register.

Design Rules – Edge-Triggered always Block

1. Only clock-edge and reset-edge events can be included in the edge-triggered *always* block. Listing other active signal edges in the trigger list might work for simulation purposes, but it is not permitted for practical hardware circuits.
2. It is highly recommended to segregate the design of combinational circuits and the descriptions of sequential registers into separate blocks. When working within edge-triggered *always* blocks, focus solely on the register description.

3.5.2.3 Simulation results: level trigger and edge trigger behaviors

This section discusses the simulation differences between a level-triggered *always* block and an edge-triggered *always* block. Below is a simple design code example, where the first line shows a level-triggered *always* block, and the second line shows an edge-triggered *always* block.

```
1  always @(en, d) if(en) q <= d;   // D Latch
2  always @(posedge clk)  q <= d;   // D Register
```

In the first design example, the trigger list includes level events of the enable input "en" and the data input "d", which can trigger the assignment "q <= d". The assignment is conditioned by the "if (en)" which executes only when "en" is TRUE. This synthesizable design creates a latch, and the simulation result is depicted in Figure 3.4(a):

- At 10 ns, the input "en" switches from binary 1'b0 to 1'b1, triggering the assignment "if(en) q <= d;". Since "if(en)" is TRUE at this moment, the output "q" follows the input "d" and becomes binary 1'b0.
- At 58 ns, the input "d" changes from binary 1'b0 to 1'b1, triggering the assignment "if(en) q <= d;". As "if(en)" is TRUE at this moment, "q" follows "d" and becomes binary 1'b1.
- At 88 ns, although the input "d" changes from binary 1'b1 to 1'b0 and triggers the *always* block, "q" remains binary 1'b1 because the "if(en)" statement is FALSE.
- The output "q" sustains binary 1'b1 until 90 ns when the input "en" changes, and the assignment is triggered again. Thus, "q" follows "d" and becomes binary 1'b0 at this moment.

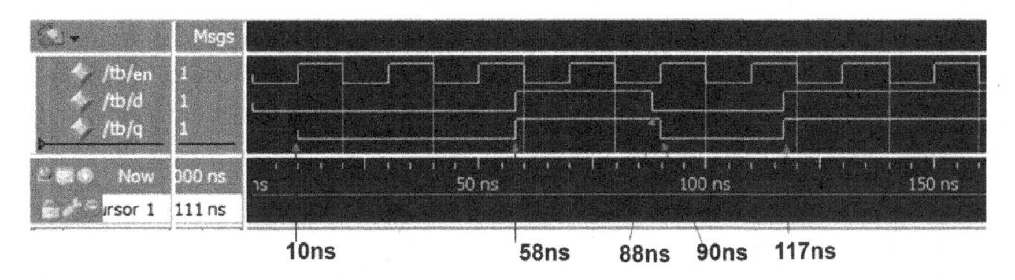

(a) Simulation Waveform with Level-Triggered Design

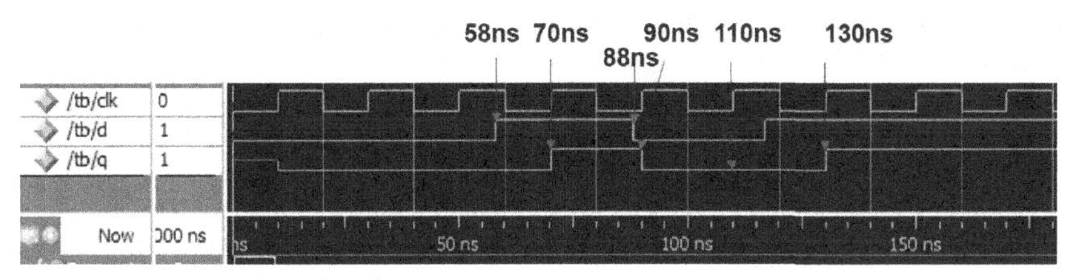

(b) Simulation Waveform with Edge-Triggered Design

FIGURE 3.4
Comparison between level-triggered and edge-triggered always blocks.

- At 117 ns, the output "q" follows the input "d" and becomes binary 1'b1, because the condition "if(en)" is TRUE. The change of the input "d" triggers the assignment.

Returning to the second design example, an edge-triggered *always* block is employed. Only "(*posedge* clk)" is listed in the trigger list, indicating that every rising clock edge can trigger the assignment "q <= d;". This synthesizable design creates a register, and the simulation results are illustrated in Figure 3.4(b):

- At 58 ns, the input "d" switches from binary 1'b0 to 1'b1. However, the assignment "q <= d;" can be triggered until the following rising clock edge occurring at 70 ns. After 70 ns, the output "q" remains 1'b1 within the clock cycle.
- At 88 ns, the input "d" toggles from binary 1'b1 to 1'b0. At the following rising clock edge occurring at 90 ns, the output "q" takes the binary value 1'b0 and further retains this value until the next rising clock edge at 110 ns.
- At the rising clock edge occurring at 110 ns, since the input "d" remains binary 1'b0, the register retains its previous value, and "q" remains binary 1'b0 within the clock cycle.
- At the rising clock edge occurring at 130 ns, the output "q" transitions from 1'b0 to 1'b1 because the current value of "d" is binary 1'b1.

These simulation results demonstrate the behavior of the edge-triggered register design. The register captures the value of "d" at each rising clock edge, holding it within the current clock cycle. Understanding this behavior is vital to ensure correct functionality in register-transfer designs.

3.5.3 Unsynthesizable block – initial

An unsynthesizable behavioral block – *initial* can be used exclusively in a testbench. It executes only once, starting at time zero (beginning of simulation), and terminates when it reaches the end. For an *initial* block:

- The LHS signals must be declared as *reg*.
- It can be used in parallel with other *assign*, *always*, and *initial* blocks.
- It cannot be "nested".

Serving as unsynthesizable Verilog design, the *initial* block is primarily employed in testbenches to initialize variables and furnish stimuli for simulations. To illustrate this with a simple example involving an *initial* block, lines 3–6 initiate the process by setting "rst = 1'b0;", "clk1 = 1'b0;", and "clk2 = 1'b0;" to initialize the clock and reset signals. Subsequently, after a duration of 10 ns (assuming a timescale unit of 1 ns), "rst = 1'b1;" is triggered to terminate the reset operation. It's crucial to note that within the *initial* block, LHS signals, "rst", "clk1", and "clk2", need to be declared as the *reg* data type, as indicated in the second line.

```
1 // An Example of initial Block
2 reg  rst, clk1, clk2;
3 initial begin
4   rst = 1'b0; clk1 = 1'b0; clk2 = 1'b0;
5   #10 rst = 1'b1;
6 end
```

Exercises

Problem 3.1. What are the differences between level-sensitive and edge-sensitive *always* blocks?

Problem 3.2. What are the differences between *reg* and *wire* data types?

Problem 3.3. The following examples demonstrate the use of signal assignment with mismatched bit widths. Assuming an 8-bit signal "wire [7:0] a = 8'h89;", what values will the signals "b" and "c" have in each example?

```
1 wire [7 :0] a = 8'h89;
2 wire [15:0] b = a    ;
3 wire [3 :0] c = a    ;
```

Problem 3.4. Assuming we have two 4-bit signals "a" and "b" with the values of binary 4'b0101 and 4'b1100, respectively, what are the resulting values of signals "c", "d", "e", "f", and "g" when performing bitwise operations on them?

```
1 wire [3:0] a = 4'b0101 ;
2 wire [3:0] b = 4'b1100 ;
3
4 wire [3:0] c = a & b   ; // AND
5 wire [3:0] d = a | b   ; // OR
6 wire [3:0] e = a ^ b   ; // XOR
7 wire [3:0] f = ~(a & b); // NAND
8 wire [3:0] g = ~(a | b); // NOR
```

Problem 3.5. Assuming we have a 4-bit signal "a" with the value of binary 4'b0011, what are the resulting values of signals "b", "c", "d", "e", and "f" when using reduction operations?

```
1 wire [3:0] a = 4'b0011;
2
3 wire b = &a ;     // Reduction AND
4 wire c = |a ;     // Reduction OR
5 wire d = ^a ;     // Reduction XOR
6 wire e = ~&a;     // NOT Reduction AND
7 wire f = ~|a;     // NOT Reduction OR
```

Problem 3.6. Assuming we have a 128-bit signal "a", how can we use a reduction operation in Verilog to determine the statement "if(a==128'hffff_ffff_ffff_ffff_ffff_ffff_ffff_ffff)", which checks if all the 128 bits of the signal "a" are all binary ones. Similarly, express the statement "if(a==128'h0)" with a reduction operation, which checks if all the 128 bits are all binary zeros.

Additionally, use reduction operations to express the following two statements: "if(a!=128'h0)" and "if(a!=128'hffff_ffff_ffff_ffff_ffff_ffff_ffff_ffff)", which check if not all the 128 bits are binary zeros and ones, respectively.

Problem 3.7. Assume that we have two 8-bit signals "a" and "b" with the hexadecimal values 8'h12 and 8'h34, respectively. What are the resulting values of signals "vec1", "vec2", "vec3", "vec4", and "vec5" when using concatenation operations below?

```
1  wire [7:0] a = 8'h12;
2  wire [7:0] b = 8'h34;
3
4  wire [15:0] vec1 = {a, b}              ;
5  wire [7:0]  vec2 = {a[3:0], b[7:4]} ;
6  wire [15:0] vec3 = {4{a[3:0]}}         ;
7  wire [15:0] vec4 = {8'h0, a}           ;
8  wire [15:0] vec5 = {b, 8'h0}           ;
```

Problem 3.8. Assume that we have an 8-bit signal "a" with the hexadecimal values 8'h17. What are the values of 10-bit signals "lls2" and "lrs2" when performing logical left and right shift operations below?

```
1  wire [7:0] a = 8'h17   ;
2  wire [9:0] lls2 = a<<2;
3  wire [9:0] lrs2 = a>>2;
```

Training: Vim, Simulation Env, Siemens ModelSim

Prior to commencing the projects, it is advisable to peruse **Appendix A** and **Appendix B**. These sections provide essential guidance on tasks such as editing and inserting Verilog code using the Vim editor, simulating Verilog designs using Siemens Modelsim, and setting up a fundamental simulation

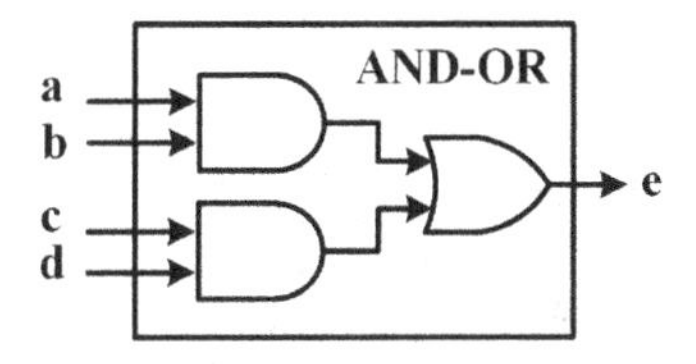

FIGURE 3.5
Design structure of AND-OR circuit.

environment through the creation of a TCL script, a filelist, and a testbench. Following this preparation, you can proceed to gain hands-on experience by working on the projects outlined below.

PBL 1: AND-OR Combinational Circuit

1) Design a combinational circuit using a concurrent *assign* block. Refer to Figure 3.5 for the design structure, which includes two AND gates and one OR gate. The circuit should have single-bit inputs "a", "b", "c", and "d", and a single-bit output "e".
2) Create a basic testbench with the following test cases.

```
1  initial begin
2    a = 1'b0; b = 1'b0; c = 1'b0; d = 1'b0;
3    #10; a = 1'b1; b = 1'b1; c = 1'b0; d = 1'b1;
4    #10; a = 1'b1; b = 1'b0; c = 1'b1; d = 1'b1;
5    #10; a = 1'b1; b = 1'b0; c = 1'b1; d = 1'b0;
6  end
```

PBL 2: AND-MUX Combinational Circuit

1) Design a combinational circuit using a conditional *assign* block. Refer to Figure 3.6 for the design structure, which includes two AND gates and one multiplexer. The circuit should have single-bit

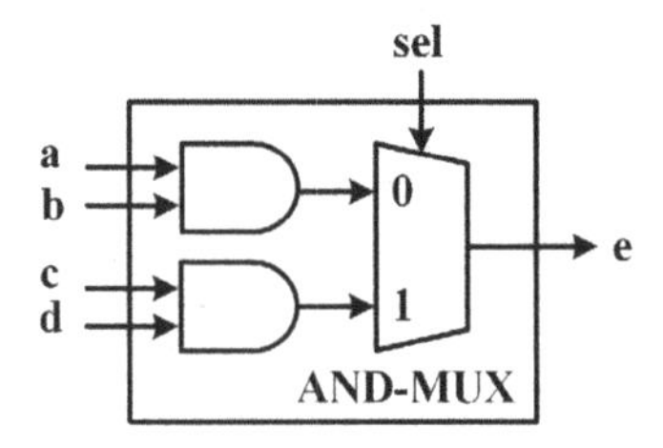

FIGURE 3.6
Design structure of AND-MUX circuit.

inputs "a", "b", "c", and "d", a single-bit select input "sel", and a single-bit output "e".

2) Create a testbench featuring the following test cases and verify the design functionality.

```
1  initial begin
2    sel = 1'b0; a = 1'b0; b = 1'b0; c = 1'b0; d = 1'b0;
3    #10; sel = 1'b0; a = 1'b1; b = 1'b1; c = 1'b0; d = 1'b1;
4    #10; sel = 1'b1; a = 1'b1; b = 1'b1; c = 1'b0; d = 1'b1;
5    #10; sel = 1'b0; a = 1'b0; b = 1'b1; c = 1'b1; d = 1'b1;
6    #10; sel = 1'b1; a = 1'b0; b = 1'b1; c = 1'b1; d = 1'b1;
7  end
```

PBL 3: AND-OR-Tri Combinational Circuit

1) Design a combinational circuit using a conditional *assign* block. Refer to Figure 3.7 for the design structure, which includes two AND gates, one OR gate, and one tri-state buffer. The circuit should have single-bit inputs "a", "b", "c", and "d", a single-bit enable input "en", and a single-bit output "f".

2) Create a testbench featuring the following test cases and verify the design functionality.

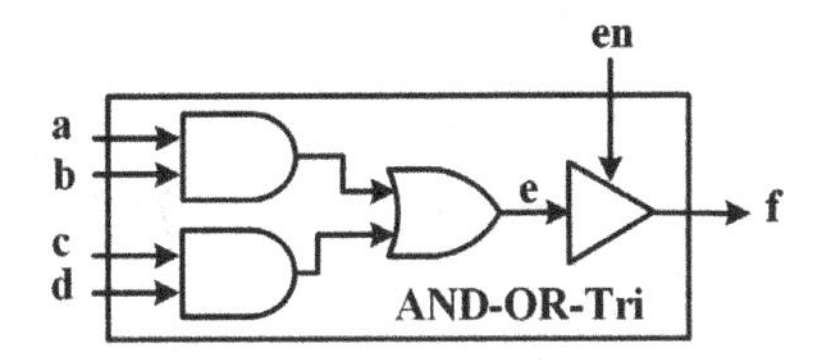

FIGURE 3.7
Design structure of AND-OR-Tri circuit.

```
1  initial begin
2    en = 1'b0; a = 1'b0; b = 1'b0; c = 1'b0; d = 1'b0;
3    #10; en = 1'b0; a = 1'b1; b = 1'b1; c = 1'b1; d = 1'b1;
4    #10; en = 1'b1; a = 1'b1; b = 1'b1; c = 1'b0; d = 1'b1;
5    #10; en = 1'b1; a = 1'b0; b = 1'b1; c = 1'b1; d = 1'b0;
6    #10; en = 1'b1; a = 1'b1; b = 1'b0; c = 1'b1; d = 1'b1;
7    #10; en = 1'b1; a = 1'b0; b = 1'b0; c = 1'b1; d = 1'b1;
8  end
```

4

RTL Design with Verilog HDL

Together with Chapter 3, Chapter 4 delves into the fundamentals of Verilog design methodologies for describing basic hardware. It emphasizes the essential knowledge of RTL designs using Verilog, encompassing topics such as blocking and non-blocking designs, asynchronous and synchronous reset, hierarchical design, and instantiation. PBL 4–5 demonstrate the design and simulation of basic sequential circuits.

4.1 Design Statements: if-else, case, for/while/repeat/-forever loop

Verilog design statements encompass conditional structures like *if-else* and *case*, as well as loop constructs including *for*, *while*, *repeat*, and *forever*. These statements find application within both *always* and *initial* blocks within Verilog modules.

4.1.1 if-else statement

4.1.1.1 Missing-else design on a latch

The *if-else* statement in Verilog is used to conditionally execute a code block. The syntax for this structure is as follows. This structure can include multiple *if-else* pairs and may omit the final *else*. Similar to various programming languages, Verilog's *if-else* statements can also be nested for handling more complex conditions.

```
1  if (expression1) begin
2    statement1;
3  end else if (expression2) begin
4    statement2;
5  end else begin //can be omitted if it is assigned to itself
6    statement3;
7  end
```

DOI: 10.1201/9781003187080-4

The following is an example of using an *if-else* statement in a level-triggered *always* block. In this example, the LHS signal "q" in the *always* block must be declared as *reg* (line 2). Additionally, all signals that can trigger the computation must be included in the trigger list, including not only "a" and "b" on the RHS of the assignment, but also the "rst" and "en" signals in the conditional statements.

In the *if-else* statement (lines 5–11), the behavior of the signal "q" is described based on three conditions. When "∼rst" (the negation of reset signal) is TRUE, "q" is reset to binary 1'b0 (line 6). This initialization sets the output to a known state when the circuit is reset. When "∼rst" is FALSE and the enable signal "en" is TRUE, the value of "q" is updated with the result of "a AND b" (line 8). This operation computes the AND gate output and assigns it to "q". Consequently, the output follows the logic operation "a AND b" when the enable signal is active. When both "∼rst" and "en" are FALSE, meaning that both the reset and enable signals are inactive, "q" retains its previous value (line 10). This behavior can be implemented using a latch, allowing "q" to retain its previous state until new valid inputs are received.

```
1  //Ex 1: A Latch Design Using if-else Statement
2  module example_latch (input      rst, en, a, b,
3                        output reg q            );
4  always @(rst, en, a, b) begin
5    if(~rst) begin
6      q <= 1'b0 ;
7    end else if(en) begin
8      q <= a&b  ;
9    end else begin
10     q <= q    ;
11   end
12 end
13 endmodule
14
15 //Recommended Design for a Latch
16 module example_rec_latch (input      rst, en, a, b,
17                           output reg q            );
18 wire nxt_q = a&b; //combinational AND gate
19 always @(rst, en, nxt_q) begin
20   if(~rst) begin
21     q <= 1'b0 ;
22   end else if(en) begin
23     q <= nxt_q;
24   end //last else (q<=q;) is omitted
25 end
26 endmodule
```

A recommended coding style for designing the same circuit is shown in lines 15–26. First, the circuit design is divided into a combinational AND logic using an *assign* block (line 18) and a sequential latch using an *always* block (lines 19–25). Additionally, a missing-else approach is employed in the *if-else* statement, which is equivalent to the first design example.

4.1.1.2 Missing-else design on a register

As a comparison, consider the following example that uses an *if-else* statement within an edge-triggered *always* block. The sensitivity list of the *always* block lists two events, "*negedge* rst" and "*posedge* clk", indicating a design involving a register.

In the *if-else* statement in lines 5–11, the signal "q" is initialized to binary 1'b0 when "~rst" (the negation of reset signal) is TRUE, indicating that the reset is asserted. When "~rst" is FALSE and the enable signal "en" is TRUE, "q" is assigned the result of "a AND b" at rising clock edges (line 8). This allows the register output to follow the logic operation of the "a AND b" when the enable signal is active at the rising clock edge. When both "~rst" and "en" are FALSE, meaning that both reset and the enable signals are inactive, "q" retains its previous value at rising clock edges (line 10). This behavior can be implemented using a register, allowing "q" to retain its previous state until the following clock edges.

```
1  //Ex 2: A Register Design Using if-else Statement
2  module example_reg (input        rst, clk, en, a, b,
3                      output reg q                   );
4  always @(negedge rst, posedge clk) begin
5    if(~rst) begin
6      q <= 1'b0 ;
7    end else if(en) begin
8      q <= a&b  ;
9    end else begin
10     q <= q    ;
11   end
12 end
13 endmodule
14
15 //Recommended Design for a Register
16 module example_rec_reg (input        rst, clk, en, a, b,
17                         output reg q                   );
18 wire nxt_q = a&b; //combinational AND gate
19 always @(negedge rst, posedge clk) begin
20   if(~rst) begin
21     q <= 1'b0 ;
22   end else if(en) begin
23     q <= nxt_q;
```

```
24    end //last else (q<=q;) is omitted
25  end
26  endmodule
```

A recommended coding style for designing the same circuit is shown in lines 15–26. First, the circuit design is divided into a combinational AND logic using an *assign* block (line 18) and a sequential register using an *always* block (lines 19–25). Additionally, a missing-else approach is employed in the *if-else* statement, which is equivalent to the first design example.

4.1.1.3 Using logic operators in if-else statement

The third example illustrates the use of reduction operators within an *if-else* statement. In line 5, the statement "$if(\sim|a)$" is written as an equivalent expression "$if(a == 32'h0)$" to check whether all bits in the signal "a" are set to binary zeros. Similarly, in line 7, the statement "$if(\&a)$" is used to express an identical function of "$if(a == 32'hffff\_ffff)$". The last *else* is omitted in this *if-else* statement, meaning that the signal "b" will sustain its previous value if both "$\sim|a$" and "$\&a$" are FALSE. Therefore, a latch is needed for the design circuit, except for the NAND and OR gates.

```
1   //Ex 3: Using Logic Operators in if-else Statement
2   module example_logic_op_ifelse (input [31:0] a,
3                                   output reg   b);
4   always @(a) begin
5     if(~|a) begin          // if(a==32'h0)
6       b <= 1'b0;
7     end else if(&a) begin // if(a==32'hffff_ffff)
8       b <= 1'b1;
9     end
10  end
11  endmodule
12
13  //Recommended Design for EX3
14  module example_rec_ex3 (input [31:0] a,
15                          output       b);
16  assign b = ~|a ? 1'b0 :
17                   &a ? 1'b1 : b ;
18  endmodule
```

An optimal design approach for this instance is illustrated in lines 14–18. It is essential to declare the output "b" as a *wire* type since it serves as the LHS signal within a nested conditional *assign* block. While there exist alternative Verilog representations, it is advisable to employ *assign* blocks for designing combinational circuits.

4.1.1.4 Nested if-else statement

Take a look at the example below, which involves a nested *if-else* statement. Initially, it may seem like the hardware resembles a 2-stage multiplexer with two selection signals, "a" and "b". However, upon closer inspection, it reveals itself as a straightforward AND gate. This is because "c" is set to binary 1'b1 only when both "a" and "b" are TRUE. To represent such straightforward logic, we recommend using a concurrent *assign* block, as demonstrated thoughtfully in lines 18–21.

```
1  //Ex 4: Nested if-else Statement
2  module example_nested_ifelse (input      a, b,
3                                output reg c   );
4  always @(a,b) begin
5    if(a) begin
6      if(b) begin
7        c <= 1'b1;
8      end else begin
9        c <= 1'b0;
10     end
11   end else begin
12     c <= 1'b0;
13   end
14 end
15 endmodule
16
17 //Recommended Design for EX4
18 module example_rec_ex4 (input  a, b,
19                         output c   );
20 assign c = a&b ;
21 endmodule
```

Design Rule – if-else Statement

- It is allowed to omit the last *else* in an *if-else* statement. However, designers must keep in mind that omitting the last *else* in a level-triggered *always* block will generate latches to sustain the value for the LHS signals, and omitting the last *else* in an edge-triggered *always* block will generate registers to preserve the value for the LHS signals.

- Within a level-triggered *always* block, confine your designs solely to the associated signals and logic. Any designs pertaining to unrelated signals and logic should be segregated into separate level-triggered *always* blocks or arranged within *assign* blocks.
- When using edge-triggered *always* blocks, it is advisable to use them exclusively for describing registers. Combinational circuits, on the other hand, should be defined within other edge-triggered *always* blocks or in *assign* blocks.

4.1.2 case statement

In Verilog, the *case* statement is used for conditional branching based on the value of an expression. The syntax of a *case* statement is as follows: it begins with the keyword *case* and ends with the keyword *endcase*. The expression within the following parenthesis is compared with the items listed in order. The corresponding item statements are executed when a match is found. The *default* item is necessary to handle any possible values of the expression that do not match any specific items.

```
1  case(expression)
2    item1     : begin item statement1    end
3    item2     : begin item statement2    end
4    default   : begin DEFAULT statement end
5  endcase
```

4.1.2.1 A 4-to-1 multiplexer design using case-endcase statement

The first example below illustrates the design of a 4-to-1 multiplexer. It utilizes a 2-bit signal named "sel", which can take on values represented by hexadecimal 2'h0, 2'h1, and 2'h2, as listed in the three items. The *default* branch handles any other possible value, including hexadecimal 2'h3, as well as combinations of high-Z and/or unknown states. In cases where "sel" is equal to 2'h3 or exhibits a high-Z or unknown state, the simulation will report "Mismatch: sel = 2'h3, X, or Z" to highlight the mismatched scenario. While the Verilog standard defines X and Z values, real-world circuits typically adhere to binary zeros and ones. The use of high-Z values is limited to situations involving the description of tri-state buffers.

```
1  // Ex1: A Design Example Using case-endcase Statement
2  module example_case1 (input      [1:0] a, b, c, sel,
3                        output reg [1:0] d);
4  always @ (a, b, c, sel) begin
5    case (sel)
6      2'h0    : d <= a;
7      2'h1    : d <= b;
8      2'h2    : d <= c;
9      default : begin
10                 d <= 2'b0;
11                 $display("Mismatch: sel = 2'h3, X, or Z");
12               end
13   endcase
14 end
15 endmodule
16
17 // An Equivalent Design to Ex1
18 module example_case2 (input   [1:0] a, b, c, sel,
19                       output  [1:0] d);
20 assign d = (sel==2'h0) ? a :
21                (sel==2'h1) ? b :
22                   (sel==2'h2) ? c : 2'h0;
23 endmodule
```

An equivalent design using a conditional *assign* block is shown in lines 17–23. Note that the output data type in the first example is a *reg*, while in the alternative design it must be a *wire*.

4.1.2.2 Missing default in case-endcase statement

The second example below illustrates a design featuring a *case-endcase* statement without a *default* branch. The 2-bit "sel" signal can take on three possible values: hexadecimal 2'h0, 2'h1, and 2'h2, which are listed in lines 7–9. However, there is no matching item for the value 2'h3. Consequently, an unintended latch will be generated to retain the value of the signal "d" when unmatched items occur. To tackle this concern, it is advisable to incorporate a *default* item to ensure the comprehensiveness of the *case-endcase* statement.

```
1  //Ex 2: Missing default Causes Unwanted Latch!
2  module example_case_missing_item (input      [1:0] a, b, c,
3                                    input      [1:0] sel,
```

```
4                                       output reg [1:0] d);
5  always @ (a, b, c, sel) begin
6    case (sel)
7      2'h0 : d <= a;
8      2'h1 : d <= b;
9      2'h2 : d <= c;
10   endcase
11 end
12 endmodule
```

4.1.2.3 X and Z in case-endcase statement

The *X* and *Z* data types are applicable for bit-wise matching within the *case* expression. For instance, in the provided *case-endcase* statement, four entries are enumerated: 4'b0000, 4'b0zzz, 4'b0xxx, and 4'b1010. Any remaining values of the 4-bit signal "sel" are routed to the *default* item.

```
1  //Ex 3: An Example Using X and Z in case-endcase Statement
2  module example_case1_x_z (input  [3:0]  sel);
3  //--- Design Code---//
4  always @(sel) begin
5    case (sel)
6      4'b0000 :   $display ("item0 matches");
7      4'b0zzz :   $display ("item1 matches");
8      4'b0xxx :   $display ("item2 matches");
9      4'b1010 :   $display ("item3 matches");
10     default :   $display ("nothing matches");
11   endcase
12 end
13
14 //--- Testbench ---//
15 initial begin
16   sel = 4'b0000;          // Printed log: # item0 matches
17   #100 sel = 4'b0xxx;     // Printed log: # item2 matches
18   #100 sel = 4'b0zzz;     // Printed log: # item1 matches
19   #100 sel = 4'b1111;     // Printed log: # nothing matches
20 end
21 endmodule
```

In the testbench, the signal "sel" is initialized with the binary value 4'b0000 (line 16), resulting in a successful match with the first item (line 6). This successful match is indicated in the printed-out log as "item0 matches", as noted in line 16. Subsequently, following a delay of 100 ns (assuming a timescale unit of 1 ns), the signal "sel" is assigned the binary value 4'b0xxx (line 17). Consequently, the third item (line 8) is matched, depicted by the printed-out log as "item2 matches". In a similar manner, after an additional 100 ns interval, the execution of line 18 triggers the printing of "item1 matches". This arises from a bit-wise match being performed with the second item in the *case-endcase* statement (line 7). Lastly, since "sel=4'b1111" fails to match any of the listed entries, it defaults to the *default* item (line 10), prompting the output of "nothing matches".

Below is another example of a *case-endcase* statement for item matching. The Verilog system task *$time* is used to print out the simulation time in the timescale unit of 1 ns. In the design module, the *case-endcase* statement lists the possible values for a 2-bit "sel" signal – 2'b00, 2'b01, 2'b1x, and 2'b1z – with all remaining values directed to the *default* item in line 10.

```
1  //Ex 4: An Example Using X and Z in case-endcase Statement
2  module example_case2_x_z (input [1:0] sel);
3  //---Design Code---//
4  always @(sel) begin
5   case (sel)
6     2'b00: $display("%d ns, sel=%b, 2'b00 sel", $time,sel);
7     2'b01: $display("%d ns, sel=%b, 2'b01 sel", $time,sel);
8     2'b1x: $display("%d ns, sel=%b, 2'b1x sel", $time,sel);
9     2'b1z: $display("%d ns, sel=%b, 2'b1z sel", $time,sel);
10    default: $display("%d ns, sel=%b, def sel", $time,sel);
11  endcase
12 end
13
14 //--- Testbench ---//
15 initial begin
16   sel = 2'b00;      //Printed log: # 0 ns, sel=00, 2'b00 sel
17   #10 sel = 2'b01; //Printed log: # 10 ns, sel=01, 2'b01 sel
18   #10 sel = 2'b10; //Printed log: # 20 ns, sel=10, def sel
19   #10 sel = 2'b1x; //Printed log: # 30 ns, sel=1x, 2'b1x sel
20   #10 sel = 2'b1z; //Printed log: # 40 ns, sel=1z, 2'b1z sel
21 end
22 endmodule
```

In the testbench, the initial value of "sel=2'b00" matches the first item, so the printed log is "# 0 ns, sel=00, 2'b00 sel" (line 16). When "sel=2'b01", the second item is matched, and the log displays "# 10 ns, sel=01, 2'b01 sel" (line 17). Similarly, when "sel=2'b10", the *default* item is matched since there is no listed item that matches this value, and the log shows "# 20 ns, sel=10, def sel" (line 18). The last two cases match the third and fourth items, with "sel=2'b1x" and "sel=2'b1z", respectively. The corresponding logs demonstrate "# 30 ns, sel=1x, 2'b1x sel" (line 19) and "# 40 ns, sel=1z, 2'b1z sel" (line 20).

Design Rule – case-endcase Statement

- Using *case-endcase* statements is highly recommended over a long *if-else* statement in level-triggered *always* blocks. This approach enhances the clarity and efficiency of the design code, especially when dealing with a large number of listing items.
- It is crucial to ensure that all items in a *case-endcase* statement are completely listed. In cases where a *default* item is missing, the design may inadvertently generate unwanted latches or registers. This can lead to unintended behavior and potential simulation failures.

4.1.3 Loop statements: for, while, repeat, and forever

Similar to other programming languages, Verilog HDL supports loop statements in *always* and *initial* blocks. These loop statements, including *for*, *while*, *repeat*, and *forever*, offer a way to control the execution of a statement zero, one, or more times. Loop statements can be efficient for simulation models and testbenches, but they are generally not recommended for use in synthesizable RTL designs. This is because loop statements can lead to uncertain hardware synthesis and may introduce potential performance and timing issues during the implementation process.

Design Rule – for/while/repeat/forever Loop Statement

- The loop statement serves as an efficient approach for simulation and testbenches. However, it is generally not recommended in synthesizable designs due to uncertain hardware implementations and performance.
- To ensure reliable hardware implementation, consider using alternative constructs such as sequential circuits for the iterative control, which are more appropriate for synthesis purposes.

4.1.3.1 for loop

A *for* loop is particularly beneficial when dealing with situations that involve repetitive operations. The Verilog syntax is illustrated below. It enables the control of statement execution through a 3-step process: 1) Initialization: a variable that controls the number of loop iterations initialized. 2) Evaluation: an expression is evaluated, and the statement within the loop's scope is executed when the expression is TRUE. The loop terminates if the expression is FALSE. 3) Step Assignment: the value of the loop-control variable is modified to control the execution of subsequent iterations.

```
1  for (Initial assignment; expression; step assignment) begin
2    statement;
3  end
```

An instance where a *for* loop finds practical application is the refreshing of memory blocks or register arrays in a testbench. In the subsequent example, the first line declares a register array with a depth of 16, labeled as "mem[0:15]". Noteworthy is the specification of the range, spanning from the minimum value of 0 to the maximum value of 15. Each individual memory cell has a width of one byte, defined by "reg [7:0]". This method conforms to the customary approach of defining a register array, complete with width and depth declarations.

```
1  reg [7:0] mem[0:15];
2  integer i;
3  initial begin
4    for (i=0; i<16; i=i+1) begin
5      mem[i] = i;
6      $display ("%d register is initialized as %h", i, i);
7    end
8  end
```

The *for* loop in lines 4–7 is included in an *initial* block. By sweeping the index "i", the loop initializes the memory array with the value of i. For example, "mem[0]" is initialized as integer 0, "mem[1]" is initialized as integer 1, "mem[2]" is initialized as integer 2, and so on, up to "mem[15]", which is initialized as integer 15. Note that the memory width is in bytes, so only the least significant eight bits of the integer can be assigned to the memory array.

4.1.3.2 while loop

A *while* loop repeatedly executes the statement as long as the specified *expression* remains TRUE. After each round execution, the loop returns to the beginning and re-evaluates the *expression*. If the *expression* remains TRUE, the *statement* is executed again, and this process continues until the *expression* becomes FALSE. Once the *expression* is FALSE, the loop terminates.

```
1  while (expression) begin
2    statement;
3  end
```

Here is an example demonstrating the use of a *while* loop within an *initial* block. In line 3, the variable "a" is initialized to zero. The following *while* loop, located between lines 4 and 6, is then executed. During simulation, every 10 ns (assuming the timescale unit is 1 ns), the *while* loop increments the variable "a" by one until the condition "$a < 4$" becomes FALSE (when the variable "a" reaches the integer 4).

```
1  integer a;
2  initial begin
3    a=0;
4    while (a<4) begin
5      #10 a = a + 1;
6    end
7  end
```

4.1.3.3 repeat loop

A *repeat* loop executes the specified statement a fixed number of times, as determined by the value of the expression within the parentheses. It provides a straightforward way to perform repetitive tasks for a known number of iterations during simulation.

```
1  repeat (number) begin
2    statement;
3  end
```

Here is an example utilizing the *repeat* loop. In line 3, the variable "a" is initialized to binary 1'b0 within the *initial* block. The following *repeat* loop, located between lines 4 and 6, toggles the value of "a" every 10 ns delay (assuming the timescale unit is 1 ns). The loop statement will be executed four times in total.

```
1  reg a;
2  initial begin
3    a=1'b0;
4    repeat (4) begin
5      #10 a = ~a;
6    end
7  end
```

Below is an additional example illustrating the use of a *repeat* loop for memory block refreshing. In the first line, the memory array's depth is defined using a parameter called "MEM_SIZE", which in this instance is assigned a value of 16. Consequently, the memory access address is defined as a 4-bit *reg* named "addr[3:0]", covering a range from hexadecimal 4'h0 to 4'hf. The width of the memory array is set to eight bits, corresponding to a single byte, as indicated in line 3.

Within the *initial* block, the memory address begins with the initialization of hexadecimal 4'h0 in line 6. Subsequently, a *repeat* loop is employed between lines 7 and 10 to facilitate memory refreshing. This loop iterates 16 times, which corresponds to the value established by the "MEM_SIZE" parameter. During each iteration, the "addr" signal undergoes incrementation by one through the use of the arithmetic operator "+". This process ensures sequential memory access to the succeeding location, as portrayed in line 9.

```
1  parameter MEM_SIZE = 16;
2  reg [3:0] addr;
3  reg [7:0] mem[0:MEM_SIZE-1];
4
5  initial  begin
6    addr = 4'h0;
7    repeat (MEM_SIZE) begin
8      mem [addr] = 8'h0;
9      addr = addr + 4'h1;
10   end
11 end
```

4.1.3.4 forever loop

Unlike other loop statements that have a specific number of iterations, the *forever* loop ensures the continuous execution of the enclosed statement until the simulation completes or is manually terminated. This loop is commonly used for creating continuous or infinitely running processes in a Verilog testbench.

```
1 forever begin
2   statement;
3 end
```

The presented example emulates a 100 MHz clock using a *forever* loop. In line 3, a signal named "clk" is initialized to binary 1'b0 within an *initial* block. Inside the following *forever* loop, the "clk" signal toggles its value every 5 ns (assuming a timescale unit of 1 ns). This toggling effectively simulates a clock period of 10 ns, which corresponds to a 100 MHz clock frequency. However, it's crucial to acknowledge that this design is non-synthesizable and is intended solely for simulation purposes. Generating a clock in this manner does not lead to a valid hardware implementation.

```
1 reg clk;
2 initial begin
3   clk = 1'b0;
4   forever begin
5     #5 clk = ~clk;
6   end
7 end
```

4.2 Blocking and Non-Blocking

Verilog HDL incorporates two types of procedural assignments: blocking and non-blocking. These two assignment types differ in their execution order and their ability to model hardware behavior.

In Verilog, the blocking assignment uses the equals (=) character, while the non-blocking assignment uses the less-than-equals (<=) character pair.

In an *assign* block, only blocking assignments are allowed. In contrast, non-blocking assignments are typically used in an *always* block. Both non-blocking and blocking assignments can be used in an *initial* block. The key distinction lies in the fact that non-blocking assignments can be executed in parallel, whereas blocking assignments must be carried out sequentially.

4.2.1 Examples of blocking designs

Two examples using blocking assignments are provided below. In the first example, the assignments commence with "a=b" (line 9), followed by "b=c" (line 10), then "c=d" (line 11), and finally "d=e" (line 12). Figure 4.1(a) shows the simulation waveform by pulling up and down the input "e" at 21 ns and 31 ns, respectively.

```
1  // An Example of Blocking Design #1
2  module example_blk1 (input       e, clk, rst,
3                       output reg  a           );
4  reg b, c, d;
5  always @ (posedge clk, negedge rst) begin
6    if (~rst) begin
7      a = 1'b0; b = 1'b0; c = 1'b0; d = 1'b0;
8    end else begin
9      a = b;
10     b = c;
11     c = d;
12     d = e;
13   end
14 end
15 endmodule
16
17 // Another Example of Blocking Design #2
18 module example_blk2 (input       e, clk, rst,
19                      output reg  a           );
20 reg b, c, d;
21 always @ (posedge clk, negedge rst) begin
22   if (~rst) begin
23     a = 1'b0; b = 1'b0; c = 1'b0; d = 1'b0;
24   end else begin
25     d = e;
26     c = d;
27     b = c;
28     a = b;
29   end
30 end
31 endmodule
```

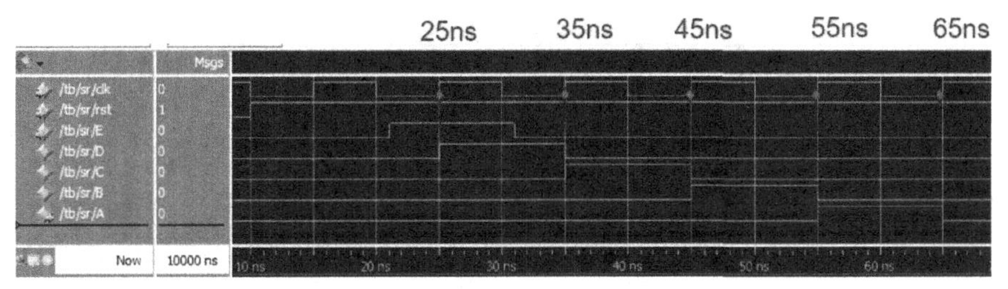

(a) Simulation Result of Blocking Design #1

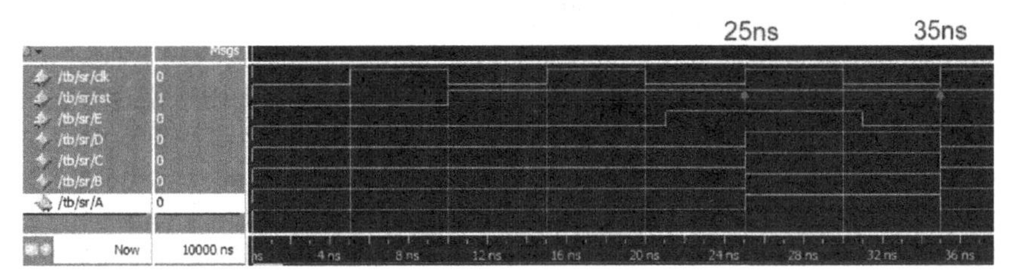

(b) Simulation Result of Blocking Design #2

FIGURE 4.1
Simulation results of blocking designs.

Due to their order-sensitive nature, during the initial rising clock edge at 25 ns, the sequence of execution proceeds as follows: first, the LHS signal "a=b=0" is processed, followed by "b=c=0", then "c=d=0", and finally "d=e=1". Similarly, during the subsequent rising clock edge at 35 ns, the sequence unfolds as "b=0", "c=0", "d=1", and "e=0", resulting in "a=b=0", "b=c=0", "c=d=1", and "d=e=0". At the third rising clock edge, the sequence updates to "a=b=0", "b=c=1", "c=d=0", and "d=e=0". The fourth rising clock edge, occurring at 55 ns, triggers a sequence update of "a=b=1", "b=c=0", "c=d=0", and "d=e=0". Finally, during the last rising clock edge at 65 ns, all RHS signals "b-e" take on binary zero values, and the corresponding LHS signals "a-d" update in sequence as follows: "a=b=0", "b=c=0", "c=d=0", and "d=e=0".

The second example is a similar blocking design but with a different assignment order: "d=e" in line 25, "c=b" in line 26, "b=c" in line 27, and finally "a=b" in line 28. As depicted in Figure 4.1(b), after the RHS signal "e" transitions to binary one, the first rising clock edge occurs at 25 ns. Since the RHS signal "e=1", the LHS signal "d=e=1" follows. After that, "c=d" is executed with the updated "d=1", resulting in "c=d=1". Similarly, "b=c=1" and then "a=b=1" are executed subsequently.

In the synthesis analysis depicted in Figure 4.2, the first blocking design seems to represent a shift register, utilizing four registers to transition the input "e" to the output "a" over four clock cycles. On the other hand, the second blocking design appears to describe a single registers with the input

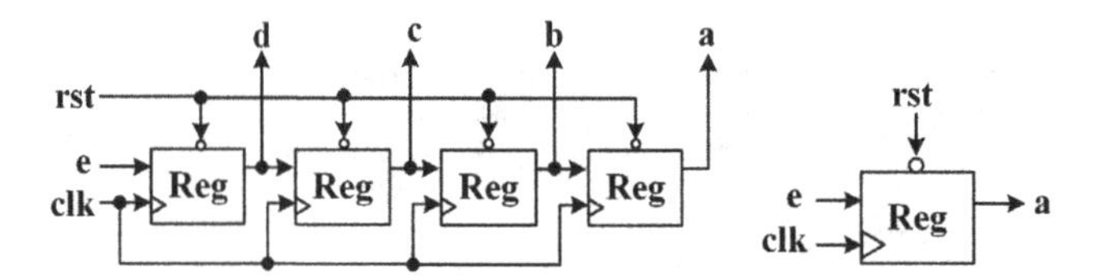

(a) Hardware Results of a 4-Stage Shifter (b) Hardware Results of a Register

FIGURE 4.2
Hardware results of blocking designs.

"e" and the output "a", exhibiting behavior to transition the input to the output over one clock cycle. The internal *reg* variables "b-d" are optimized in the second design, as they do not contribute to any intermediate logic functionality.

However, it is essential to note that neither of these blocking designs is recommended for describing a shift register or register in actual hardware. The reason is that blocking assignments in an *always* block are order-sensitive, which diverges from the behavior of real hardware. In RTL description, it is generally preferred to use non-blocking assignments in an *always* block to accurately model concurrent behavior and ensure sequential updates, which aligns with practical hardware operations.

4.2.2 Examples of non-blocking designs

The non-blocking designs provided below are recommended for creating a 4-stage shift register and a register in Figure 4.2. By utilizing descriptions in the first example, the order of execution for the statements "a <= b;" (line 9), "b <= c;" (line 10), "c <= d;" (line 11), and "d <= e;" (line 12) is insensitive in the *always* block. These assignments can be executed in parallel as soon as a rising clock edge or a falling reset edge occurs, precisely emulating the real hardware behavior of a shift register. At each rising clock edge, the simulator accesses the current values of all the RHS signals "b-e" and assigns their values to the LHS signals "a-d" concurrently, achieving the desired behavior of a shift register as found in actual hardware.

The second non-blocking design is a typical description of a register. The input to the register is "e", and the output is "a". The assignment activates on every rising clock edge.

```
1  // Non-Blocking Design on a 4-Stage Shift Register
2  module example_nonblk1 (input       e, clk, rst,
3                         output reg a          );
4  reg b, c, d;
```

```
5  always @ (posedge clk, negedge rst) begin
6    if (~rst) begin
7      a <= 1'b0; b <= 1'b0; c <= 1'b0; d <= 1'b0;
8    end else begin
9      a <= b;
10     b <= c;
11     c <= d;
12     d <= e;
13   end
14 end
15 endmodule
16
17 // Non-Blocking Design on a Register
18 module example_nonblk2 (input      e, clk, rst,
19                         output reg a           );
20 always @ (posedge clk, negedge rst) begin
21   if (~rst) begin
22     a <= 1'b0;
23   end else begin
24     a <= e    ;
25   end
26 end
27 endmodule
```

4.3 Asynchronous and Synchronous Reset

In RTL design, a global reset is necessary to initialize the entire circuit into a known state. There are two types of resets: asynchronous and synchronous. An asynchronous reset becomes active as soon as the reset signal is asserted, while a synchronous reset activates on the active clock edge when the reset signal is asserted.

In many ASIC designs, an asynchronous reset is commonly used because it can occur independently of the clock signal. However, certain FPGA blocks may offer support for a synchronous reset, allowing it to synchronize with the clock signal. The choice of reset type depends on the specific requirements and constraints of the design and the target FPGA platform.

4.3.1 Verilog design examples of asynchronous and synchronous reset

To illustrate the difference between asynchronous and synchronous reset, let's consider the following examples. The first example (lines 1–11) demonstrates an asynchronous reset. The trigger list comprises both the falling reset edge and the rising clock edge. As soon as the reset signal goes low, the reset activates in line 6: "q <= 1'b0;" to initialize the "q" output. In contrast, if the reset is binary one, the normal operation in line 7: "q <= d;" activates on every rising clock edge, since the event "*posedgeclk*" is also included in the trigger list. This corresponds to a register description with an asynchronous reset.

```
1  // A Design Example of Asynchronous Reset
2  module example_async_reset (input       d, clk, rst,
3                             output reg q           );
4  always @ (posedge clk, negedge rst) begin
5    if (~rst) begin
6      q <= 1'b0;
7    end else begin
8      q <= d    ;
9    end
10 end
11 endmodule
12
13 // A Design Example of Synchronous Reset
14 module example_sync_reset (input       d, clk, rst,
15                            output reg q           );
16 always @ (posedge clk) begin
17   if (~rst) begin
18     q <= 1'b0;
19   end else begin
20     q <= d    ;
21   end
22 end
23 endmodule
```

The second example (lines 13–23) illustrates a synchronous reset. In this case, the trigger list exclusively encompasses the rising clock edge, indicating that the *always* block can be activated solely on every rising clock edge. If the reset signal is binary zero, the output "q <= 1'b0;" can be carried out once the rising clock edge occurs. On the other hand, during normal operation when the reset is high, "q <= d;" activates on every rising clock edge. This description represents a register with synchronous reset.

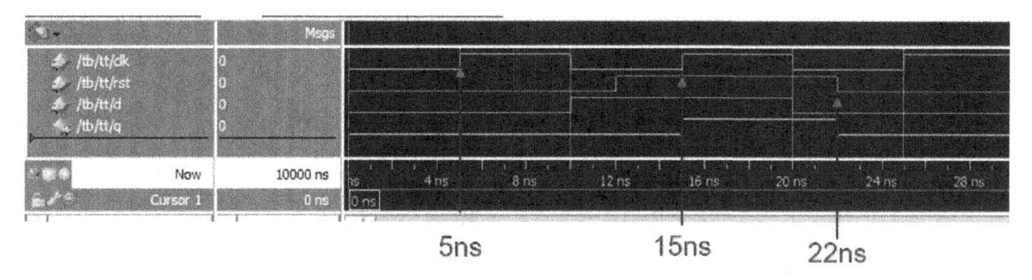

(a) Simulation Result of Asynchronous Reset

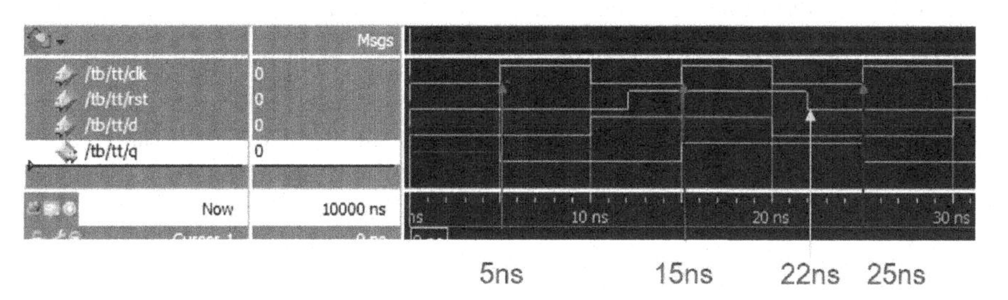

(b) Simulation Result of Synchronous Reset

FIGURE 4.3
Simulation results of asynchronous and synchronous reset.

4.3.2 Simulation and synthesis of asynchronous and synchronous reset

Figure 4.3 illustrates the simulation waveform for both design examples. The testbench applies the same input stimuli, including "clk", "rst", and "d". At the beginning of the simulation, the initial reset sets the output "q" to zero for the asynchronous reset simulation. However, for the synchronous reset simulation, the reset occurs when the rising clock edge happens at 5 ns. At the second rising clock edge occurring at 15 ns, the output "q" follows the input "d" because "rst=1". This behavior is observed in both designs.

The distinction between the two simulations becomes evident when a falling reset edge occurs at 22 ns. In Figure 4.3(a), illustrating the behavior of an asynchronous reset, the falling reset edge immediately triggers the reset operation as "q<=1'b0;". In contrast, in the case of a synchronous reset, as demonstrated in Figure 4.3(b), the reset operation must wait until the next rising clock edge at 25 ns to take effect. This simulation demonstrates the fundamental difference between asynchronous and synchronous reset in terms of their responsiveness to the reset signal. The asynchronous reset responds immediately to the reset signal, whereas the synchronous reset waits for the active clock edge to perform the reset operation.

The design incorporating an asynchronous reset is depicted in Figure 4.4(a), featuring a reset input directly initializing the register. In contrast,

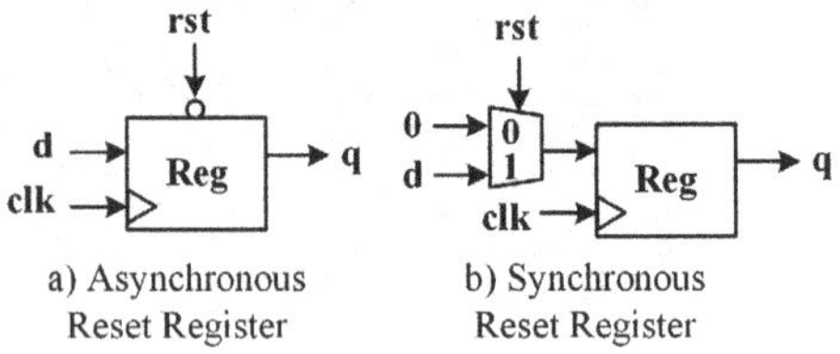

FIGURE 4.4
Hardware results of asynchronous and synchronous reset.

the design employing a synchronous reset is illustrated in Figure 4.4(b), incorporating a multiplexer to choose the register input during each rising clock edge. When the reset signal is low, the register input remains at 0; however, when the reset signal is high, the "d" input can be selected as the input to the register.

4.4 Hierarchical Design and Instantiation

In the realm of RTL design, intricate circuits can be constructed by integrating multiple modules in a hierarchical fashion. The module being integrated is commonly denoted as a "submodule", whereas the module responsible for integrating these submodules is recognized as the "parent module". Submodules are instantiated within parent modules, and the inputs/outputs of these instances can be connected with other signals within the parent module.

This hierarchical design strategy can encompass several instantiation tiers, comprising upper-tier parent modules and lower-tier submodules. The uppermost design module holds the designation of the "top module" or "level-one module", while subsequent tiers are labeled as the "level-two module", "level-three module", and so forth. Employing this hierarchical design methodology comes strongly endorsed in hardware description practices. It enables individual submodules to be independently designed and subsequently instantiated multiple times with distinct instances within the upper-level modules. Such an approach effectively reduce the overall design intricacy and fosters a comprehensible and maintainable design structure.

4.4.1 Verilog design example of hierarchical design

Here's an example that demonstrates module instantiation. The first module (lines 1–7) illustrates a half adder design, including an AND gate and an XOR gate. The module name is "half_adder", and it takes single-bit inputs "a" and "b", producing single-bit outputs "sum" and "c_out". The hardware result is depicted in Figure 4.5(a).

It is important to note that the RTL description of a half adder design is succinctly demonstrated in the comment line 4, which utilizes the arithmetic operator "+" instead of gate-level constructions. In addition, this approach uses a concatenation operator to combine the carry-out result (denoted as "c_out") and the sum result (denoted as "sum").

```
1  module half_adder (output sum, c_out,
2                     input  a, b       );
3  // Behavioral model level design:
4  // assign {c_out,sum} = a + b;
5  assign sum   = a ^ b;
6  assign c_out = a & b;
7  endmodule
8
9  // Behavioral model level design:
10 // assign {c_out,sum} = a + b + c_in;
11 module full_adder (output sum, c_out,
12                    input  a, b, c_in);
13 wire w1, w2, w3;
14
15 half_adder u1_half_adder (.sum  (w1),
16                           .c_out(w2),
17                           .a    (a ),
18                           .b    (b ));
19
20 half_adder u2_half_adder (.sum  (sum ),
21                           .c_out(w3  ),
22                           .a    (w1  ),
23                           .b    (c_in));
24
25 assign c_out = w2 | w3;
26 endmodule
```

In what follows, the half adder design serves as a submodule that is instantiated in the parent module named "full_adder". In lines 11–12, the inputs ("a", "b", and "c_in") and outputs ("sum" and "c_out") are declared within the parent module. It is important to note that the IO ports in the parent module have the same names as those in the submodule, but they represent different signals within distinct designs. Three internal signals, "w1", "w2", and "w3", are declared as the *wire* data type in line 13 and will be used to connect different circuits.

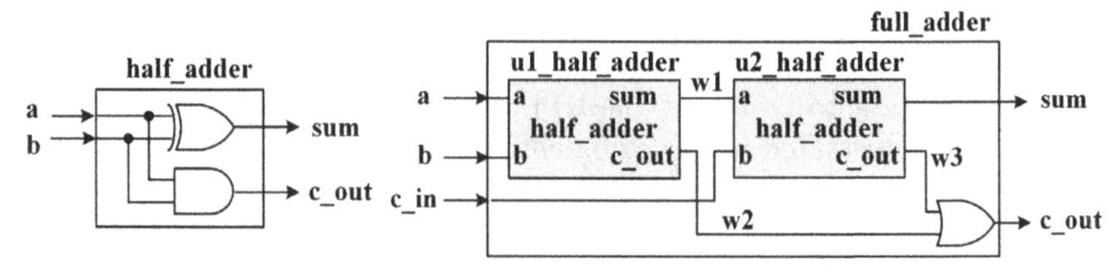

(a) Submodule Design on Half Adder (b) Top Module Instantiation into a Full Adder

FIGURE 4.5
Hierarchical design and instantiation.

The block diagram in Figure 4.5(b) depicts the parent module design, which includes two instantiations of the submodule and an additional OR gate. The first instantiation, named "u1_half_adder", and the second instantiation, named "u2_half_adder", are specified following the submodule name "half_adder" in lines 15 and 20, respectively. Each instantiation must have a unique name to avoid confusion. In this figure, the submodule's names and IOs inside the gray boxes belong to the submodules, while the signals and instantiation names outside the gray boxes belong to the parent module. More specifically, for the first instance, "u1_half_adder", its inputs "a" and "b" are connected to the parent module inputs "a" and "b", respectively. Its outputs "sum" and "c_out" are connected to the parent module's wires "w1" and "w2", respectively.

Following the block diagram, in Verilog, line 15 specifies the submodule's name ("half_adder") and assigns it a unique instance name ("u1_half_adder"). The connections for each IO port are listed in the following parenthesis, where each port is preceded by a dot to indicate the submodule's IOs. In lines 15–18, the submodule's output "sum" is connected to the internal wire "w1" in the parent module, while the submodule's output "c_out" is connected to the internal wire "w2". The submodule's inputs "a" and "b" are connected to the parent module's inputs "a" and "b", respectively.

For the second instance, "u2_half_adder", its inputs "a" and "b" are connected to the internal wire "w1" and the parent module's input "c_in", respectively. Its outputs, "sum" and "c_out", are connected to the parent module's output "sum" and internal wire "w3", respectively. Finally, line 25 describes the OR gate, which takes the two internal wires "w2" and "w3" as inputs and assigns the result to the parent module output "c_out".

It is worth noting that in the RTL description, the full adder can be succinctly described as "assign {c_out,sum} = a + b + c_in;" using the arithmetic operator "+" and the concatenation operators, as mentioned in the comment line 10.

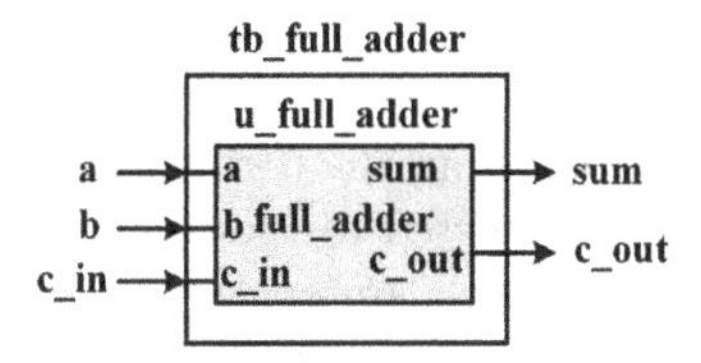

FIGURE 4.6
Design instantiation in testbench.

4.4.2 Verilog testbench example of instantiation

Within RTL designs, the utilization of hierarchical structures and instantiation can notably streamline design complexity, resulting in concise and readable design code. Another key purpose of design instantiation is to incorporate the design-under-test seamlessly into a testbench environment.

Displayed in Figure 4.6, the top design module labeled as "full_adder" and its IOs, including "a", "b", "c_in", "sum", and "c_out", are listed within the gray box. In the corresponding Verilog description, the top design module "full_adder" is utilized in line 5, and its IOs are listed with a dot prefix in lines 5–9 as ".a", ".b", ".c_in", ".sum", and ".c_out". This notation signifies the top design module and all its IOs are being connected and referenced in the testbench.

Outside the gray box, the design-under-test is instantiated within the testbench and assigned the identifier "u_full_adder". All signals within the testbench, including "a", "b", "c_in", "sum", and "c_out", interface with the corresponding IOs of the design module. In the Verilog code, the specific design instance name "u_full_adder" is provided after the design module's name in line 5, and all the IO connections are detailed within the parentheses, denoted as "(a)", "(b)", "(c_in)", "(sum)", and "(c_out)", following the design IOs being connected, in lines 5–9.

```
1  module tb_full_adder ();
2  reg  a, b, c_in;
3  wire sum, c_out;
4  // Design-Under-Test Instantiation
5  full_adder u_full_adder (.sum  (sum  ),
6                           .c_out(c_out),
7                           .a    (a    ),
8                           .b    (b    ),
9                           .c_in (c_in ));
10 endmodule
```

With this instantiation and connection, the project hierarchy is structured as follows: tb_full_adder -> u_full_adder -> u1_half_adder and u2_half_adder. During simulation, the testbench is capable of driving signals "a", "b", "c_in" (which are connected to the design's inputs) while monitoring signals "sum" and "c_out" (which are connected to the design's outputs) to verify the design's functionality. The process of design instantiation within a testbench will be further elaborated in subsequent sections.

4.5 RTL Design Rules with Verilog HDL

In this chapter, we have covered the basic Verilog syntax for describing synthesizable hardware. To summarize, here are some recommended design rules for using Verilog HDL:

Verilog Design Rules

1. **RTL Programming:**
   1) Avoid creating **design loops** within combinational circuits that can lead to unpredictable behavior in hardware.
   2) Since RTL designs rely on clock-edge-triggered registers, generating half-cycle signals is not feasible.
   3) Align the bit numbers of signals within signal assignments to avoid bit-mismatching issues.
   4) In edge-triggered *always* blocks, include only clock edge and reset edge in the sensitivity list. In level-triggered *always* blocks, include all signals that can trigger the LHS signals inside the sensitivity list.
   5) Separate combinational and sequential circuit descriptions into different design blocks, in order to promote clarity in design code and simplify debugging and modifications.
   6) Avoid using **deep logic** to maintain concise and readable design code.
   7) Organize your code by grouping functionally related design statements within the same *always* block, and place functionally unrelated statements in separate design blocks.
   8) For synthesizable RTL designs, avoid unknown statuses for signals and IOs during design and simulation. Initialize all inputs of the design-under-test in the test cases.

2. **RTL Design:**

   1) Use the **hierarchical structure** to divide complex circuit designs into different levels of submodules, promoting independent design and simulation.
   2) Implement a **global reset** for the entire system to enhance system stability and predictability during startup or error conditions.
   3) Minimize the use of **multiple clocks** to avoid complexity and clock domain crossing issues.
   4) For data transfer between different clock domains, use a **data buffer** or **synchronizer** to ensure proper synchronization.
   5) Follow the **register-in register-out** rule for RTL designs to create a clear pipeline between registers and avoid unnecessary timing issues.
   6) Establish consistent register designs across various scenarios involving low/high valid reset, synchronous/asynchronous reset, and rising/falling clock edge.
   7) Avoid using the division operator provided by Verilog HDL for both ASIC and FPGA designs. Use arithmetic IP modules instead.

3. **FPGA Design and Verification:**

   1) For FPGA verification within the ASIC design flow, strive to maximize the resemblance between the FPGA design code and the ASIC design code.
   2) FPGAs may automatically reset all signals when powered on, which may not apply to ASICs. In ASIC designs, ensure that all signals are reset in the initial stage.

4. **Design for Testing and Remediation Circuitry:**

   1) Integrate testing circuits to verify proper circuit functionality.
   2) Incorporate Build-in-Self-Test (BIST) modules for memory blocks, enabling the identification and containment of errors.
   3) Develop remediation circuits to rectify identified errors.

5. **Consideration of Timing:**

   1) When developing RTL code, it's crucial to take timing considerations into account. Practical outcomes might be different from RTL simulation results due to timing issues.

2) During design specification, identify the anticipated operational clock frequency and the longest critical path.

6. **Memory:** Avoid reading memory before writing to it. Some FPGAs may reset all memory blocks to zeros when powered up.

Exercises

Problem 4.1. What are the differences between asynchronous and synchronous reset?

Problem 4.2. What are the differences between non-blocking and blocking assignments in Verilog HDL designs?

Problem 4.3. Identify the ten bugs present in the Verilog code below, which include seven different types of design simulation warnings/errors: 1) signal declaration errors, 2) bit mismatch warnings, 3) incomplete trigger lists, 4) incomplete *case-endcase* statements, 5) errors in non-blocking and blocking assignments, 6) incomplete *begin – end* statements, and 7) incomplete *module – endmodule* blocks.

```
1  module design_bugs (input       [7:0] a, b, c,
2                      input       [1:0] sel,
3                      input             rst, clk,
4                      output reg        d       );
5  assign d = a & b | c;
6
7  wire [7:0] e;
8  always @ (a, b) begin
9    case (sel)
10     2'h0 : e <= a;
11     2'h1 : e <= b;
12   endcase
13 end
14
15 wire [7:0] f;
16 always @ (posedge clk, negedge rst) begin
17   if(~rst) begin
18     f = 8'h0;
19   end else begin
20     f = e   ;
21   end
```

```
22
23 wire [7:0] g;
24 initial begin
25   g = 4'h0;
26   #10; g = f;
27 end
```

Problem 4.4. Consider the following two designs, labeled as Design #1 and Design #2, both using *always* blocks.

1) Which one of the designs is level sensitive, and which one is edge sensitive?

2) Do they use non-blocking or blocking assignments?

3) In Design #1, is the data type of "c" a *reg* or *wire*? In Design #2, is the data type of "c" a *reg* or *wire*?

```
1 // Design #1
2 always (a, b) begin
3     c <= a & b;
4 end
5
6 // Design #2
7 always (posedge clk) begin
8     c <= a & b;
9 end
```

Problem 4.5. In designs #1 and #2 below, which design uses asynchronous reset, and which design uses synchronous reset?

```
1  // Design #1
2  always (posedge clk, negedge rst) begin
3    if(~rst) begin
4      c <= 1'b0 ;
5    end else begin
6      c <= a & b;
7    end
8  end
9
10 // Design #2
11 always (posedge clk) begin
```

```
12     if(~rst) begin
13       c <= 1'b0 ;
14     end else begin
15       c <= a & b;
16     end
17   end
```

Problem 4.6. Figure 4.7 illustrates a Verilog design representing a 4-bit 2-to-1 multiplexer. The design comprises two 4-bit inputs labeled as "bus_a[3:0]" and "bus_b[3:0]", along with a 4-bit output labeled as "bus_c[3:0]". A single-bit signal denoted as "sel" serves as the selection input. When "sel" is set to 0, the output "bus_c[3:0]" is driven by the input "bus_a[3:0]". Conversely, when "sel" is set to 1, the output "bus_c[3:0]" is determined by the input "bus_b[3:0]".

The design utilizes four individual single-bit 2-to-1 multiplexers. These submodules' IOs connections are established with the top module's IOs, progressing from MSB to LSB. Specifically, the IOs of the first submodule, designated as "a-c", are interconnected with the MSB signals of "bus_a[3]", "bus_b[3]", and "bus_c[3]" within the top module. The second submodule's IOs are correspondingly linked to the intermediate bits "bus_a[2]", "bus_b[2]", and "bus_c[2]" of the main module, and this pattern continues until the LSB is reached. Furthermore, the selection signals labeled as "s" in the submodules are integrated together with the top module's selection signal "sel".

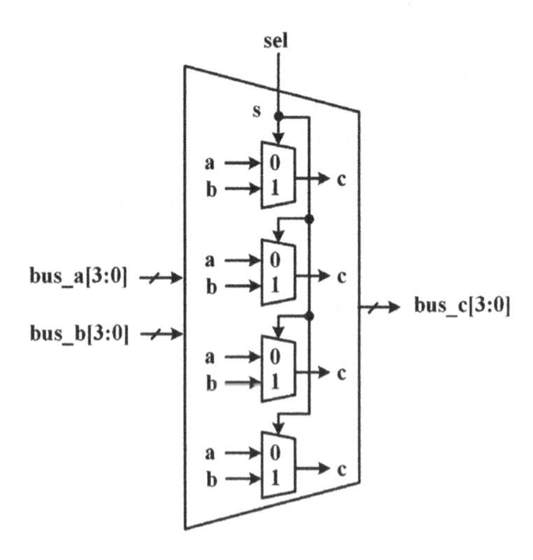

FIGURE 4.7
Design instantiation (four single-bit mux into a 4-bit mux).

The Verilog design of the submodule is presented in lines 1–5, followed by the declaration of the top module name and IOs in lines 7–16. Write the instantiation code below the comments (lines 12–14).

```
1  // Submodule Design: Single-Bit 2-to-1 Mux
2  module mux2to1_1bit (input  a, b, s,
3                       output c      );
4  assign c = ~s ? a : b;
5  endmodule
6
7  // Top Module Instantiation: 4-bit 2-to-1 Mux
8  module mux2to1_4bit (input  [3:0] bus_a, bus_b;
9                       input        sel         ,
10                      output [3:0] bus_c       );
11 //-------- Instantiation -----------------
12
13 endmodule
```

PBL 4: AND-OR-Latch Sequential Circuit

1) Design a sequential circuit according to the illustration in Figure 4.8. The combinational section, housing two AND gates and one OR gate, can be succinctly described using a concurrent *assign* block. For the sequential aspect, incorporating a latch, can be succinctly described employing a conditional *assign* block or a level-triggered *always* block.

   The circuit has six single-bit IO ports: four inputs labeled "a-d", one input "en" signal (for latch enable), and one output "f". An internal signal "e" is used to connect the combinational and sequential parts.

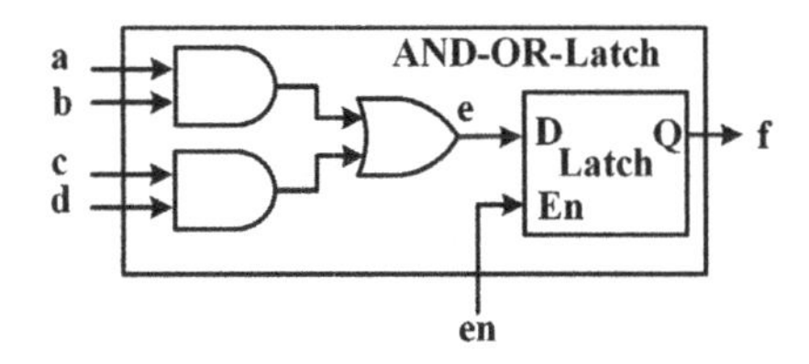

FIGURE 4.8
Design structure of AND-OR-latch circuit.

2) Create a testbench featuring the following test cases and verify the design functionality.

```
1  initial begin
2    en = 1'b0; a = 1'b0; b = 1'b0; c = 1'b0; d = 1'b0;
3    #10; en = 1'b1; a = 1'b1; b = 1'b1; c = 1'b1; d = 1'b0;
4    #10; en = 1'b0; a = 1'b1; b = 1'b0; c = 1'b0; d = 1'b1;
5    #10; en = 1'b1; a = 1'b0; b = 1'b1; c = 1'b1; d = 1'b1;
6    #10; en = 1'b0; a = 1'b1; b = 1'b1; c = 1'b0; d = 1'b1;
7  end
```

PBL 5: AND-OR-Reg Sequential Circuit

1) Design a sequential circuit according to the illustration in Figure 4.9. The combinational section, housing two AND gates and one OR gate, can be succinctly described using a concurrent *assign* block. For the sequential aspect, incorporating a register, can be succinctly described employing an edge-triggered *always* block. It's important to note that the register responds to both a falling-edge reset (indicated as "negedge rst") and a rising-edge clock (indicated as "posedge clk").

   The IO ports are all single-bit signals, including four inputs "a-d", one input "clk", one input "rst" signal, and one output "f". The internal signal "e" is used for connecting the combinational and sequential parts.

2) Create a testbench featuring the following test cases and verify the design functionality.

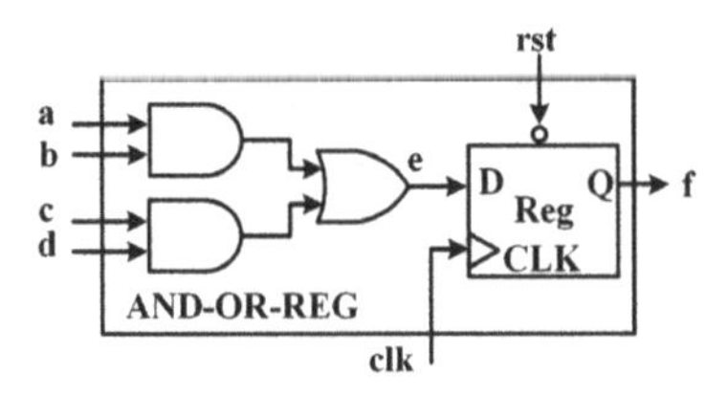

FIGURE 4.9
Design structure of AND-OR-REG circuit.

```
1  initial begin
2    rst=1'b0; clk=1'b0;
3    a = 1'b0; b = 1'b0; c = 1'b0; d = 1'b0;
4    #10 rst=1'b1;
5    #10; a = 1'b1; b = 1'b1; c = 1'b0; d = 1'b0;
6    #10; a = 1'b0; b = 1'b1; c = 1'b0; d = 1'b1;
7    #10; a = 1'b0; b = 1'b1; c = 1'b1; d = 1'b1;
8    #10; a = 1'b0; b = 1'b1; c = 1'b0; d = 1'b1;
9  end
10
11 always #5 clk=~clk;
```

5

Design Simulation with Verilog HDL

Chapter 5 provides an introduction to Verilog designs for building a standard testbench, covering elements such as system tasks, compiler directives, functions and tasks, and delay controls. It concludes with a demonstration of an automated simulation environment, showcasing a structured project directory, an automated testbench utilizing bus functional models and monitors, the creation of test cases and test plans, and the analysis of simulation logs. Design simulation and verification are pivotal procedures within industry IC design flow. An automated simulation environment is indispensable for verifying and covering design functionalities, as well as identifying corner cases through a multitude of randomized tests.

PBL 6–8 focus on establishing an automated design simulation environment, involving tasks such as developing a testbench, utilizing TCL scripts, and creating filelists. PBL 9 illustrates design instantiation through hierarchical design methodologies.

5.1 System Tasks

Verilog provides several system tasks that primarily for simulation and verification purposes. Some of the commonly used system tasks include \$*display*, \$*monitor*, \$*fwrite*, \$*time*, \$*finish*, \$*stop*, \$*dumpfile*, \$*dumpvars*, \$*readmemh*, and \$*writememh*. It's important to note that these system tasks are meant for simulation and debugging purposes only and are not intended to be used for synthesizable hardware design. Synthesis tools will disregard these system tasks when generating hardware.

5.1.1 \$display, \$monitor, and \$fwrite

In Verilog, the \$*display*, \$*monitor*, and \$*fwrite* system tasks are commonly used to generate informational and debugging messages, which can be recorded in log files or the transcript window, facilitating the tracking of simulations.

The \$display task serves as a simple printing mechanism to display variables in the code. Its syntax is shown in the first line below. The debug messages are printed out with specific data formats for the variables specified

DOI: 10.1201/9781003187080-5

within the "string" argument. For instance, %b prints variables in binary, %d prints variables in decimal, and %h prints variables in hexadecimal. The second line demonstrates an example where the variables "a", "b", and "c" are printed in binary, decimal, and hexadecimal, respectively, when the $display task is executed.

On the contrary, the *$monitor* task exclusively prints its variables when they undergo changes. The syntax for *$monitor* is depicted in line 4 below, and its argument in the form of a "string" follows identical syntax and format guidelines as the *$display* task. In line 5, an illustrative instance is provided, demonstrating that signals "a", "b", and "c" are printed in binary, decimal, and hexadecimal formats, respectively, every time any of the three signals experiences a change.

```
1  Syntax: $display("string", variable1, variable2, etc.);
2  $display("display signals a=%b, b=%d, c=%h", a, b, c);
3
4  Syntax: $monitor("string", variable1, variable2, etc.);
5  $monitor("monitor signals a=%b, b=%d, c=%h", a, b, c);
```

The *$display* and *$monitor* tasks are commonly used for quick debugging and simulations to print out data in the simulation console. On the other hand, the *$fwrite* task can be employed to log simulation results, generate reports, or create output files that can be read by other programs. Unlike *$display* and *$monitor*, which output data to the console, *$fwrite* writes the formatted data to a log file.

The syntax for the *$fwrite* task is shown in the first line below. The "file_id" is an integer that represents the file to which the data will be written. This file ID can be obtained using the *$fopen* function. The "string" argument specifies the format of the data, following the conventions of employing the *$display* and *$monitor* system tasks.

```
1  Syntax: $fwrite(file_id, "string", variable1, etc.);
2
3  initial begin
4    file_id = $fopen("Output.log", "w");
5    if(file_id) begin
6      $fwrite(file_id, "An example of fwrite System Task!");
7      $fclose(file_id);
8    end else begin
9       $display("Error opening file!");
10   end
11 end
```

In the example starting in the third line, the *$fwrite* task is demonstrated within an *initial* block. The file "output.log" is opened for writing (line 4), the string "An example of fwrite system task!" is written to it (line 6) if the log file is successfully opened (line 5), and then the file is closed (line 7). If *$fopen* fails (e.g., if the file cannot be created), writing to the file should be avoided. The error message "Error opening file!" will be displayed in the simulation console.

5.1.2 $time

The *$time* task is a valuable tool for tracing signals in a simulation waveform. It allows printing out the specific simulation time when used within *$display*, *$monitor*, and/or *$fwrite* tasks. This aids in precisely locating when and where bugs occur in the simulation waveform.

Below is an example demonstrating the use of the *$time* within a *$display* task. In the *$display* string, the format of the time is specified as decimal (%d in the first line). The example log in line 3 illustrates how the *display* task prints out the simulation time and the values of the signal "a" in different formats.

```
1  $display("ERROR occurs at %d ns: a=%d, a=%b, a=%h",
2                                  $time, a, a, a);
3  //Printed log: ERROR occurs at 10 ns: a=16, a=10000, a=10
```

5.1.3 $finish and $stop

The *$finish* task is used to terminate the simulation, while the *$stop* task is utilized to pause the simulation. Both tasks can be employed in situations where the simulation needs to be stopped.

In the example below, an *initial* block compares the "rtl_result" with the "golden_result" in the second line. If the data checking fails, a debug message is printed using the *$display* task in line 3. The simulation is then paused using the *$stop* task with a delay of 100 ns (assuming the timescale unit is 1 ns) in line 4. It is recommended to include the delay information to observe what happens following the test failure, as otherwise, the simulation waveform stops exactly at the time when the data checking fails.

```
1  initial begin
2    if(rtl_result != golden_result) begin
```

```
3       $display("ERROR occurs at %d ns!", $time);
4       #100; $stop;
5     end
6   end
```

5.1.4 $dumpfile and $dumpvars

The *$dumpfile* task serves the purpose of generating a waveform file that records all signal changes during simulation. To specify which signals should be included in the waveform file, you can make use of the *$dumpvars* task. Sharing this waveform file within project groups can significantly enhance the efficiency of debugging and analyzing simulation results.

The syntax for *$dumpfile* and *$dumpvars* is provided below in lines 1–2, respectively. Keep in mind that the format of the dump file depends on the simulator being used. For example, Siemens ModelSim produces waveform files in *.vcd* format, while Debussy/Verdi generate files in *.fsdb* format. The *$dumpvars* task, shown in the second line, accepts two arguments: the first argument determines the hierarchy level to be included in the waveform. Using "0" includes all signals in the current module and any lower-level instantiated modules, while "1" includes only signals in the current module. The second argument specifies the name of the module to be included in the waveform.

```
1   Syntax: $dumpfile("dump_file.vcd");
2   Syntax: $dumpvars(level, Module name);
3
4   $dumpfile("dut.vcd");
5   $dumpvars(0, tb.dut);
```

In line 4, an example of utilizing *$dumpfile* to generate a waveform file named "dut.vcd" is demonstrated. Meanwhile, line 5 showcases the *$dumpvars* task, specifying the inclusion of all signals in the module "tb.dut" and its lower-level instantiated modules in the waveform. Here, "tb" represents the testbench, and "dut" is the top-level design module instantiated within the testbench. The hierarchical path effectively defines the module and any sub-modules to be incorporated into the waveform.

5.1.5 $readmemh and $readmemb

As mentioned earlier, a memory block can be defined as an array of registers with a width specified by the *reg* declaration and a depth corresponding to the number of *reg* arrays. In the first line below, an example is presented to declare a register array. It is 8 bits wide ($reg[7:0]$), representing bytes, and has a depth of 256 ($mem[0:255]$), giving rise to an array with indices ranging from 0 to 255.

The memory array can be loaded by reading data from a text file using the system tasks *$readmemh* (for hexadecimal data) or *$readmemb* (for binary data). The syntax for *$readmemh* is shown in the second line. The first argument of the *$readmemh* task specifies the file name, while the second argument indicates the memory array to be loaded.

```
reg [7:0] mem[0:255];
Syntax: $readmemh("file name", mem);

$readmemh("array.txt", mem);
```

For instance, in line 4, it demonstrates the loading of all data from a file named "array.txt" into the memory array. Assuming that the file "array.txt" contains 256 hexadecimal data entries in the following format:
00
01
02
...
ff
Each line in the file corresponds to a single hexadecimal byte. Specifically, the task is to load the hexadecimal value 8'h00 into memory location "mem[0]", 8'h01 into "mem[1]", 8'h02 into "mem[2]", and so on, until the final value 8'hff is loaded into "mem[255]". If you choose to use *$readmemb*, the data in the text file should be in the binary format.

5.1.6 $random and $urandom

$random and *$urandom* are system tasks utilized for generating random numbers: *$random* can generate both signed and unsigned random numbers, whereas *$urandom* specifically generates unsigned random numbers. In both cases below, "seed" is an optional argument that specifies the seed for the random number generator. If provided, it initializes the random number generator to a specific state. If omitted, the random number generator is seeded based on the current simulation time. Both system tasks return a 32-bit random integer value.

```
Syntax: $random(seed)
Syntax: $urandom(seed)
```

Typically, the *$random* and *$urandom* system tasks are utilized in combination with *initial* or *always* blocks within Verilog testbench. In the example below, both tasks are called 100 times within an *initial* block to generate 200 random numbers, which are then displayed using *$display*. A seed of "02142024" is used in both tasks to ensure that the random sequence can be repeated each time running the simulation. This is particularly helpful when debugging or verifying the design.

```
initial begin
  repeat(100) begin
    $display("Random Number: %0d", $random(02142024));
    $display("Unsigned Random Number: %0d",
              $urandom(02142024));
  end
end
```

5.1.7 A testbench example using system tasks

This section illustrates a simulation example using a single-bit full adder design. The full adder takes three single-bit inputs: "a", "b", and "c_in" (the carry-in bit), and produces two single-bit outputs: "sum" and "c_out" (the carry-out bit).

```
module full_adder (output sum   ,
                   output c_out ,
                   input  a     ,
                   input  b     ,
                   input  c_in  );
assign {c_out, sum} = a + b + c_in;
endmodule
```

Below is a comprehensive testbench designed to verify the functionality of the given design. It comprises four fundamental components: the design instantiation, waveform dumping, bus functional model, and the monitor. The testbench commences with the compiler directive *$timescale* on the first line, which sets the time unit and the time precision for the modules. The time unit represents the measurement unit for simulation time and delay values. Further details about this directive will be elaborated in the subsequent section.

```
1  `timescale 1ns/1ns //reference time/resolution
2  module tb_full_adder();
3  reg  a, b, c_in;  // LHS signals in an initial block
4  wire sum, c_out;  // wire connection between tb and dut
5
6  // --------  DUT Instantiation ------------
7  full_adder u_full_adder (.sum   (sum  ),
8                           .c_out (c_out),
9                           .a     (a    ),
10                          .b     (b    ),
11                          .c_in  (c_in));
12
13 // --------  Dump VCD Waveform ------------
14 initial begin
15   $dumpfile("full_adder.vcd");
16   $dumpvars(0,tb_full_adder.u_full_adder);
17 end
18
19 // ------ Bus Function Model (BFM) ---------
20 initial begin
21   a=1'b0;  b=1'b0; c_in=1'b0;             // golden result: 00
22   #10;  a=1'b0;  b=1'b0; c_in=1'b1;  // golden result: 01
23   #10;  a=1'b0;  b=1'b1; c_in=1'b0;  // golden result: 01
24   #10;  a=1'b0;  b=1'b1; c_in=1'b1;  // golden result: 10
25   #10;  a=1'b1;  b=1'b0; c_in=1'b0;  // golden result: 01
26   #10;  a=1'b1;  b=1'b0; c_in=1'b1;  // golden result: 10
27   #10;  a=1'b1;  b=1'b1; c_in=1'b0;  // golden result: 10
28   #10;  a=1'b1;  b=1'b1; c_in=1'b1;  // golden result: 11
29 end
30
31 // ------------  Monitor -----------------
32 reg [1:0] mem [0:7];
33 $readmemb("../monitor/GoldenModel.txt", mem);
34
35 integer i;
36 initial begin
37   for (i=0; i<8; i=i+1) begin
38     #5;
```

```
39      if({cout,sum}==mem[i]) begin
40        $display ("Data Comparison Passes!");
41      end else begin
42        $display ("ERROR at: %d ns, golden c_out=%b, sum=%b,
43      but DUT c_out=%b, sum=%b when a=%b, b=%b, c_in=%b.",
44      $time, mem0[i][1], mem0[i][0], c_out, sum, a, b, c_in);
45        $stop;
46      end
47      #5;
48    end
49  end
50  endmodule
```

A. Design Instantiation

The testbench is named "tb_full_adder" within the *module – endmodule* encapsulation. Unlike hardware descriptions, a typical testbench does not have IO ports as it is specifically created for simulation purposes. In lines 3–4, the testbench declares several signals: "a", "b", and "c_in" are declared as *reg* because they will be assigned in an *initial* block, while "sum" and "c_out" are declared as *wire* since they are used solely for connecting to the IO ports of the design. All five signals are local in the testbench.

In lines 6–11 of the code, an instance of the design module "full_adder" is created using the unique instance name "u_full_adder". During this module instantiation, the five signals that were previously declared in the testbench are connected to the corresponding IOs of the design module. For the sake of simplicity, the same signal names are used for both the signals in the testbench and the IOs of the design module. The IO ports of the design module are indicated by a dot notation preceding their names, while the signals enclosed in parentheses refer to the signals declared in the testbench. This establishes a clear association between the testbench signals and the design module's IOs.

When analyzing signals in the waveform, one should pay attention to the hierarchical path of the signals to differentiate between the testbench and the design. If the testbench signals are assigned correctly but the design IOs appear to be different or unknown, it is crucial to review the code related to the instantiation connections.

B. Waveform Dumping

The first *initial* block in lines 13–17 uses the *$dumpfile* task to capture signal changes and create a waveform file named "full_adder.vcd". The *$dumpvars* task is utilized to specify that all the signal changes within the hierarchical module "tb_full_adder.u_full_adder" will be included in the ".vcd" file.

C. Bus Functional Model

Following that, a bus functional model is employed as the bus/interface driver to provide stimulus for the design module. In this case, the possible combinations of the three single-bit inputs are binary 3'b000 to 3'b111, listed in lines 21–28. Corresponding golden output results are provided in the comments and pre-stored in a "txt" file named "GoldenModel.txt". A 10 ns delay is inserted between each two stimulus inputs to allow observing the results in the waveform.

D. Monitor

A monitor is established to compare the design outputs with the pre-stored golden results. In line 33, the golden results from the "GoldenModel.txt" file are loaded into a memory array named "mem[0:7]" using the *$readmemb* task. The register array width is set to two bits, where the MSB corresponds to the golden result for the design output "c_out" and the LSB corresponds to the golden result of the design output "sum". For example, the binary data in the first line of the "GoldenModel.txt" is 00, which corresponds to "c_out=0" and "sum=0". These are the expected golden results for the first stimulus in line 21: "a=1'b0;", "b=1'b0;", and "c_in=1'b0;".

E. Golden Results and Golden Models

Complex designs often require the creation of golden models, which are implemented using high-level programming languages such as C, Scala, MATLAB, etc. These golden models are capable of producing anticipated outcomes, which are then recorded in files like "GoldenModel.txt". These files play a crucial role as reference data, enabling the validation of the accuracy of the design's outputs.

Alternatively, the golden model can be incorporated into the Verilog testbench through the utilization of the direct programming interface (DPI) offered by simulators. This integration empowers real-time simulation, enabling the assessment of design outputs against the golden results while the simulation is in progress. By harnessing the capabilities of the DPI, it provides enhanced versatility for conducting intricate simulations and validating the functionality of complex designs.

F. Simulation Log

In the testbench, various C-style data types, such as *integer*, *time*, and *real*, can be utilized. In this case, for instance, the index i is of type *integer*, and it sweeps through the memory depth from 0 to 7. Inside the *initial* block, a *for* loop is employed to compare each design output "{c_out, sum}" with the corresponding golden result loaded in the memory array. If the comparison passes, the testbench displays the string "Data Comparison Passes!" (in line 40). However, if the comparison fails, a debug message is printed out, providing information about the simulation time, the golden results, and the design outputs (in line 42–44).

Please take note that the *for* loop initiates with a delay of 5 ns (as shown in line 38) to instigate the data comparison right in the middle of each data output interval. To elaborate, the initial data output transpires between 0 and 10 ns and the comparison is precisely executed at the midpoint of 5 ns. After concluding the comparison for a specific data input, another delay of 5 ns (as seen in line 47) is introduced to synchronize with the forthcoming data input.

5.2 Compiler Directives

Verilog compiler directives are denoted by a backtick ` and take effect as soon as the compiler encounters them. Some common compiler directives include `define, `ifdef-`else-`endif, `include, and `timescale. These directives play a crucial role in controlling the compilation behavior and processing of the Verilog code.

Moreover, compiler directives in one Verilog file can influence the compilation behavior and processing across multiple files. By using these directives strategically, Verilog designers can efficiently manage code reuse, conditional compilation, and timescale specification throughout their projects.

5.2.1 `define vs. parameter

A `define can be utilized to create a "global" text macro that applies to all modules being compiled, similar to the C-language *#define* directive. In contrast, a *parameter* is primarily used inside a module to parameterize an attribute specific to that module alone. The main distinction between `define and *parameter* lies in the fact that `define is a text substitution mechanism, while parameters are actual variables capable of being assigned different values during module instantiation.

In the first example provided below, the second line defines "WIDTH" as an integer with a value of 8 using `define WIDTH 8. This compiler directive is then utilized in lines 3–5 to declare IOs "a", "b", and "s" with a bus width of 8 bits. It's important to note that `define is placed outside the *module – endmodule* block and can function as a global constant accessible to all files compiled together. When the Verilog compiler encounters the bus WIDTH statement in the code, it replaces all occurrences of the defined WIDTH with the integer 8 substitution. This process allows for efficient text-based manipulation of code during compilation.

```
1  // An Example Using 'define
2  `define WIDTH 8
3  module adder (input  [`WIDTH-1:0] a,
4                input  [`WIDTH-1:0] b,
5                output [`WIDTH-1:0] s);
6  assign s = a+b;
7  endmodule
8
9  // An Example Using parameter
10 module adder #(parameter WIDTH=8) (input  [WIDTH-1:0] a,
11                                     input  [WIDTH-1:0] b,
12                                     output [WIDTH-1:0] s);
13 assign s = a+b;
14 endmodule
```

In the second example, line 10 introduces the *parameter* "WIDTH" as a constant with a value of 8 within the *module* – *endmodule* block. Subsequently, in the port declaration, all the IOs "a", "b", and "c" are parameterized with a byte width. By using *parameter*, the design module becomes more flexible and reusable, enabling designers to alter the values of the *parameter* during the instantiation of the design module. This adaptability allows for easy customization of the module's behavior, making it suitable for various applications and scenarios.

5.2.2 `ifdef-`else/`elsif-`endif

Conditional compiler directives in Verilog HDL provide the capability to selectively include certain lines of code during compilation. The `ifdef directive checks for the existence of a defined text macro name, and if the name is defined, the lines following the `ifdef are included. If the name is not defined and an `else directive is present, the source following the `else is compiled. Additionally, the `ifndef directive can be used to verify the absence of a text macro name definition, and if the name is not defined, the lines following the `ifndef directive are included.

To specify multiple options, the `elsif directive can be utilized. Similar to the *if – else if* statement in Verilog, the `elsif directive allows the compiler to check for the existence of the text macro name, and if it exists, the lines following the `elsif are included.

Below is an example of using these conditional compiler directives in Verilog HDL. Suppose we have three floating-point designs with the same functionality but different precision: 32 bits, 64 bits, and 128 bits. You can implement all the Verilog code in the same module and differentiate between different interfaces using `ifdef-`elsif-`endif. In particular, the `ifdef statement checks if the macro PROJECT_32BIT is defined. If it is defined, the module expects 32-bit input data ("i32_data" in line 5) and produces 32-bit output data ("o32_data" in line 6). If PROJECT_64BIT is defined instead, the module expects 64-bit input data ("i64_data" in line 8) and produces 64-bit output data ("o64_data" in line 9). Similarly, if PROJECT_128BIT is defined, the module expects 128-bit input data ("i128_data" in line 11) and produces 128-bit output data ("o128_data" in line 12). By defining the appropriate macros during compilation, you can select the desired precision and customize the module's behavior accordingly.

```
1  // An Example Using `ifdef-`elsif-`endif
2  module FP_adder (input                clk       ,
3                   input                rst       ,
4  `ifdef PROJECT_32BIT
5                   input  [31:0]  i32_data ,
6                   output [31:0]  o32_data
7  `elsif PROJECT_64BIT
8                   input  [63:0]  i64_data ,
9                   output [63:0]  o64_data
10 `elsif PROJECT_128BIT
11                  input  [127:0] i128_data,
12                  output [127:0] o128_data
13 `endif
14                                           );
15 endmodule
```

You have two interface selection options during compilation. The first approach involves using the `define PROJECT_64BIT (an instance to select lines 8–9) within the design module to specify the source code. Alternatively, you can use a TCL command during compilation to globally select the appropriate source code. For instance, in the TCL script in line 5, the directive "+define+PROJECT_64BIT" is utilized to define the text macro name PROJECT_64BIT. Consequently, the lines 8–9 in the design module, which follow the `elsif PROJECT_64BIT directive in line 7, will be selected.

```
1  // Select the source code with a TCL script
2  vlog +define+PROJECT_64BIT \
3  -v ../dut/FP_adder.v
```

5.2.3 `include

The Verilog compiler directive `include enables the complete content of a source file to be integrated into another file during the Verilog compilation process. This results in the same outcome as if the source code were manually copied and pasted into the location where the `include directive is situated. Leveraging `include offers several benefits, such as incorporating frequently utilized designs and functions without the necessity to duplicate code across various locations. This capability contributes to code maintenance and elevates the overall organization of Verilog HDL files.

As an example, provided below is a Verilog file named "load_mem.v", which embodies a very common functionality of loading two dense matrices into register arrays. Specifically, two matrices are initially stored in the "dmx_ieee754.txt" and "tri_ieee754.txt" files in a row-wise or column-wise fashion and are subsequently loaded into two register files, "dmx_ram" and "tri_ram".

```
1  initial begin
2       $display("%d ns, Load Matrices into Reg Files",$time);
3       $readmemh("../golden/dmx_ieee754.txt", dmx_ram);
4       $readmemh("../golden/tri_ieee754.txt", tri_ram);
5  end
```

In the testbench below, the size of the register files is specified. The width and depth of the "dmx_ram" array are determined using the `DMX_MX_WIDTH and `DMX_MX_DPTH directives, while the width and depth of the "tri_ram" array is defined using the `TRI_MX_WIDTH and `TRI_MX_DPTH directives. In line 5, the `include directive is employed to incorporate the contents of the "load_mem.v" file into the testbench, which loads the matrices into the two register arrays.

```
1  module tb ();
2  reg [`DMX_MX_WIDTH*32-1:0] dmx_ram[0:`DMX_MX_DPTH-1];
3  reg [`TRI_MX_WIDTH*32-1:0] tri_ram[0:`TRI_MX_DPTH-1];
4
5  `include "load_mem.v"
6  endmodule
```

5.2.4 `timescale

The `timescale directive is utilized to define the time unit and time precision of the modules that follow it. The statement is typically in the format: `timescale unit/precision. The time unit represents the measurement unit for time values, such as the simulation time and delay values.

For instance, the directive `timescale 1ns/1ps specifies a time unit of 1 ns and a simulation precision of 1 ps. The Verilog code below illustrates a register with an input "d" and an output "q". Two *initial* blocks are used to simulate the same design: the first *initial* block specifies a 30.1 ns delay after the initialization, and the second *initial* block utilizes a 31 ns delay.

```
1  `timescale 1ns/1ps
2  always @(posedge clk) begin
3    q <= d;
4  end
5
6  always #10 clk = ~clk;
7
8  initial begin
9    d = 0;
10   clk = 0;
11   #30.1 d = 1;
12   #20 d = 0;
13 end
14
15 initial begin
16   d = 0;
17   clk = 0;
18   #31 d = 1;
19   #20 d = 0;
20 end
```

The simulation results are presented in Figure 5.1. In Figure 5.1(a), theoretically, the input "d" switches from binary zero to one at 30.1 ns, as described in the first *initial* block. However, since the time unit is 1 ns, the 0.1 ns delay specified in the *initial* block is not considered by the simulator. As a result, the register detects the change in the signal "d" at exactly 30 ns. Following the rising clock edge at 30 ns, "q" follows "d" and becomes binary one.

In Figure 5.1(b), the input "d" switches from binary zero to one at 31 ns, as described in the second *initial* block. Due to the time unit being 1 ns, the change in the signal "d" is not recognized at the rising clock edge at 30 ns. Consequently, during the clock cycle from 30 ns to 50 ns, the output "q" follows the previous value of "d" and remains binary zero. It is only at the

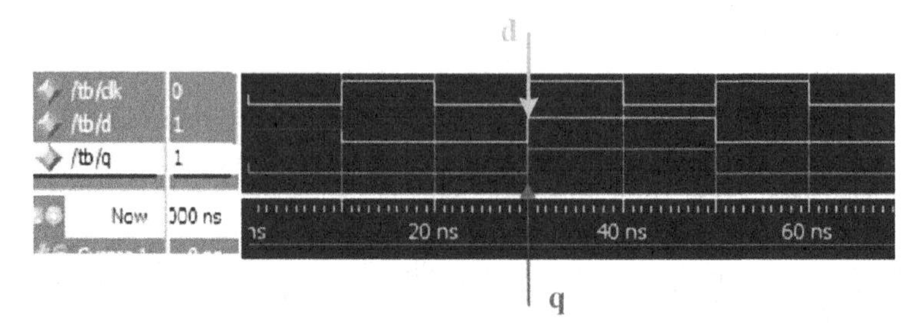

(a) Simulation Results of 30.1 ns Delay

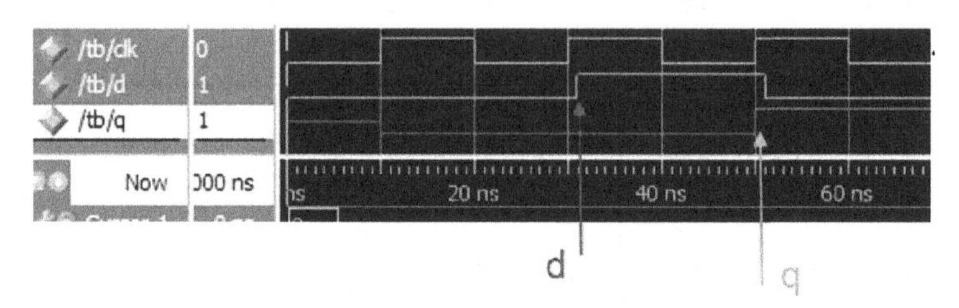

(b) Simulation Results of 31 ns Delay.

FIGURE 5.1
ModelSim simulation results with timescale unit.

next rising clock edge at 50 ns that "q" follows the updated value of "d" and becomes binary one.

It is essential to acknowledge that real circuits exhibit delays, both for combinational circuits and sequential circuits. In practical scenarios, the second timing behavior, where there is a delay between signal changes and the rising clock edge, is more commonly encountered than in the first case. This delay consideration ensures proper synchronization between input signals and clock edges in register-transfer design circuits.

5.3 Functions and Tasks

Functions and Tasks are valuable tools for dividing large procedures into smaller, more manageable units, improving the readability and ease of debugging in the Verilog code. The key distinction between functions and tasks lies in their execution behavior. Functions are blocking, and they must complete their execution before the remaining code can proceed. Conversely, tasks are invoked in a non-blocking manner, meaning that their execution does not hinder the progress of the rest of the code. In this section, we will provide typical examples that demonstrate the use of functions and tasks, highlighting their differences and discussing how to create and invoke them effectively.

5.3.1 Functions

In Verilog, a *function* is a subprogram designed to encapsulate frequently performed operations within the Verilog code. The syntax for a *function* definition is as follows:

```
1 function [<data_type>] <function_name> (<inputs>);
2 <variable declration>
3 <statements>
4 <return result>
5 endfunction
```

The *function* definition begins with the keyword *function*, followed by the data type of the return value, the name of the function, and the input declarations. After that, a semicolon is used, and then the function body is enclosed between *function* and *endfunction*. Inside the *function*, you can declare variables, write statements to perform operations, and finally, use the return operation to specify the value that the *function* should return.

Functions are more limited than tasks and must adhere to the following design rules:

- A *function* cannot include delay control statements, such as *@posedge*, *@negedge*, # delay, or *wait*(). In other words, a *function* must be executed with zero time delay.
- A *function* can have one or more input arguments but cannot have any argument declared as output or inout. A *function* is designed to return a single value to the calling *function*.
- A *function* cannot enable *tasks*. In other words, *functions* cannot be used to trigger or initiate *tasks*.

In the following example, we present a design illustration of a *function* that converts data from the IEEE 754 format to a real data type. Additionally, we demonstrate how to invoke this *function* within a testbench.

A. IEEE 754 Format

In digital circuit design, a single-precision floating-point (FP) number is commonly represented following the IEEE 754 standard, which is widely used for FP arithmetic in digital systems. For instance, a single-precision FP number,

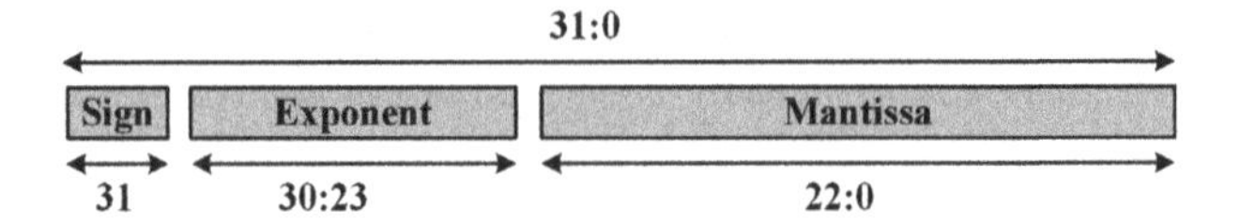

FIGURE 5.2
IEEE 754 Standard (Single Precision): $(-1)^S \times (1.0 + M) \times 2^{(E-127)}$.

depicted in Figure 5.2, can be represented using 32 bits divided into three sections: the sign (S), exponent (E), and mantissa (M), also known as the significand. The MSB denotes the sign bit, indicating whether the number is positive (zero) or negative (one). The subsequent eight bits represent the exponent, which is biased by 127 as per the IEEE 754 standard. This bias implies that to determine the actual power of two, one needs to subtract 127 from the exponent. The remaining 23 bits make up the mantissa, representing the fraction forming the significant digits of the number.

To convert data from the IEEE 754 format to a single-precision FP format, we can utilize the formula $(-1)^S \times (1.0+M) \times 2^{(E-127)}$. This conversion process is suitable for regular FP numbers but does not address special cases such as zero, infinity, denormalized numbers, or NaN (Not a Number) representations. These special cases have unique rules defined in the IEEE 754 standard and require additional handling in the conversion process.

B. Function of IEEE 754 to FP Conversion

As previously mentioned, the provided illustration exemplifies the process of crafting the *function* and invoking it within a Verilog testbench. The designated *function* is titled "ieee754_to_fp", and it accepts a 32-bit input referred to as "ieee754_data", signifying an IEEE 754 formatted floating-point number. Notably, the returned value from the *function* is of the *real* data type.

The sections "sign", "exponent", and "mantissa" are declared in lines 2–4 and assigned in lines 11–13. In line 16, the integer "int_exp" is used to store the actual exponent value by subtracting the bias, which is typical for IEEE 754 format. Additionally, in lines 17–18, "mantissa_val" and "fp_output" are real numbers used to compute the final FP value.

The last line, line 20, of the function is "return fp_output;" which indicates that "fp_output" is the result that the function will return after execution. It is important to note that "fp_output" is declared as a *real* data type at the start of the *function* and is calculated based on the IEEE 754 single-precision FP formula in the body of the *function*.

```
1 function real ieee754_to_fp (input [31:0] ieee754_data);
2 reg          sign    ;
```

```
3  reg [7:0]  exponent;
4  reg [22:0] mantissa;
5
6  integer int_exp      ;
7  real    mantissa_val ; // Divide by 2^23
8  real    fp_output    ;
9
10 // Extracting sign, exponent, and mantissa bits
11 sign     = ieee754_data[31]    ;
12 exponent = ieee754_data[30:23];
13 mantissa = ieee754_data[22:0]  ;
14
15 // Calculating floating-point value
16 int_exp      = exponent-127;
17 mantissa_val = 1.0+(mantissa/8388608.0); // Divide by 2^23
18 fp_output    = (sign?-1:1)*mantissa_val*(2.0**int_exp);
19
20 return fp_output;
21 endfunction
```

If a *function* does not have a *return* statement, it implicitly returns the value of the function itself at the end of its execution. For example, if line 20 is omitted in the "ieee754_to_fp" *function*, the *function* will implicitly return the value of "fp_output" because the last operation of the *function* is assigning a value to "fp_output". However, it is generally considered good practice to include an explicit *return* statement in a *function* for clarity. This makes it unambiguous which value is being returned from the *function*.

C. Call the Function in Testbench

Presented below is a testbench that facilitates the comparison between the output of the design-under-test (referred to as "dut_result") and the expected outcome (referred to as "golden_result"). In the realm of IC design simulation, where outcomes are expressed in IEEE 754 format, the utilization of the "ieee754_to_fp" *function* is imperative. This *function* is employed to convert hexadecimal representations into the *real* data type (lines 10–11), thereby permitting a meaningful and insightful juxtaposition of results. Subsequently, the testbench computes the percentage error (labeled as "error_percent" in line 13) and logs the test outcome into a designated file termed "report.log" (in lines 14–24).

```
1 module tb();
2 reg [31:0] golden_result, dut_result;
3 real       golden_real, dut_real, diff_real;
```

```
4  real        error_percent;
5  integer     log;
6
7  initial begin
8    log=$fopen("./report.log", "w");
9    wait(data_check_en);
10   golden_real   = ieee754_to_fp(golden_result)      ;
11   dut_real      = ieee754_to_fp(dut_result)         ;
12   diff_real     = golden_result_real-dut_result_real;
13   error_percent = diff_real/golden_real*100         ;
14   if(error_percent<5) begin
15     $fwrite (log, "Test Pass!");
16     $fwrite (log, "Error percent: %f%%", error_percent);
17   end else begin
18     $fwrite (log, "Test FAIL!");
19     $fwrite (log, "Error percent: %f%%", error_percent);
20     $fwrite (log, "Golden real: %f", golden_real);
21     $fwrite (log, "DUT real: %f\n", dut_real);
22     $fwrite (log, "Golden Hex: %h", golden_result);
23     $fwrite (log, "DUT Hex: %h\n", dut_result);
24   end
25   $fclose(report);
26 end
27 endmodule
```

In the *initial* block, the log file named "report.log" is opened using the *$fopen* system function, and the file identifier is stored in the integer variable "log" (line 8). The testbench waits for the signal "data_check_en", which acts as a trigger to start the data comparison process.

In lines 10–11, the 32-bit "golden_result" and "dut_result" data are converted from IEEE 754 format to single-precision FP format by calling the ieee754_to_fp *function*. The results are then stored in the "golden_real" and "dut_real" variables, both of which are of the *real* data type. The difference between the golden and design-under-test results is calculated and stored in the variable "diff_real" (line 12). Additionally, the error percentage is computed by dividing the difference by the golden result and multiplying by 100 (line 13).

Based on the computed error percentage, the testbench determines whether the test has passed or failed. If the error percentage is less than 5%, it considers the test as passed (line 14). On the other hand, if the error is 5% or more, it marks the test as failed (line 17). In case of a test failure, the testbench writes "Test FAIL!" to the log file and also includes the error percentage, golden result, and design-under-test result in both real and hexadecimal formats (lines 18–23). This information can be helpful for debugging purposes. Finally, after completing the test, the report log file is closed using the *$fclose* system function (line 25).

5.3.2 Tasks

In Verilog, a *task* is similar to a *function*, but it allows for more complex operations that cannot be accomplished within *functions*. The syntax for defining a task is as follows:

```
1  task <task_name>(<inputs>  ,
2                   <outputs>);
3  <variable declrations>
4  <statements>
5  endtask
```

The *task* definition starts with the keyword *task*, followed by a name for the *task*, a *task* port list, and ends with the keyword *endtask*. Between the *task* and *endtask* keywords, you can include variables declarations and statements, enabling the *task* to perform various operations and tasks.

The following are design rules for *tasks*:

- A *task* can include delay control, such as *@posedge*, *@negedge*, *#delay*, and/or *wait*().
- A *task* can have any number of inputs and outputs.
- A *task* can call other *tasks* or *functions*.

In the following example, we present a design illustration of two *tasks* that emulate the data write and read operations utilizing a control bus. Additionally, we demonstrate how to invoke the *tasks* within a testbench.

A. Tasks of Bus Control Operations

In SoC architectures, the control bus is responsible for managing fundamental bus control operations related to functional register configurations. In the initial stages of SoC development, simulation models can be employed to emulate single burst data transfers on the control bus.

The Verilog *task* below describes a basic write operation using a straightforward control bus. As illustrated in lines 1–13, it consists of two clock phases: in the first clock cycle, the bus commands including the enable signal "ce", data write indicator "wr", register address "addr", and write data "wrdata" are driven on the bus. In the second clock cycle, the write data will be written into the register, and all the commands are reset to binary zeros.

```
//%%%%%%%%%%%%%%%%%%%%%%%%%%%%%%%%%%%%%%%%%%%%%%%%%%%%%%%%%%%%
//               _____         _____
//   clk  _____|     |_____|     |_____
//               ___________
//   ce   _____|           |___________
//               ___________
//   wr   _____|           |___________
//
//addr  -----|  tk_addr  |---8'h00---|
//
//wrdata-----| tk_wrdata |---8'h00---|
//
//%%%%%%%%%%%%%%%%%%%%%%%%%%%%%%%%%%%%%%%%%%%%%%%%%%%%%%%%%%%%
task task_cpu_write; // a task to mimic cpu write
input [7:0] tk_addr    ;
input [7:0] tk_wrdata  ;
begin
  $display ("%d CPU Write address: %h, data: %h",
            $time, tk_addr, tk_wrdata);
  $display ("%d -> Driving ce, wr, wrdata, addr", $time);
  @(posedge clk)       ;
  ce     = 1           ;
  wr     = 1           ;
  addr   = tk_addr     ;
  wrdata = tk_wrdata   ;
  @(posedge clk)       ;
  ce     = 0           ;
  wr     = 0           ;
  addr   = 0           ;
  wrdata = 0           ;
  $display ("=====================");
end
endtask
```

The Verilog code is presented within lines 14–33, encapsulated within a task. It incorporates two inputs: the bus address "tk_addr" and the corresponding write data "tk_wrdata". Lines 21–25 delineate the operations executed during the first clock stage, while lines 26–30 illustrate the behaviors within the second clock stage.

Likewise, the *task* below describes a basic read operation utilizing the control bus. In lines 1–13, the timing diagram for the bus read operation is illustrated. It consists of two clock phases: in the first clock cycle, the bus commands including the enable signal "ce", data read indicator "rd", and register address "addr" are driven on the bus. In the second clock cycle, all the commands are reset to binary zeros. At the falling clock edge within the same clock cycle, the read data is transmitted as a valid and stable data output.

The Verilog code is provided in lines 14–33 within the task description, comprising an input for the bus address "tk_addr" and an output for the read data "tk_rddata". Lines 20–23 represent the operations in the first clock stage, while lines 24–29 describe the behaviors in the second clock stage.

```
1  //%%%%%%%%%%%%%%%%%%%%%%%%%%%%%%%%%%%%%%%%%%%%%%%%%%%%%%%%%%
2  //            -----       -----
3  //  clk _____|     |_____|     |_____
4  //            -----------
5  //  ce  _____|           |___________
6  //            -----------
7  //  rd  _____|           |___________
8  //
9  //addr -----|  tk_addr  |---8'h00---|
10 //
11 //rddata-----------------------|  rddata  |
12 //
13 //%%%%%%%%%%%%%%%%%%%%%%%%%%%%%%%%%%%%%%%%%%%%%%%%%%%%%%%%%%
14 task task_cpu_read;
15 input  [7:0] tk_addr  ;
16 output [7:0] tk_rddata;
17 begin
18   $display ("%d CPU Read address: %h!", $time, tk_addr)
19   $display ("%d -> Driving ce, rd, addr", $time);
20   @(posedge clk)       ;
21   ce        = 1        ;
22   rd        = 1        ;
23   addr      = tk_addr  ;
24   @(posedge clk)       ;
25   ce        = 0        ;
26   rd        = 0        ;
27   addr      = 0        ;
28   @(negedge clk)       ;
29   tk_rddata = rddata   ;
30   $display ("%g CPU Read data : %h", $time, tk_rddata);
31   $display ("=====================");
32 end
33 endtask
```

B. Call the Tasks in Testbench

The Verilog testbench presented below utilizes two tasks for conducting bus write and read operations. Lines 3–5 declare the signals employed in the testbench. The LHS signals declared within the *initial* block are of the data type *reg*, while the "rddata" signal is of data type *wire*. In lines 6–7, we include two tasks, "cpu_write" and "cpu_read", which are then invoked within the *initial* block in lines 19–22.

```
1  // Call cpu_write and cpu_read tasks
2  module task_cpu_wr_rd();
3  reg          ce, rd, wr   ;
4  reg   [7:0] addr, wrdata;
5  wire  [7:0] rddata       ;
6  `include "task_cpu_write.v"
7  `include "task_cpu_read.v"
8
9  initial clk = 0;
10 always #5 clk = ~clk;
11
12 initial begin
13   ce     = 1'b0;
14   wr     = 1'b0;
15   rd     = 1'b0;
16   addr   = 8'h0;
17   wrdata = 8'h0;
18
19   task_cpu_write(8'h01, 8'h10    ); //Call the write task
20   task_cpu_read (8'h01, rddata   ); //Call the read task
21   #5 task_cpu_write(8'h00, 8'h85 );
22      task_cpu_read (8'h00, rddata);
23 end
24 endmodule
```

In the *initial* block, all declared *reg* signals are initialized. In line 19, the "cpu_write" *task* is called with the hexadecimal value 8'h01 as the first argument "tk_addr" and 8'h10 as the second argument "tk_wrdata". This bus operation configures the functional register at address 8'h01 with the data value 8'h10. The data written to the functional register is then read out by calling the "cpu_read" *task* in line 20. The same address is used to call the *task*, and the output of the *task*, "rddata", is of data type *wire*. In lines 21–22, the two *tasks* are invoked once more to configure the register located at address 8'h00. The data being written to this register is 8'h85, and the expected result after reading should also be 8'h85.

After executing the testbench, the printed log in the simulation console appears as follows. In lines 2–3, the log displays a write operation with address 8'h01 and write data 8'h10. Subsequently, in lines 5–7, it indicates that the data read from the same address is also 8'h10. This demonstrates the successful validation of the bus write and read operations. Similarly, lines 9–14 depict the results for the second bus write and read operation. The final data read from this operation is 8'h85, matching the data previously written to the functional register.

```
1  // Printed log:
2  0 CPU Write address: 01, data : 10
3  0 -> Driving ce, wr, wdata, addr
4  ======================
5  25 CPU Read address : 01
6  25 -> Driving ce, rd, addr
7  40 CPU Read data: 10
8  ======================
9  45 CPU Write address: 00, data: 85
10 45 -> Driving ce, wr, wrdata, addr
11 ======================
12 65 CPU Read address : 00
13 65 -> Driving ce, rd, addr
14 80 CPU Read data : 85
15 ======================
```

5.4 Verilog Delay Control

Verilog provides two types of delay control: delay expression and event expression. Delay expression can be used to model the propagation delay of signals and specify timing constraints in a simulation environment. However, it is essential to note that timing descriptions that use delay expressions are not synthesizable. They are only meant for simulation purposes.

On the other hand, event expression is used to specify events that trigger a block of code in a procedural block, such as *@posedge clk*, *@negedge clk*, and *wait(en)*. Event expressions, such as edge-triggered and level-triggered *always* blocks, play a crucial role in describing combinational and sequential circuits. They are also used to synchronize events in testbenches, enabling efficient simulation and accurate modeling of hardware behaviors. The *wait()* event is primarily employed within the testbench to synchronize various simulation events and behaviors.

5.4.1 Delay expression

Timing delay in Verilog is used to introduce a delay before executing the following statements in a procedural block. Its syntax is # DELAY, where the DELAY value specifies the amount of time to wait before executing the subsequent statements. The DELAY value for each statement is based on the time unit defined in the `timescale directive.

For example, in the code below, the signals "rst" and "clk" are initialized to binary 1'b0 in an *initial* block. After a delay of 10 ns (assuming `timescale 1ns/1ns), the "rst" signal is disabled to activate the entire circuit.

```
1  initial begin
2    rst=1'b0; clk=1'b0;
3    #10; rst=1'b1;
4  end
5
6  always #5 clk=~clk;
```

It's crucial to grasp that the # DELAY statement serves a specific purpose solely in simulation. It aids in modeling timing behaviors and introduces time delays between events during the simulation process. However, it's vital to understand that this type of delay is non-synthesizable and should never be included in RTL code intended for synthesis. In real hardware, delays manifest naturally due to the physical properties of hardware components, rather than relying on explicit timing delay statements like the # DELAY used in Verilog simulations.

5.4.2 Event expression

The event expression in Verilog is used to schedule a statement to execute when a certain expression evaluates to TRUE. It allows for timing control events, including edge-trigger and level-trigger events.

A. Posedge and Negedge Events

In synthesizable designs, edge-trigger events should be used with caution and primarily included in the trigger list of *always* blocks to describe the behavior of registers. In testbenches, on the other hand, edge-trigger events can be used more flexibly and are not limited to just the clock and reset signals. In the code snippet below, the signal "a" is initialized to binary 1'b0 (line 2) in the *initial* block, and then at the rising clock edge (line 3), it is set to binary 1'b1 (line 4). This illustrates how the signal "a" can be synchronized within the clock domain.

```
initial begin
  a=1'b0;
  @(posedge clk);
  a=1'b1;
end
```

B. Level Events

In synthesizable designs, level-trigger events are indeed included in the trigger list of *always* blocks, and they are commonly used to describe combinational circuits and latches, where the outputs are dependent on the current inputs and do not have any clock dependency.

In testbenches, the *wait* statement can be used to introduce a delay control until a specific condition becomes TRUE. This is often employed to synchronize signals across different blocks or modules during simulation. In the following example, the signal "b" is initialized to binary 1'b0 in an *initial* block (line 2). Then, the code waits until signal "a" becomes TRUE (line 3) before setting signal "b" to binary 1'b1 (line 4). It's important to note that the *wait* statement is used in testbenches for simulation purposes only and is not synthesizable.

```
initial begin
  b=1'b0;
  wait(a);
  b=1'b1;
end
```

5.5 Automated Simulation Environment and Verilog Testbench

Verilog HDL is a versatile language that enables us not only to describe hardware modules but also to create testbenches for the purpose of verifying the functionality of those design modules. The primary objective of a testbench is to simulate the behavior of the design module under specific input stimuli in order to validate its correctness. Typically, this simulation is carried out

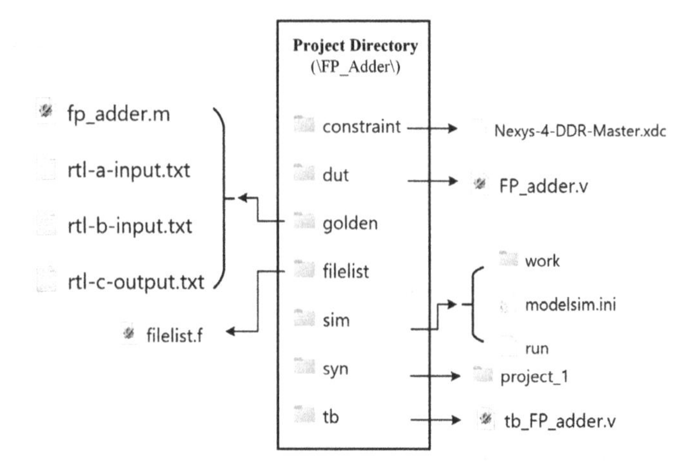

FIGURE 5.3
Structured project directory.

using EDA tools such as Synopsys VCS®, Cadence NC-Verilog®, and Siemens EDA Modelsim®, and others. The simulation results can be further analyzed using waveform viewers to visualize signal behavior and ensure that the design functions as intended.

In this section, we set up an automated simulation environment and illustrate the process of creating a typical Verilog testbench to verify the design functionalities. By establishing a well-structured simulation environment, designers can greatly enhance the verification process and identify potential bugs or issues at an early stage of the IC design cycle. Furthermore, well-designed testbenches promote code reusability, simplifying the validation of the design under various test scenarios.

5.5.1 Structured project directory

The project directory can grow in complexity, encompassing a multitude of files such as design code, testbenches, constraint files, scripts, golden models, and various files generated during simulation and synthesis. Furthermore, simulation and debugging often involve repetitive tasks. Therefore, it is vital to embrace a structured project directory approach, as exemplified in Figure 5.3, in order to efficiently manage projects and the extensive array of files and tools involved.

This directory structure includes key folders for constraint files (located in the "constraint" folder), the design under test (in the "dut"), filelists (in the "filelist"), golden models/results (in the "golden"), simulation (in the "sim"), synthesis (in the "syn"), and testbenches (in the "tb").

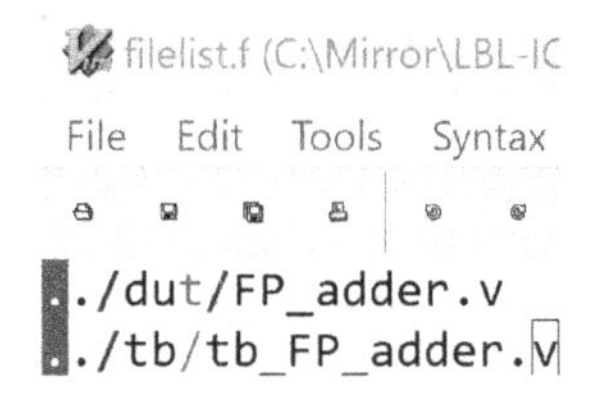

FIGURE 5.4
Filelist example.

A. dut, tb, and filelist

The project directory is carefully crafted to maintain organization and facilitate effortless access to pertinent files. The "dut" folder contains all the design files with the extension ".v", while the testbench files are located in the "tb" folder. To ensure proper handling of all the design files and the testbench, a "filelist.f" file is placed in the "filelist" folder. During simulation, this filelist is utilized to sequentially load all the ".v" files.

It is essential to note that the simulation process commences with the execution of TCL scripts present in the "sim" folder. Therefore, to accurately locate the design files in the "dut" folder and the testbench files in the "tb" folder, it is necessary to navigate up to the parent folder using the ".." notation. For example, as illustrated in Figure 5.4, the "filelist" includes two specific files: the design file (named "FP_adder.v") and its corresponding testbench (named "tb_FP_adder.v"). The hierarchical paths to access these files are as follows: "../dut/FP_adder.v" and "../tb/tb_FP_adder.v". For a comprehensive understanding of Linux/Unix related commands and hierarchical paths, refer **Appendix A.2**.

B. sim and syn

The TCL script, which automates simulations and manages the simulator, is situated in the "sim" folder. TCL is a versatile scripting language that can effectively handle repetitive simulations and streamline the simulator's operation. A detailed tutorial on employing a TCL template to automate simulations with Siemens Modelsim is provided in **Appendix B.2**.

All generated simulation files will be saved in the "sim" folder, whereas all generated synthesis files will be stored in the designated "syn" folder. This clear distinction ensures that all the relevant files are neatly organized and readily accessible for their respective tasks.

C. constraint

The constraint file plays a crucial role in providing essential information to the software regarding the specific resources to be utilized or connected to the Verilog design within the FPGA. These resources can include physical pins, switches, buttons, VGA interfaces, LEDs, and more. For example, the constraint file for the AMD/Xilinx FPGA is typically in the ".xdc" format, which stands for Xilinx Design Constraints. By configuring the constraint file effectively, FPGA users can accurately map the RTL design's functionalities to the target FPGA resources, enabling seamless communication and interaction between the Verilog design and the FPGA platform.

AMD/Xilinx Vivado serves as the tool for programming the final bitstream onto the FPGA and for loading the constraint file to ensure the proper configuration and utilization of the designated resources. For a comprehensive guide on using AMD Vivado, please refer to **Appendix C** for a detailed tutorial.

D. golden

The "golden" folder serves as a repository for all files related to the algorithm design and programming, encompassing the golden models and corresponding golden results. These golden files are typically created and generated through various software programs, providing a benchmark against which the hardware results can be compared during RTL verification.

The fundamental concept behind RTL verification revolves around the comparison of hardware results with the golden results while providing the same or similar inputs. In this context, the RTL results are considered accurate if they match the golden results or fall within an acceptable error tolerance specified in the design specifications and/or verification plans. In contrast, test cases are marked as failed if the RTL results deviate beyond the specified tolerance, necessitating debugging and further refinement of the design.

Figure 5.5 illustrates of the utilization of golden files in an FP adder simulation. In this example, MATLAB is employed to generate $2 \times 1,000,000$ single-precision data in IEEE 754 format. These data are subsequently written into two separate files: "rtl-a-input.txt" and "rtl-b-input.txt", with each file containing $1,000,000$ data. These files will serve as the input datasets for the RTL design. Using MATLAB, the resulting FP summation is calculated and recorded in a designated golden result file, labeled as "rtl-s-output.txt". This file will serve as the reference or golden result against which the RTL design's output will be compared.

Specifically, the first row of the input files exhibits the IEEE 754 format for the first design inputs "a" and "b", represented as 0xE0E22B56 and 0x618C06A1 in IEEE 754 format, corresponding to the decimal values of $-1.30377713877e+20$ and $3.22877729169e+20$, respectively. MATLAB's golden result for this example is represented as 0x6126F797 in IEEE 754 format, which translates to a decimal value of $1.92500015293e+20$. This value represents the summation of the decimal numbers $-1.30377713877e+20$ and $3.22877729169e+20$, and it is written in the first row of the golden output file.

In the testbench, the data stored in the ".txt" files can be loaded into three memory arrays: two for input and one for output. These arrays are essential for

Line	rtl-a-input.txt	rtl-b-input.txt	rtl-s-output.txt
1	E0E22B56	618C06A1	6126f797
2	60E16AA1	612C5B15	618e8833
3	6037947A	612C4711	615a2c30
4	60993A58	61771933	61a1db30
5	60CB6E29	615BA156	61a0ac35
6	61798C06	E137D7B2	608368a8
7	616B5DE2	60CE9B71	61a955cd
8	60DCC0E3	60B6563A	61498b8e
9	601A4989	615C4BB8	61816f0d
10	60CED70B	608208A3	61286fd7
11	6148901D	61822B36	61e67344
12	E1490F72	E1881FDD	e1eca796
13	DDD1BDB5	6095EECC	6092a7d5
14	6191DEA0	E0D7A0D3	6137ecd6
15	DF7AC383	6143E5DE	613439a6
16	E0572DFD	DF39B250	e082cd48
17	E05533F5	6163C775	612e7a78
18	6169B8B5	E12908C4	60815fe2
19	5F4DAB4E	60EE9A47	610427d8
20	613F9FAC	E14D2787	df587db0
21	5E4111D3	5F6C4561	5f8e44eb
22	E0134694	60805E08	5fdaeaf8
23	5FABA0E0	E18537F4	e174fbcc
24	6035EA9C	E17D58A0	e14fddf9

FIGURE 5.5
Input stimulus and golden result files.

providing input data to the design-under-test and for comparing the output data with the golden results. The Verilog design for the testbench will be presented in the following section.

5.5.2 Automated simulation testbench utilizing BFM and monitor

As shown in Figure 5.6, a typical testbench contains the instantiation of the design-under-test (DUT), bus functional models (BFM) to drive and respond to the DUT, and a monitor to automatically check the results. It's important to note that the testbench is usually not synthesizable code, which means that any Verilog system tasks and timing controls can be used when creating a testbench.

The figure illustrates the fundamental concept behind verifying the design's functionalities. It involves feeding random data into the DUT utilizing the BFM and checking the DUT's output using the monitor. The input stimulus is pre-stored in the "rtl-a-input.txt" and "rtl-b-input.txt" files, while the reference output is pre-stored in the "rtl-s-output.txt" file.

An alternative approach involves the integration of golden models, typically implemented in languages like C or MATLAB, directly into the Verilog testbench using the DPI. DPI provides a standardized interface for

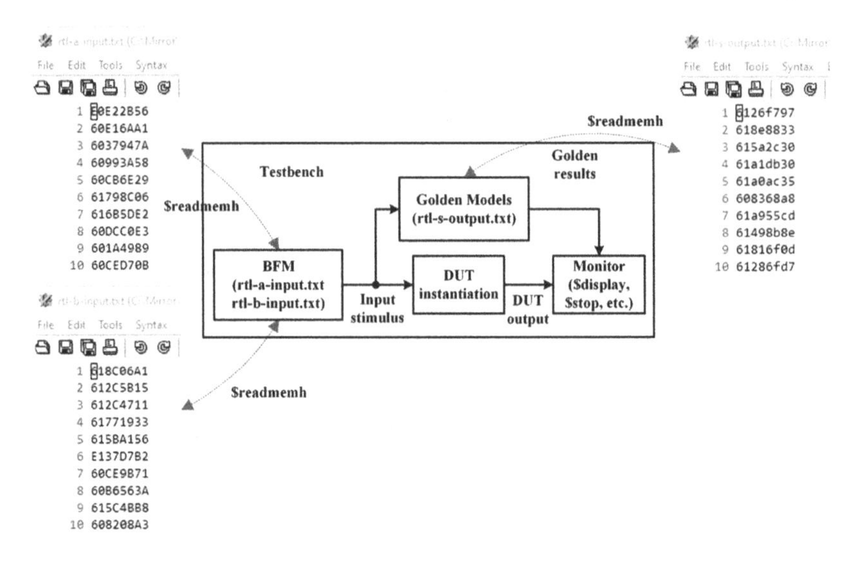

FIGURE 5.6
Automated simulation testbench utilizing BFM and monitor.

embedding code written in languages other than Verilog within a Verilog simulation environment. Within this setup, the BFM can be utilized to randomize input stimuli, which are then concurrently applied to both the DUT and the golden model. The golden model's outputs serve as the expected results, allowing for a comparison with the DUT's outcomes to determine the success or failure of the test.

5.5.3 Verilog design on automated simulation testbench

Below is the Verilog code for the testbench. The DUT represents a single-precision adder, with each addition operation taking one clock cycle, as illustrated in lines 1–15. The *parameter* specifying the number of test cases is defined in line 18.

Signal declarations for connecting and driving the DUT's input and output ports are presented in lines 19–23, following the standard Verilog declaration conventions. Specifically, the *reg* data type is used for LHS signals within *initial* blocks, while the *wire* data type is employed for DUT instantiation and connections. It's essential to note that these signals are specific to the testbench.

```
1  /* -----------------------------------------------------
2   * This source file contains a simulation testbench for a FP
3   * Adder generated by the Chisel HCL.
```

```
4   * Design-under-test: FP_adder
5   * Testbench         : tb_FP_adder
6   * Design IOs        : inputs - clock, reset,
7   *                              io_in_a[31:0], io_in_b[31:0]
8   *                     outputs- io_out_s[31:0]
9   * Latency           : One cycle per FP addition
10  * Precision         : Single precision
11  * Resource Required: 16 Binary Adders (32 x 32 bits)
12  *
13  * Author            : Xiaokun Yang
14  * Date              : June 2022
15 *///------------------------------------------------------------
16 `timescale 1ns/1ns
17 module tb_FP_adder();
18 parameter TEST_SIZE = 1_000_000;
19 reg          clock    ;
20 reg          reset    ;
21 reg   [31:0] io_in_a  ;
22 reg   [31:0] io_in_b  ;
23 wire  [31:0] io_out_s ;
24
25 // --------------------------------
26 // Stimulus Files and Golden File
27 // --------------------------------
28 reg [31:0]  input_a[TEST_SIZE-1:0] ;
29 reg [31:0]  input_b[TEST_SIZE-1:0] ;
30 reg [31:0]  output_c[TEST_SIZE-1:0];
31
32 initial begin
33    $readmemh("../golden/rtl-a-input.txt",input_a)  ;
34    $readmemh("../golden/rtl-b-input.txt",input_b)  ;
35    $readmemh("../golden/rtl-s-output.txt",output_s);
36 end
37
38 // --------------------------------
39 // --- DUT Instantiation ---------
40 // --------------------------------
41 FP_adder u_FP_adder (
42                       .clock   (clock   ),
43                       .reset   (reset   ),
44                       .io_in_a (io_in_a ),
45                       .io_in_b (io_in_b ),
46                       .io_out_s(io_out_s));
47
48 // --------------------------------
49 // --- Bus Functional Models -----
50 // --------------------------------
51 integer i;
52 initial begin
```

```
53      reset = 1'b0;
54      clock = 1'b0;
55      #100;
56      reset = 1'b1;
57      @(posedge clock);
58
59    for (i=0; i < TEST_SIZE; i = i+1) begin
60      io_in_a <= input_a[i];
61      io_in_b <= input_b[i];
62      @(posedge clock);
63    end
64  end
65
66  always #10 clock = ~clock;
67
68  // ------------------------------
69  // ----------Monitor ------------
70  // ------------------------------
71  integer j, m, n, o, p, q, r, s;
72  wire sign_check = (io_out_s[31]==output_s[j][31])       ;
73  wire exp_check  = (io_out_s[30:23]==output_s[j][30:23]);
74  wire mant_check = (io_out_s[22:0]==output_s[j][22:0])   ;
75  wire mant_22_4  = (io_out_s[22:4]==output_s[j][22:4])   ;
76  wire mant_22_8  = (io_out_s[22:8]==output_s[j][22:8])   ;
77  wire mant_22_16 = (io_out_s[22:16]==output_s[j][22:16]);
78  wire mant_3_0   = (io_out_s[3:0]==output_s[j][3:0])     ;
79  wire mant_7_0   = (io_out_s[7:0]==output_s[j][7:0])     ;
80  wire mant_15_0  = (io_out_s[15:0]==output_s[j][15:0])   ;
81
82  initial begin
83    m = 0 ;
84    n = 0 ;
85    o = 0 ;
86    p = 0 ;
87    q = 0 ;
88    r = 0 ;
89    s = 0 ;
90    wait (reset);
91    repeat (2) @(negedge clock);
92    for (j=0; j < TEST_SIZE; j = j+1) begin
93      $display("%d ns, a=%h, b=%h, golden s=%h, dut s=%h",
94  $time, input_a[i-1], input_b[i-1], output_s[j], io_out_s);
95      case({sign_check, exp_check, mant_check})
96      3'b111: begin
97                m=m+1;
98              end
99      3'b110: begin
100               if(mant_22_4 & ~mant_3_0) begin
101                 n=n+1;
```

```
102             end else if(mant_22_8 & ~mant_7_0) begin
103               o=o+1;
104             end else if(mant_22_16 & ~mant_15_0) begin
105               p=p+1;
106             end
107             end
108     default: begin
109             if(~sign_check) begin
110               $display("Sign bit different!, j=%d", j);
111               q=q+1;
112             end else if(~exp_30_23) begin
113               $display("Exponent are different!");
114               r=r+1;
115             end else if(~mant_22_16) begin
116               $display("Matanssa[22:16] are different!");
117               s=s+1;
118             end
119             end
120       endcase
121       @(negedge clock);
122     end
123
124     $display("------FP adder simulation summary-------");
125     $display("%d cases pass, %d fail!", m+n+o+p, q+r+s);
126     $display("----FP adder simulation passed cases---");
127     $display("%d cases exactly the same!", m);
128     $display("%d cases: different mantissa[3:0]!", n);
129     $display("%d cases: different mantissa[7:0]!", o);
130     $display("%d cases: different mantissa[15:0]!", p);
131     $display("----FP adder simulation failed cases---");
132     $display("%d cases: different sign bit!", q);
133     $display("%d cases: different exponent bits!", r);
134     $display("%d cases: different mantissa[22:16]!", s);
135     $display("------FP adder simulation summary-------");
136   end
137   endmodule
```

A. Load Memory Array and Instantiate Design-under-Test

In lines 25–36, the three ".txt" files are loaded into three memory arrays: "input_a" declared in line 28, "input_b" declared in line 29, and "output_c" declared in line 30. The depth of the memory arrays is the same as the number of test cases, and the width is 32 bits, representing single-precision FP data.

The DUT is instantiated in lines 38–46, with all signals from the testbench (declared in lines 19–23) connected to the design's IOs. The unique design instance name is "u_FP_adder", which extends the design module name "FP_adder". The testbench signals are placed in parentheses following the design IOs, effectively connecting them together. To simplify matters, it is common to use the same names for connected signals between the testbench and the design IOs.

B. Bus Functional Model

The bus functional model is established in lines 48–66. The clock signal is initially set to 1'b0 in line 54 and then alternates its value every 10 ns delay through an *always* block in line 66, effectively simulating a 50 MHz clock. In the *initial* block, the reset signal is asserted to 1'b1 in line 56, with a 100 ns delay after the simulation begins.

The subsequent *for* loop is responsible for supplying data to the design inputs at every rising clock edge. The initial signal assignment within the *for* loop is synchronized with the rising clock edge as shown in line 57.

C. Monitor and Test Plan

The monitor design is illustrated in the *initial* block in lines 82–136. At the beginning, initialization is carried out for all the declared integers. The *wait(reset)* statement in line 90 is employed to synchronize the executions between the monitor and the bus functional model. It ensures that the code following the *wait* statement is executed only when the circuit reset is completed.

In lines 92–122, a *for* loop is employed to tally the count of passed and failed test cases. Before delving into the results, let's recap the verification strategy used in this example. Given the disparities between the RTL design (single-precision) and the MATLAB program (double-precision), the verification plan incorporates an error tolerance approach during simulation. The sign bit and exponent bits must match exactly; for the comparison of mantissa bits, four tolerance levels are applied:

- Error Tolerance #0: All mantissa bits are identical.
- Error Tolerance #1: The least significant 22–4 bits need to match.
- Error Tolerance #2: The least significant 22–8 bits need to match.
- Error Tolerance #3: The least significant 22–16 bits need to match.

This multi-level error tolerance approach allows for a flexible assessment of the simulation results, considering various degrees of similarity between the RTL design and the MATLAB program.

With this strategy in place, the *integers* "m-p" are used to track the number of test cases that meet tolerance levels #0, #1, #2, and #3, respectively. Meanwhile, the integers "q-s" are employed to count the number of tests that fail in the sign bit comparison, the exponent bit comparison, and the mantissa bit comparison, respectively. In the Verilog code, lines 72–80 depict the comparison results between the RTL design and the golden models. Furthermore, lines 92–122 present a *for* loop responsible for tallying the number of test cases that either pass or fail based on the specified error tolerance levels.

D. Simulation Log

Lines 124–135 present the Verilog code for providing a comprehensive summary of the test outcomes. In Figure 5.7, a screenshot illustrates the simulation results. Remarkably, all 1,000,000 test cases pass the simulation, underscoring the robustness of the design. To provide a more nuanced understanding, here is a breakdown of the results:

- 659,008 test cases yield precisely the same 32-bit output as the golden results, adhering to tolerance level #0.
- 297,848 test cases match the golden results in the 31 to 4 bits of output, meeting tolerance level #1.
- 37,530 test cases align with the golden results in the 31 to 8 bits of output, satisfying tolerance level #2.
- 5,606 test cases mirror the golden results in the 31 to 16 bits of output, meeting tolerance level #3.

5.6 Guidelines for RTL Simulation and Verification

In this chapter, we have explored the fundamental Verilog syntax for creating simulation testbenches and introduced a structured project directory approach for establishing a well-organized simulation environment. To encapsulate our discussion, here are some key guidelines for RTL simulation and verification:

Guidelines for RTL Simulation and Verification

- **Functional Verification:** Functional verification (also known as RTL verification) is used to verify the RTL design features without testing timing constraints. The practical circuits have timing delays and timing requirements, which cannot be simulated during RTL verification but exist in physical chips.
- **Test Plan:** Develop a comprehensive test plan that incorporates both direct and random testing strategies. In this plan, RTL designers primarily handle direct testing to validate fundamental design features, while verification engineers concentrate on random testing to accumulate coverage data and pinpoint corner and exceptional testing scenarios. For

```
# At      19999800ns, a=e0b774e9, b=df7e60f6, expected s=e0d74108, dut s=e0d74107
# At      19999820ns, a=60881f4b, b=61021bcc, expected s=61462b72, dut s=61462b71
# At      19999840ns, a=5e823d3c, b=e02ed887, expected s=e01e90e0, dut s=e01e90e0
# At      19999860ns, a=604ffee7, b=6123b1cc, expected s=6157b186, dut s=6157b185
# At      19999880ns, a=e0fe6e88, b=e18642a3, expected s=e1c5de45, dut s=e1c5de45
# At      19999900ns, a=608b1fce, b=60b33823, expected s=611f2bf8, dut s=611f2bf8
# At      19999920ns, a=e0fb79de, b=5f70226c, expected s=e0dd7590, dut s=e0dd7591
# At      19999940ns, a=dedf091c, b=e1801d30, expected s=e1839954, dut s=e1839954
# At      19999960ns, a=e15a5517, b=e0dfc219, expected s=e1a51b12, dut s=e1a51b11
# At      19999980ns, a=e0510cbb, b=61709f39, expected s=613c5c0a, dut s=613c5c0b
# At      20000000ns, a=618ab8f0, b=60a765f7, expected s=61b4926e, dut s=61b4926d
# At      20000020ns, a=618bb7b1, b=60bee6a6, expected s=61bb715a, dut s=61bb715a
# At      20000040ns, a=60ba9f05, b=610be16b, expected s=616930ee, dut s=616930ed
# At      20000060ns, a=610c0fcf, b=e005c152, expected s=60d53ef5, dut s=60d53ef6
# At      20000080ns, a=e14e1c6f, b=601485b2, expected s=e128fb02, dut s=e128fb03
# At      20000100ns, a=5f8c0cb7, b=e0676290, expected s=e0215c34, dut s=e0215c35
# At      20000120ns, a=6114bc00, b=e0ac7466, expected s=607a0734, dut s=607a0734
# --------------------------- FP adder design simulation summary ----------------------
# For 1,000,000 test cases, there are   1,000,000 test cases pass,  0 test cases fail!
# --------------------------- FP adder design simulation passed cases -----------------
# For 1,000,000 passed test cases, 659008 test cases exactly the same!
# For 1,000,000 passed test cases, 297848 test cases with different 3-0 mantissa bits!
# For 1,000,000 passed test cases,  37530 test cases with different 7-0 mantissa bits!
# For 1,000,000 passed test cases,   5606 test cases with different 15-0 mantissa bits!
# --------------------------- FP adder design simulation fail cases -------------------
# For 1,000,000 failed test cases, 0 test cases with different sign bit!
# For 1,000,000 failed test cases, 0 test cases with different exponent bits!
# For 1,000,000 failed test cases, 0 test cases with different 22-16 mantissa bits!
# ---------------------------FP adder design simulation summary --------------------
```

FIGURE 5.7
Simulation log.

intricate design projects, uncovering exceptional cases poses a considerable challenge in terms of functional verification, and the verification team plays a critical role in addressing this challenge.

- **Reusable Testbench:** Create a testbench with reusability and scalability in mind, facilitating the straightforward inclusion of new test cases. To optimize the verification process for advanced IC designs, consider employing the SystemVerilog language in conjunction with the UVM methodology. This approach enhances efficiency, scalability, and overall productivity throughout the verification process.
- **Testbench Monitor:** Incorporate a monitor or scoreboard within the testbench to automate result verification, eliminating the need for manual checks. Monitors play a pivotal role in Verilog testbenches by enabling the real-time observation and capture of signals from the design-under-test. This automation ensures comprehensive verification of the design's functionality and behavior.
- **Bus Functional Model and Golden Models:** Integrate a bus functional model into the verification process for providing random data

inputs to RTL designs. You can achieve this by preparing data files alongside golden models. Alternatively, consider the direct integration of golden models into the Verilog testbench using the DPI. This approach offers the advantage of seamlessly incorporating a wide range of programming languages into the Verilog testbench, thereby boosting flexibility and versatility in your verification process.

- **Printed Log and Dump Waveform:** Ensure that error messages, along with related signals and precise simulation timestamps, are logged to a designated log file. Furthermore, when conducting random tests, consider dumping waveform data and specifying a unique simulation seed ID. This seed ID facilitates the reproduction of the same random data in subsequent simulations, significantly assisting in efforts related to reproducibility and debugging.
- **Signal Initialization:** Ensure that all design inputs are properly initialized, either with zeros or ones, to avoid unknown signals in the simulation waveform.
- **Regression Testing:** After making changes to the RTL code, it's essential to conduct regression tests to identify potential impacts on other design features. Regression testing should encompass comprehensive verification activities, including coverage analysis, to ensure thorough validation of the design-under-test.

PBL 6: Four-Bit Full-Adder Design and Simulation

1) Design a 4-bit full adder using the Verilog arithmetic operator "+". Figure 5.8(a) shows the design block diagram, including the 4-bit inputs "a" and "b" and the 5-bit output "c".
2) Create a testbench to simulate the design, encompassing all input combinations of "a" and "b", totaling 256 possible combinations. The testbench shown in Figure 5.8(b) connects the 4-bit signals "a" and "b" to the design inputs, and the 5-bit signal "c" to the design output to monitor the results.

Hint: Repeated loops can be used within the bus functional model to feed in data and within the monitor to automate result validation.

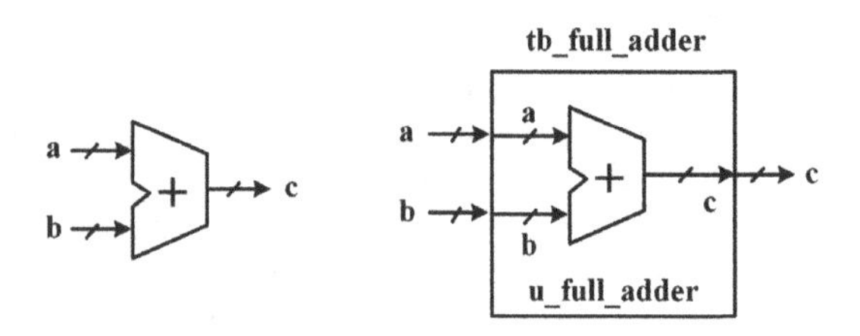

FIGURE 5.8
Design and simulation specification of 4-bit adder.

PBL 7: Combinational Circuit Design and Simulation

1) Design a combinational circuit named "com_and_or" with an AND gate and two OR gates as shown in Figure 5.9. The design should have single-bit input and output.
2) Create a testbench to simulate the design, encompassing all input combinations of "a-d", totaling 16 combinations.

PBL 8: Sequential Circuit Design and Simulation

1) Using the project presented in Figure 5.9 as a submodule, design a sequential circuit that comprises an instantiation of this submodule along with a register, as depicted in Figure 5.10. The name of the design module is "seq_and_or", and all the IOs are single-bit signals. The register can be triggered by falling reset edges and rising clock edges.
2) Create a testbench to simulate the design, encompassing all input combinations of "a-d", totaling 16 combinations. To simulate the

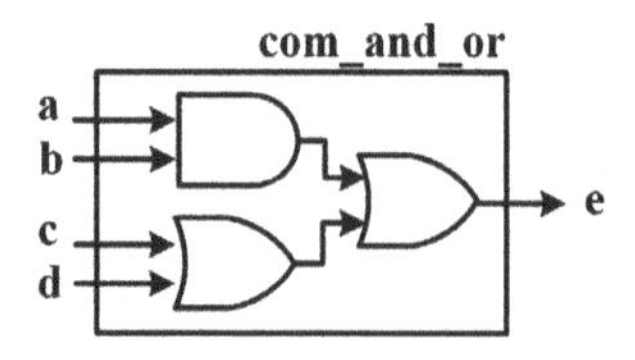

FIGURE 5.9
Combinational AND-OR circuit.

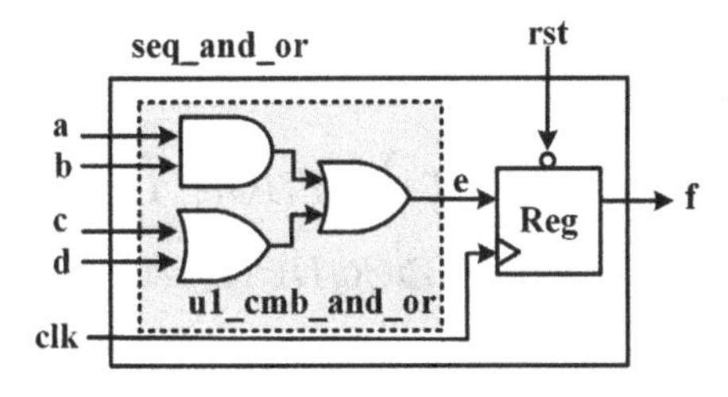

FIGURE 5.10
Sequential AND-OR circuit.

register's behavior, it's essential to have a clock and reset generator in place.

PBL 9: Sequential Circuit Instantiation and Simulation

1) Using the project in Figure 5.9 as a submodule, design a sequential circuit named "seq_inst_and_or". The sequential circuit should include two submodule instantiations (each consisting of a combinational AND-OR circuit), a single-bit adder, and three registers as depicted in Figure 5.11. All the IOs are single-bit signals. The register can be triggered by falling reset edges and rising clock edges.
2) Create a testbench to simulate the design, encompassing all input combinations of "a-g", totaling 128 combinations. To simulate the register's behavior, it's essential to have a clock and reset generator in place.

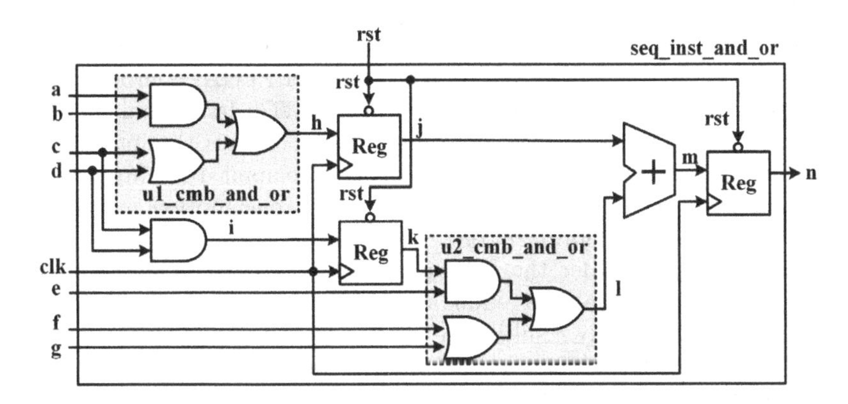

FIGURE 5.11
Sequential circuit instantiation.

6

Synthesis: Matching Verilog HDL with Basic Combinational and Sequential Circuits

Synthesis using EDA tools such as Synopsys Design Compiler (for ASIC design synthesis) or AMD Vivado (for FPGA design synthesis and implementation) enables a valuable opportunity to describe complex hardware circuits and systems at the register-transfer level (RTL), thereby significantly reducing the complexity of hardware designs. However, it poses a challenge for IC designers to establish the connection between RTL designs and hardware circuits, which is crucial for achieving high-performance and efficient hardware. Chapter 6 tackles this challenge by providing fundamental Verilog descriptions alongside the corresponding hardware circuits generated by AMD Vivado, aiming to facilitate this linkage.

PBL 10–12 feature the design synthesis of basic combinational and sequential circuits. PBL 13–16 illustrate counter and timer designs, along with their simple applications as describing timing-based interfaces and controllers.

6.1 Introduction to Synthesis

6.1.1 What is synthesis?

Synthesis is the process of converting RTL descriptions into gate-level representations suitable for implementation on ASICs or FPGAs. During synthesis, the goal is to maintain equivalence between the RTL design and the synthesized netlist, ensuring that the resulting circuit performs the intended functionality. Additionally, optimization techniques are applied to improve various design aspects, including area reduction, power consumption minimization, and timing constraint optimization.

For example, consider the following illustration that demonstrates how synthesis tools can eliminate redundant logic and optimize the utilization of internal logic, ultimately resulting in a more efficient and streamlined design. Upon initial inspection, the provided Verilog code can be translated into the hardware depicted in Figure 6.1(a), consisting of two adders and one multiplexer. The three inputs, "a-c", enter the two single-bit adders, and the results

DOI: 10.1201/9781003187080-6

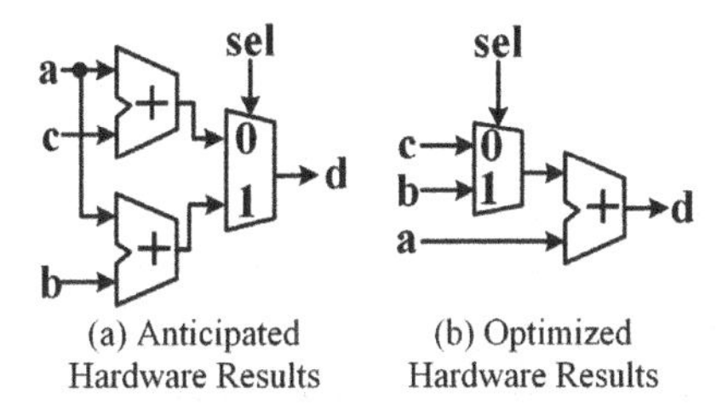

FIGURE 6.1
Synthesis results of assign d = sel ? (a+b) : (a+c);

of the two adders are further selected by the multiplexer as the final output "d".

```
1 assign d = sel ? (a+b) : (a+c);
```

However,the synthesized design, as shown in Figure 6.1(b), employs a 2-to-1 multiplexer to select signals "c" and "b" firstly. The chosen signal is then fed into an adder, where it is summed with the input "a". Despite implementing the same function, this optimization reduces the gate/slice count and minimizes power consumption within the circuit.

6.1.2 Mismatches between simulation and synthesis

There can be disparities between simulation and synthesis results when working with the same Verilog HDL design. For instance, incomplete sensitivity lists within *always* blocks may cause simulation failures since the changes of missing signals cannot trigger the associated statements. However, during synthesis, the synthesis tool can automatically analyze the design code and complete the sensitivity list.

Consider the following example where the signal "en" is missing in the sensitivity list in the first line. As a result, changes to the "en" signal will fail to trigger the *always* block during simulation. However, during synthesis, the synthesis tool can rectify this by completing the sensitivity list.

```
1 always @(sel) begin
2   if (sel & en) begin
3     a <= #1 1'b1;
```

```
4    end else if(sel==1'bx) begin
5      a <= #1 1'bx;
6    end else begin
7      a <= #1 1'b0;
8    end
9  end
```

Moreover, during synthesis, delay statements found in Verilog code, like the 1 ns delay shown in lines 3, 5, and 7 are disregarded. This can result in timing discrepancies between the simulated and synthesized outputs.

Finally, it's crucial to note that hardware can only distinguish logic levels of zeros and ones, rendering comparisons to unknown values and high impedance states in simulation irrelevant for synthesis. For instance, the expression in line 4 meant to derive the "sel" signal as unknown may work in simulation, but it lacks significance in the synthesis process.

6.1.3 Synthesizable Verilog HDL

Verilog HDL is a large and comprehensive IEEE standard that supports a wide range of design styles and modeling techniques. However, most of them are unsynthesizable, meaning they cannot be directly translated into hardware by synthesis tools. For instance, Verilog descriptions such as behavioral blocks with delays, system tasks, display statements, are intended for simulation and modeling purposes rather than direct hardware descriptions.

The subset of Verilog that is considered synthesizable is commonly referred to as RTL description or synthesizable Verilog. RTL code describes the flow of data between registers and is more hardware-oriented, making it suitable for synthesis. It focuses on the structural and behavioral aspects of a hardware design, allowing synthesis tools to generate the corresponding gate-level representation. To ensure a Verilog design is synthesizable, designers need to adhere to the subset of constructs and coding styles supported by synthesis tools. As a summary, synthesizable Verilog descriptions encompass a small subset constructs that are suitable for hardware synthesis, including concurrent *assign*, conditional *assign*, level-triggered *always* blocks, and edge-triggered *always* blocks.

- **Concurrent assignment** is commonly employed to describe combinational circuits. For instance, consider the example shown in line 2, representing a combinational circuit with an AND gate and an OR gate.

- **Conditional assignment** finds application in various combinational circuits. In line 5, a conditional *assign* block is used to illustrate a tri-state buffer, while in line 6, its implementation in a multiplexer is demonstrated.

- **Level-triggered always** blocks serve the purpose of describing both combinational and sequential circuits. Line 9 demonstrates a level-triggered *always* block that represents a latch with an omitted *else* condition. When the expression of the "en" signal evaluates to FALSE, the left-hand side signal (LHS) "c" retains its previous value through the use of a latch.
- **Edge-triggered always** blocks, on the other hand, should typically include only "posedge clock" and "posedge/negedge reset" in the sensitivity list. In line 12, a basic single-bit register with a rising active clock is demonstrated.

```
1  // Concurrent assignment
2  assign d = a & b | c; // combinational logic
3
4  // Conditional assignment
5  assign b = enable ? a : 1'bz; // tri-state buffer
6  assign c = enable ? a : b;    // multiplexer
7
8  // Level-trigger always (if-else, case, loop):
9  always @(a, b, en)  if (en) c <= a & b;  //latch
10
11 // Edge-trigger always (if-else, case, loop):
12 always @(posedge clk) b <= nxt_b;        //register
```

Design Guidelines for Synthesizable Verilog Descriptions

- For concurrent/conditional *assign* blocks, declare the data type of the LHS signals as *wire*, and utilize blocking assignment (=) for the design statements.
- For level/edge-triggered *always* blocks, declare the data type of the LHS signals as *reg*, and use non-blocking assignment (<=) for the design statements.
- Ensure that the trigger list in a level-triggered *always* block is complete, and provide complete *case* items if applicable. In $if - else$ statements, it is permissible to omit the final *else* condition when describing latches.
- Ensure that the trigger list in an edge-triggered *always* block only consists of active clock and reset edges. In $if - else$ statements, it is permissible to omit the final *else* condition when describing registers.
- To prevent multiple drivers from affecting the same LHS signals in RTL

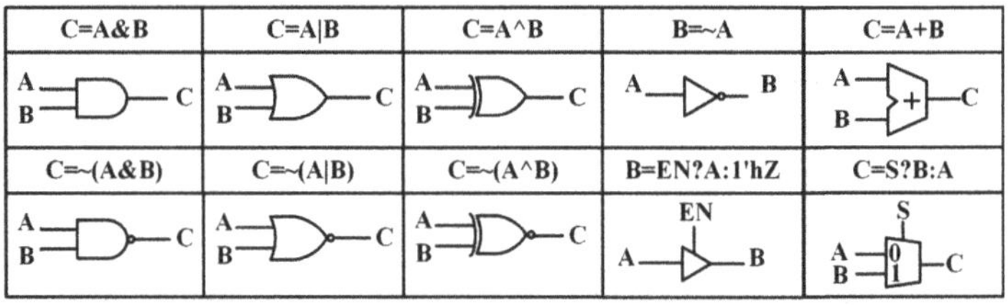

C=A&B	C=A|B	C=A^B	B=~A	C=A+B
C=~(A&B)	C=~(A|B)	C=~(A^B)	B=EN?A:1'hZ	C=S?B:A

FIGURE 6.2
Basic logic descriptions using concurrent/conditional assignments.

designs, make sure not to assign values to the same signal across different blocks. Nevertheless, it's worth noting that within the simulation testbench, it is considered acceptable to have multiple drivers for signals like the clock initialization.

6.2 Synthesis of Combinational Logic

Creating a clear correlation between the Verilog code and the targeted hardware is a critical aspect for RTL designers. This section presents the RTL analysis results for fundamental combinational circuits, demonstrating the effectiveness of the design in mapping to the intended hardware.

6.2.1 Fundamental combinational logic

In Figure 6.2, we present the fundamental logic descriptions using concurrent *assign* and conditional *assign* blocks, providing a clear demonstration of combinational logic.

Consider the first design (lines 1–5) below, utilizing a concurrent *assign* block, to describe a circuit shown in Figure 6.3(a). This circuit comprises two AND gates followed by an OR gate. Alternatively, the same circuit can be described in the second example using a level-triggered *always* block (lines 7–13). Nevertheless, it's clear that opting for a concurrent *assign* block is considerably more straightforward, and as such, it is the recommended approach for most situations.

```
// Example #1: Design AND-OR Using Concurrent Assign
module example1_and_or (output e          ,
                        input  a, b, c, d );
assign e = (a & b) | (c & d);
endmodule

```

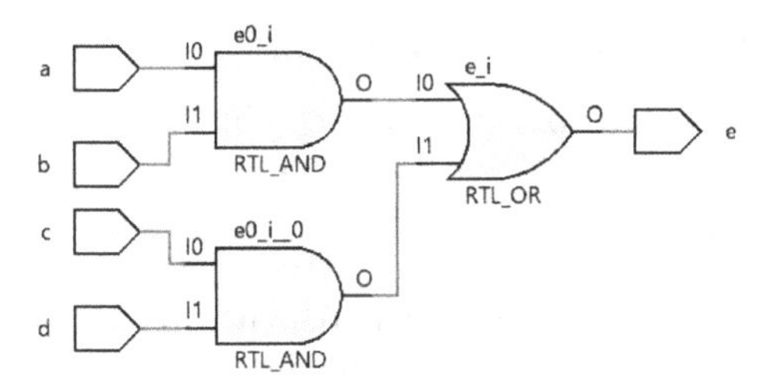

(a) Example #1 and #2: Design AND-OR Circuit Using Continuous Assign or Level-Trigger Always

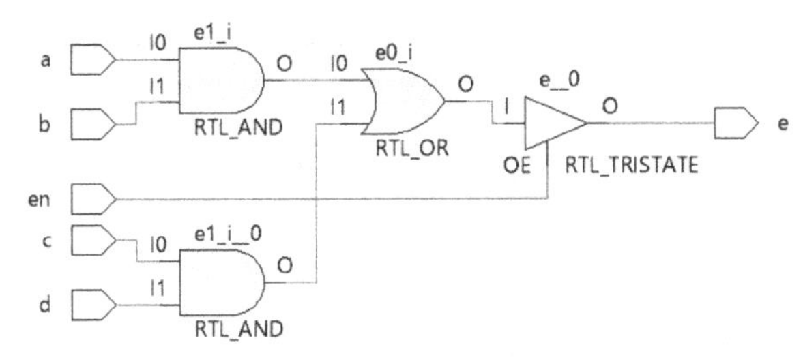

(b) Example #3: Design AND-OR Gates with Tri-State Buffer Using Conditional Assign

FIGURE 6.3
Synthesis results of AND-OR and AND-OR-tristate designs.

```
7  // Example #2: Design AND-OR Using Level-Trigger Always
8  module example2_and_or (output reg e          ,
9                          input      a, b, c, d );
10 always @ (a, b, c, d) begin
11   e = (a & b) | (c & d);
12 end
13 endmodule
14
15 // Example #3: Design AND-OR with Tri-State Buffer
16 // Using Conditional Assign
17 module example3_and_or_tri (output e              ,
18                             input  a, b, c, d, en );
19 assign e = en ? (a & b) | (c & d) : 1'bz;
20 endmodule
```

Furthermore, the third example showcases a combination of an AND-OR circuit with a tri-state buffer using a conditional *assign* block (lines 15–19). The RTL analysis results for these designs are illustrated in Figure 6.3(b).

6.2.2 Uni-directional and bi-directional buses

Below are two design examples utilizing conditional *assign* blocks: the first design represents a uni-directional bus (lines 1–6), while the second design represents a bi-directional bus (lines 8–16).

In the synthesis results depicted in Figure 6.4(a) (for the uni-directional bus), the behavior of the output "bus" depends on the values of the input "send_en". When "send_en" is TRUE, the "bus" output is driven by the input "a". Conversely, when "send_en" is FALSE, the "bus" output functions as an open circuit.

```
1  // Example #1: Design Uni-Direction Bus
2  module uni_dir_bus (output [31:0] bus         ,
3                      input  [31:0] a           ,
4                      input         send_en     );
5  assign bus = send_en ? a : 32'bz;
6  endmodule
7
8  // Example #2: Design Bi-Direction Bus
9  module bi_dir_bus (inout  [31:0] bus          ,
10                    input  [31:0] a            ,
11                    output [31:0] b            ,
12                    input         send_en      ,
13                    input         receive_en   );
14 assign b   = receive_en ? bus : 32'bz;
15 assign bus = send_en    ? a   : 32'bz;
16 endmodule
```

In the synthesis results depicted in Figure 6.4(b) (for the bi-directional bus), the behavior of the "bus" signal depends on the values of the inputs "send_en" and "receive_en". It is important to note that the direction of the "bus" signal in this case is declared as *inout*, indicating that it can function as both an *input* and an *output*. When "send_en" is TRUE and "receive_en" is FALSE, the "bus" signal acts as an *output* and is driven by the input "a". Conversely, when "send_en" is FALSE and "receive_en" is TRUE, the bus signal functions as an *input*, driving the *output* "b".

6.2.3 Multiplexer

Here are three examples illustrating Verilog descriptions of a 4-to-1 Multiplexer. The first example utilizes a *case* – *endcase* statement to specify the four channels of the multiplexer. By choosing a channel, the output "e" is assigned the corresponding input "a", "b", "c", or "d". The synthesis result is depicted in Figure 6.5(a).

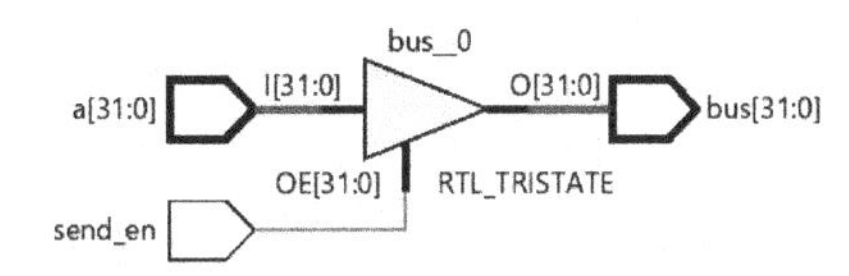

(a) Example #1 : Design Uni-Directional Bus

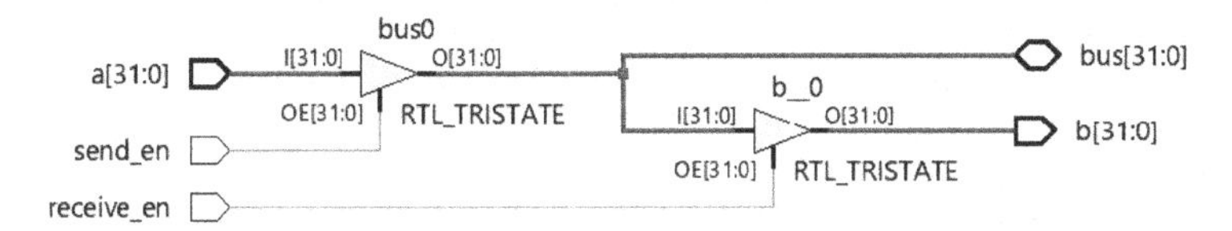

(b) Example #2 : Design Bi-Directional Bus

FIGURE 6.4
Synthesis results of uni- and bi-Directional buses designs.

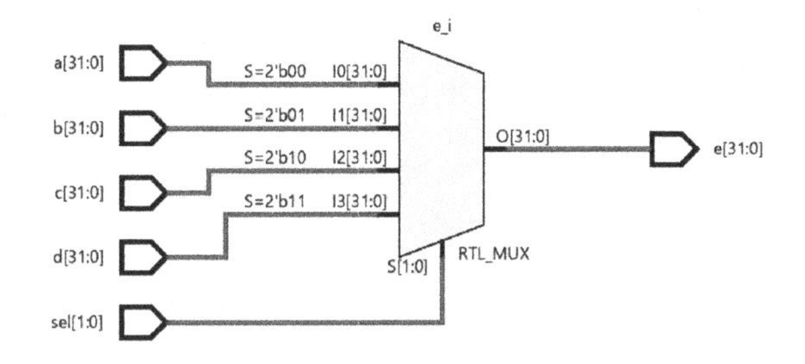

(a) Example #1: Design a Multiplexer Using Always Case

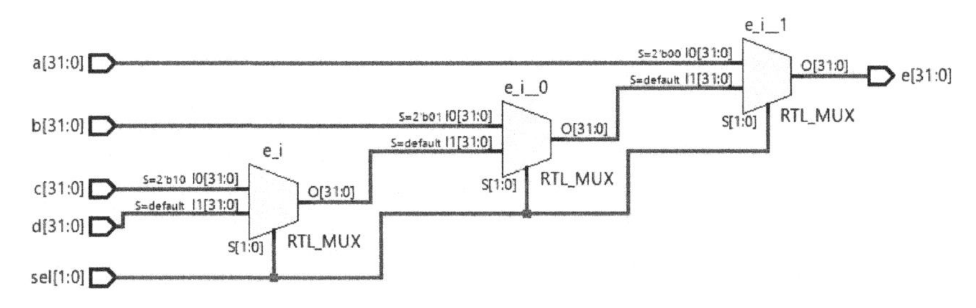

(b) Example #2: Design a Multiplexer with Different Priority Using If-Else in Always or Conditional Assign

FIGURE 6.5
Synthesis results of multiplexer designs.

```
1  // Example #1: Multiplexer Design Using case-endcase
2  module example1_mux (output reg [31: 0] e          ,
3                       input      [31: 0] a, b, c, d,
4                       input      [1 : 0] sel        );
5  always @ (a, b, c, d, sel) begin
6    case (sel)
7      2'h0    : e <= a ;
8      2'h1    : e <= b ;
9      2'h2    : e <= c ;
10     2'h3    : e <= d ;
11     default: e <= a ;
12   endcase
13 end
14 endmodule
15
16 // Example #2: Multiplexer Design Using if-else
17 module example2_mux (output reg [31: 0] e          ,
18                      input      [31: 0] a, b, c, d,
19                      input      [1 : 0] sel        );
20 always @ (a, b, c, d, sel) begin
21   if (sel == 2'h0) begin
22     e <= a;
23   end else if (sel == 2'h1) begin
24     e <= b;
25   end else if (sel == 2'h2) begin
26     e <= c;
27   end else begin
28     e <= d;
29   end
30 end
31 endmodule
32
33 // Example #3: Multiplexer Design Using Conditional Assign
34 module example3_mux (output [31: 0] e          ,
35                      input  [31: 0] a, b, c, d,
36                      input  [1 : 0] sel        );
37 assign e = (sel == 2'h0) ? a :
38            (sel == 2'h1) ? b :
39                (sel == 2'h2) ? c : d;
40 endmodule
```

In the second design, an *always* block with an *if* – *else* statement is employed to define the behavior of the multiplexer. The *if* – *else* statement, specifically lines 21–29, establishes the priority of the channels, where channel 0 is given the highest priority, and channel 3 is assigned the lowest priority.

As illustrated in Figure 6.5(b), the synthesis result comprises three 2-to-1 multiplexers, effectively implementing the channel selection function.

In the third design, a conditional *assign* block is employed, utilizing a priority-based approach. In this setup, the channels are prioritized, with channel 0 having the highest priority, followed by channels 1, 2, and 3 in order. Remarkably, the synthesis result of this priority-based multiplexer is identical to the circuit illustrated in Figure 6.5(b).

6.3 Synthesis of sequential latches

The synthesis of latches involves intentional designs as well as accidental generations. Unintended latch creation may occur due to factors such as an incomplete *case – endcase* statement and/or missing trigger signals in an *always* sensitivity list. While some synthesis tools may automatically complete the sensitivity list and the *case – endcase* statement, relying solely on this behavior can result in simulation warnings and potential functional failures. To ensure a robust design, it is advisable to provide complete statements and sensitivity lists in the design code, explicitly specifying all signals that can trigger the desired behavior.

6.3.1 Intentional latches design with Verilog

Below are two designs for implementing a 4-bit latch using a conditional *assign* block and a level-triggered *always* block. These latches store the values of the input "d" as long as the enable signal "en" is TRUE. In the first design example, the output "q" retains its value when the enable signal "en" is FALSE. In the second design example, the absence of an *else* statement results in the identical behavior where the output "q" retains its previous value when the "en" expression evaluates to FALSE. The synthesis result is depicted in Figure 6.6, illustrating a straightforward 4-bit latch implemented with these two designs.

```
1  // Example #1: Design Latches Using Conditonal Assign
2  module example1_latch_conditonal (output [3:0] q    ,
3                                    input  [3:0] d    ,
4                                    input        en   );
5  assign q = en ? d : q;
6  endmodule
7
8  // Example #2: Design Latches Using Omitted else
9  module example2_latch_always (output reg [3:0] q    ,
10                               input      [3:0] d    ,
```

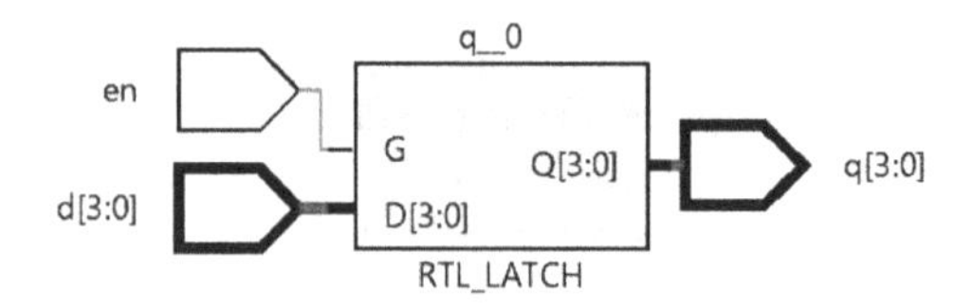

FIGURE 6.6
Synthesis result of latch design.

```
11                                     input              en    );
12  always @(en, d) begin
13    if (en) q <= d;
14  end
15  endmodule
```

Here are two more examples that demonstrate the integration of combinational logic for computations with sequential latches to preserve the results. In the first design, the combination of the combinational logic and sequential latches includes the incorporation of an omitted *else* statement. As demonstrated in lines 7–9, this design comprises three logic gates and three single-bit latches. When the input "sel" is set to TRUE, the design retains the results: "r1" from "a AND c", "r2" from "r1 XOR b", and the final output "d" from "r1 AND r2", all of which are stored using the three latches.

In contrast, the second example divides the design of combinational logic from the sequential latch. The combinational logic, outlined in lines 18–20, is articulated using concurrent *assign* blocks to implement the three logic gates. Subsequently, in line 24, the sequential latch is defined within an *always* block. This latch exclusively stores the final output "d", which is the result of "w1 AND w2".

```
1   // Example #1: Design Combinational Logic and Latches
2   module example1_comb_latch (input        a, b, c, sel,
3                               output reg d              );
4   reg r1, r2;
5   always @ (a, b, c, r1, r2, sel) begin
6     if (sel) begin
7       r1 <= a & c   ;
8       r2 <= r1 ^ b ;
9       d  <= r1 & r2;
10    end
11  end
12  endmodule
```

```
13
14 // Example #2: An Alternative Way to the Design of
15 // Combinational Logic and Latches
16 module example2_comb_latch (input       a, b, c, sel,
17                             output reg d            );
18 wire w1 = a & c;
19 wire w2 = w1 ^ b;
20 wire w3 = w1 & w2;
21
22 always @(w3, sel) begin
23   if (sel) begin
24     d  <= w3;
25   end
26 end
27 endmodule
```

The synthesis results for the two examples are showcased in Figure 6.7. As anticipated, in the first design example, three latches are produced to store the outputs from the three logic gates, as illustrated in Figure 6.7(a). In contrast, the second design example incorporates three logic gates and a single latch, as demonstrated in Figure 6.7(b).

The second design, which separates the description of combinational logic from the sequential latch, is highly recommended as a superior coding style. This approach significantly improves clarity and enforces a clear separation between different design blocks, especially in scenarios involving a combination of combinational and sequential circuits.

6.3.2 Accidental latches with Verilog

Accidental latches should be avoided when writing Verilog code. Typical causes of unintended latches are incomplete $case-endcase$ statements and/or incomplete sensitivity lists within *always* blocks.

A. Incomplete case-endcase Statement

Here, we present two design examples: one featuring an incomplete $case-endcase$ statement and the other with a fully defined $case-endcase$ statement. In the first example, the $case-endcase$ statement lacks the possible binary value 2'b11 for the 2-bit input "sel", rendering it incomplete. Consequently, this specific case won't be addressed within the $case-endcase$ statement, potentially resulting in unintended latch behavior. In contrast, the second example completes the $case-endcase$ statement by incorporating a $default$ branch. This ensures that all possible scenarios for the 2-bit input "sel" are explicitly considered, thereby establishing a comprehensive design.

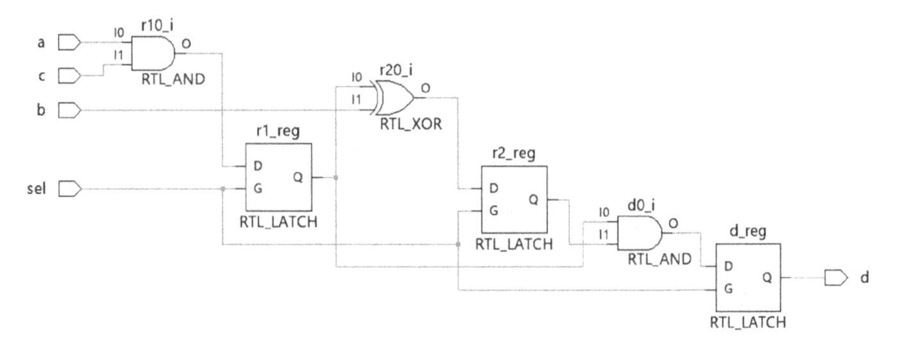

(a) Example #1: Design Combinational Logic and Latches

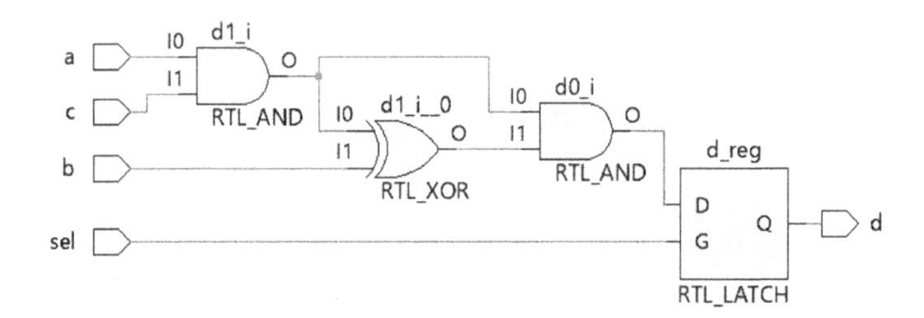

(b) Example #2: An Alternative Way to the Design of Combinational Logic and Latches

FIGURE 6.7
Synthesis results of intentional latches design.

```
1  // Example #1: Accidental Latches with Incomplete Case
2  module example1_acc_latch_co_case (input        a, b, c,
3                                     input [1:0] sel    ,
4                                     output reg  d      );
5  always @ (a, b, c, sel) begin
6    case(sel)
7      2'b00  : d <= a   ;
8      2'b01  : d <= b   ;
9      2'b10  : d <= c   ;
10   endcase // Missing default
11 end
12 endmodule
13
14 // Example #2: Design with Complete Case
15 module example2_acc_latch_co_case (input        a, b, c,
16                                    input [1:0] sel    ,
17                                    output reg  d      );
```

```
18 always @ (a, b, c, sel) begin
19   case(sel)
20     2'b00   : d <= a    ;
21     2'b01   : d <= b    ;
22     2'b10   : d <= c    ;
23     default: d <= 1'b0; // Complete case-endcase
24   endcase
25 end
26 endmodule
```

The synthesis results are depicted in Figure 6.8(a) and Figure 6.8(b), illustrating two distinct design scenarios. In the design featuring an incomplete *case – endcase* statement, as shown in Figure 6.8(a), a latch becomes essential for preserving previous values when the "sel" input is set to binary

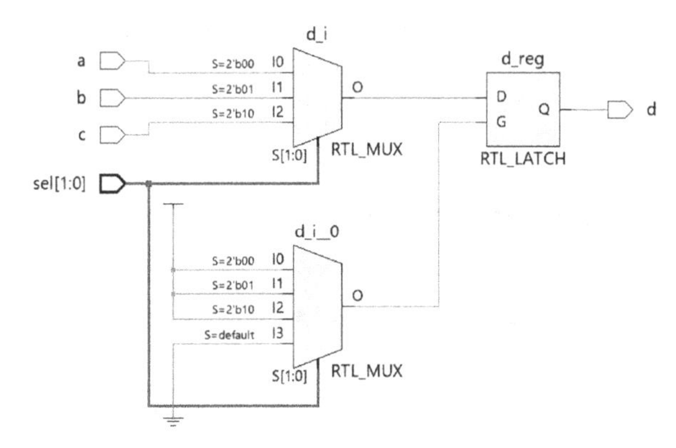

(a) Example #1: Accidental Design of Latches with Incomplete Case-Endcase

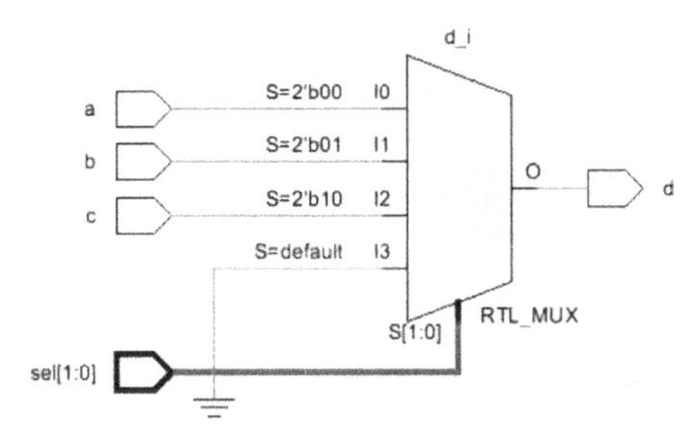

(b) Example #2: Design Results with Complete Case-Endcase

FIGURE 6.8
Synthesis results of incomplete and complete case-endcase.

2'b11. Specifically, the lower multiplexer selects the last channel to deassert the latch's enable input "G" if "sel=2'b11", ensuring that the latch retains the previously stored value. Conversely, when the "sel" signal is not 2'b11, the latch's enable signal asserts through the lower multiplexer, and the latch output follows the input from the upper multiplexer channels 2'b00-2'b10.

On the other hand, Figure 6.8(b) displays the synthesis result when the *case – endcase* statement is complete. In this scenario, the design represents a typical 4-channel multiplier without requiring a latch. The complete *case – endcase* statement ensures that all possible cases for the "sel" input are considered, and the design operates smoothly without any additional latch.

B. Incomplete Sensitivity List

Here's another design example that underscores the significance of maintaining a complete sensitivity list within an *always* block. In this instance, the sensitivity list in line 5 omits the signal "c". Consequently, any modifications to the signal "c" will not activate the *always* block during simulation. This omission can lead to the simulation failing to register changes in the signal "c", resulting in functional errors. Moreover, the simulator may issue a warning regarding the absence of certain signals in the sensitivity list, serving as an alert about potential issues.

```
1  // Example #1: Accidental Latches with Incomplete List
2  module example1_acc_latch_inc_list (input         a, b, c,
3                                      input  [1:0] sel    ,
4                                      output reg   d      );
5  always @(a, b, sel) begin // Incomplete sensitivity list
6    case(sel)
7      2'b00   : d <= a;
8      2'b01   : d <= b;
9      2'b10   : d <= c;
10     default: d <= 1'b0;
11   endcase
12 end
13 endmodule
```

From the synthesis process, the synthesis tool can automatically analyze the design code and complete the sensitivity list to generate the desired hardware. As shown in Figure 6.9, the synthesis result demonstrates the successful synthesis of a multiplexer without any unintended latches being generated. Nevertheless, it's important to emphasize that while the synthesis tool can sometimes compensate for an incomplete sensitivity list, this behavior is not guaranteed and may vary depending on the synthesis tool and the

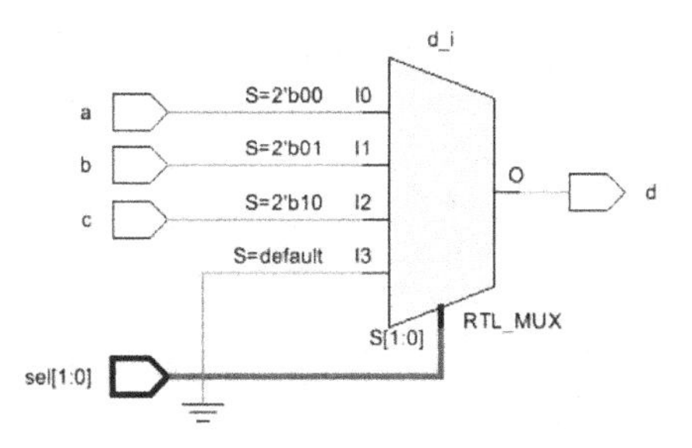

FIGURE 6.9
Synthesis result of incomplete sensitivity list.

specific design scenario. To ensure reliable and predictable behavior, it is recommended to explicitly include all relevant signals in the sensitivity list of *always* blocks. This practice not only promotes accurate simulation but also reduces the chances of encountering unexpected issues during synthesis.

6.4 Synthesis of Sequential Registers

6.4.1 Single-bit register with asynchronous and synchronous reset

Registers are essential storage elements in digital circuits. Section 4.3 discusses single-bit register design and simulation, focusing on both asynchronous and synchronous reset configurations. The synthesis results for these designs are presented in Figure 6.10(a) for the asynchronous reset and Figure 6.10(b) for the synchronous reset.

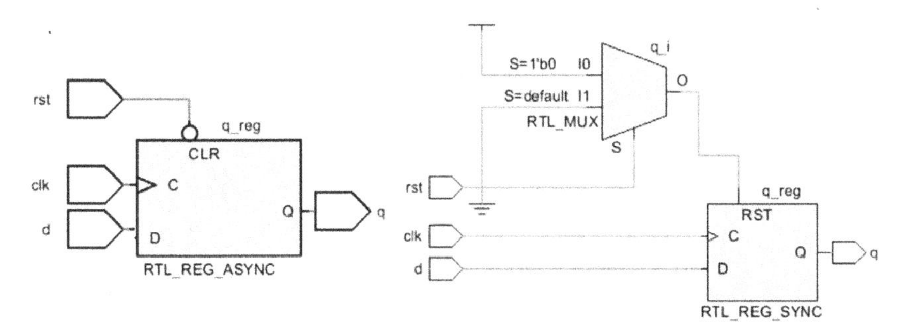

FIGURE 6.10
Synthesis results of single-bit register design with asynchronous and synchronous reset.

It's evident that a multiplexer is employed to switch the reset signal, as the Verilog design employs a zero-valid reset, whereas the synchronous reset register provided by the synthesis tool assumes a one-valid reset. This multiplexer serves to adapt the reset signal from one valid to zero valid, ensuring compatibility between the original design and the synthesized hardware.

6.4.2 Multi-bit register with asynchronous and synchronous reset

The Verilog design for multi-bit registers closely resembles that of single-bit registers. Below, we present design examples of an 8-bit register featuring both asynchronous and synchronous reset functionality. In the first example (lines 1–12), the 8-bit register utilizes an asynchronous reset signal labeled "rst" to initialize the register to the binary value 8'b0 when the reset signal is asserted. Otherwise, it retains the value of the input "d" on the positive clock edge. In contrast, the second example (lines 15–242) demonstrates an 8-bit register design that employs a synchronous reset signal "rst". When the reset "rst" signal is asserted on the positive clock edge, the register is reset to binary 8'b0. Otherwise, it stores the value of the input "d".

```
1  // Example #1: 8-bit Register with Asynchrnous Reset
2  module reg_async_rst (input              rst, clk,
3                        input        [7:0] d       ,
4                        output reg   [7:0] q       );
5  always @ (negedge rst, posedge clk) begin
6    if (~rst) begin
7      q<=8'h0;
8    end else begin
9      q<=d   ;
10   end
11 end
12 endmodule
13
14 // Example #2: 8-bit Register with Synchrnous Reset
15 module reg_sync_rst (input              rst, clk,
16                      input        [7:0] d       ,
17                      output reg   [7:0] q       );
18 always @ (posedge clk) begin
19   if (~rst) begin
20     q<=8'h0;
21   end else begin
22     q<=d   ;
23   end
24 end
25 endmodule
```

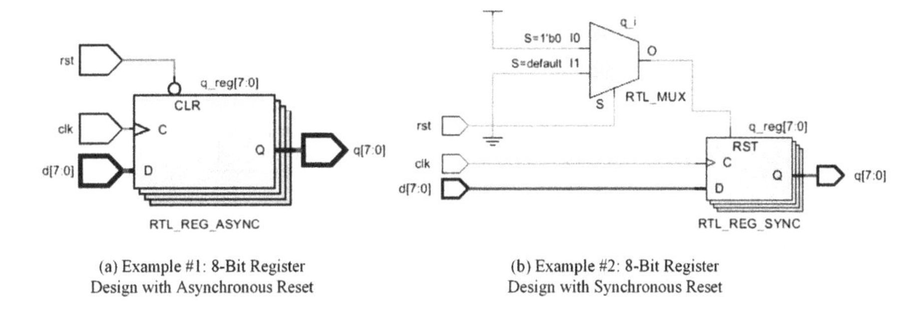

(a) Example #1: 8-Bit Register Design with Asynchronous Reset

(b) Example #2: 8-Bit Register Design with Synchronous Reset

FIGURE 6.11
Synthesis results of 8-bit register design with asynchronous and synchronous reset.

The synthesis outcome of the asynchronous reset register is depicted in Figure 6.11(a), while Figure 6.11(b) illustrates the synthesis result of the synchronous reset register. In the context of FPGA synthesis, these results demonstrate that the functionality of multi-bit registers is achieved by utilizing multiple single-bit registers. Each individual bit within the multi-bit register is implemented using a dedicated single-bit register. Similar to the single-bit register implementation, the synchronous reset register incorporates an additional multiplexer to adapt the reset signal to fit the synchronous reset register provided by the synthesis tool.

It's important to recognize that the synthesis results can exhibit variations contingent on the chosen FPGA device. Diverse FPGA families come equipped with distinct resources and features that can impact the synthesis outcome. In contrast, for ASIC designs, registers with asynchronous reset functionality are frequently employed and readily accessible within most synthesis libraries.

6.4.3 Shift register

A shift register plays a crucial role in various applications involving serial data reception and the introduction of clock cycle delays for data processing. Typically, the shift register design allows for the propagation of input data over multiple clock cycles. The number of clock cycle delays is determined by the number of registers connected in series.

In addition to this, the design of the serial data receiver offers two distinct modes: big-endian input and little-endian input. In the big-endian input mode, the data stream begins with the input of the most-significant bit (MSB), followed by the successive single-bit data in sequential order until reaching the least significant bit (LSB) of the data stream. Conversely, in the little-endian input mode, the process starts with the input of the LSB, followed by the input of the remaining data in sequential order.

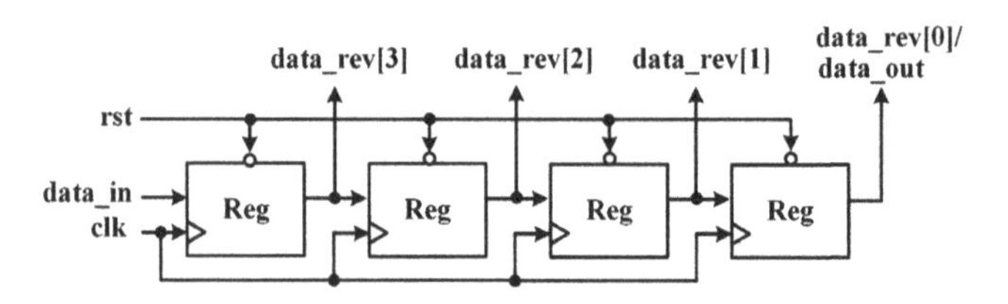

FIGURE 6.12
Block diagram of shift register (little-endian input).

A. Design Specification

To illustrate as a case study, Figure 6.12 depicts a 4-bit shift register, which encompasses two primary design features: a 4-bit data receiver operating in little-endian mode and a 4-cycle delay for data input. This shift register consists of four individual single-bit registers that are connected in series.

The input signal, denoted as "data_in", represents a single-bit signal serving as the input to the shift register. On the other hand, the single-bit output, labeled as "data_out", represents the output of the shift register across four clock cycles. The 4-bit output, labeled as "data_rev", captures a 4-bit data stream over the course of four clock cycles, starting from the LSB and concluding with the MSB, following a little-endian data collection approach.

B. Verilog Design and Synthesis

Based on the design specification, the Verilog code for the 4-bit shift register can be written as follows. In this code, the concatenation operator "{}" is utilized in line 13 to push the single-bit input "data_in" into the 4-bit shift register from the LSB to the MSB. Over four clock cycles, the 4-bit serial inputs can be collected by the 4-bit output "data_rev[3:0]" and the LSB "data_rev[0]" is specified as the single-bit output "data_out" output. This is accomplished by assigning the most-significant three bits, "data_rev[3:1]" (on the right hand side), to the least-significant three bits, "data_rev[2:0]" (on the left hand side) over clock cycles, as shown in line 13. Simultaneously, the MSB of "data_rev[3]" (on the left-hand side) is updated with the input data from "data_in".

```
1  // Shift Register Design
2  module shift_reg (input              rst      ,
3                    input              clk      ,
4                    input              data_in  ,
5                    output             data_out ,
6                    output reg [3:0]   data_rev);
7
```

```
8  assign data_out = data_rev[0];
9  always @ (negedge rst, posedge clk) begin
10   if (~rst) begin
11     data_rev <= 4'h0                    ;
12   end else begin
13     data_rev <= {data_in, data_rev[3:1]};
14   end
15 end
16 endmodule
```

Figure 6.13 showcases the synthesis results, confirming the successful implementation of the 4-bit shift register as anticipated. This implementation employs four separate single-bit registers to accumulate the 4-bit data output labeled as "data_rev" over the span of four clock cycles. Additionally, it's worth noting that the LSB of the "data_rev" output, denoted as "data_rev[0]", is routed to the single-bit output labeled as "data_out".

6.5 Synthesis of Counter and Timer

6.5.1 Counter 0–f

The register-transfer design primarily relies on data processing across clock cycles. As such, the counter plays a crucial role in managing timing and sequencing, not only for serial buses like I2C, SPI, SDIO, GPIO, UART, and similar interfaces but also for numerical designs that involve extensive datapaths such as vector-vector multiplications and matrix-matrix operations.

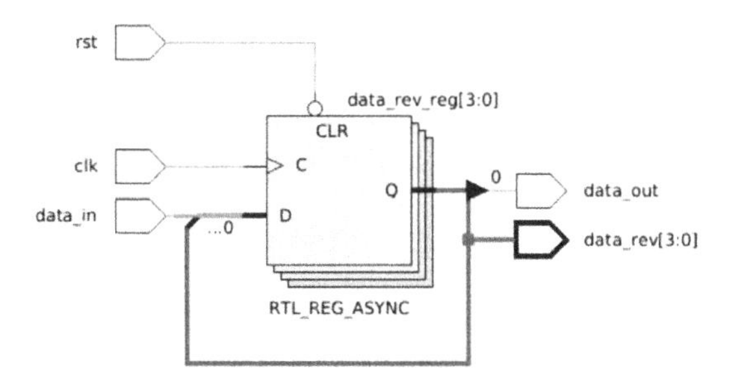

FIGURE 6.13
Synthesis result of Shift register.

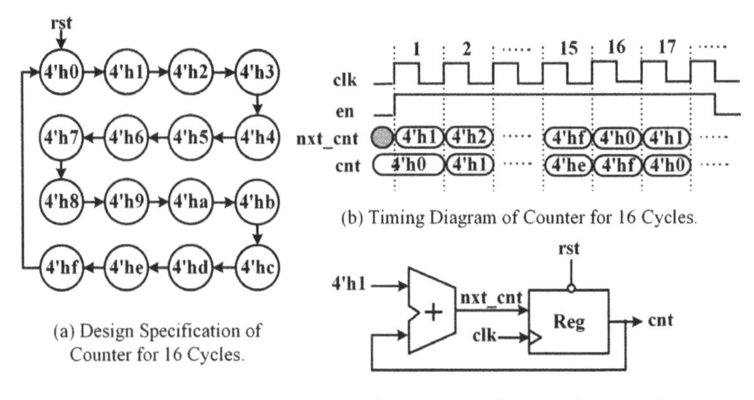

(a) Design Specification of Counter for 16 Cycles.

(b) Timing Diagram of Counter for 16 Cycles.

(c) Block Diagram of Counter for 16 Cycles.

FIGURE 6.14
Design specification of counter for 16 cycles.

A. Design Specification

Figure 6.14 presents a design specification for a 4-bit counter. It includes the design requirements, block diagram, and timing diagram to provide a comprehensive overview of the design details. In Figure 6.14(a), the design specification outlines the desired behavior of the counter, specifying a counting loop from hexadecimal 4'h0 to 4'hf. To provide a more detailed description of the design functions, Table 6.1 provides a summary of the inputs and outputs of the 4-bit counter design. These include the counter enable signal ("en"), the counter output ("cnt"), as well as the clock signal ("clk") and the asynchronous reset ("rst").

Figure 6.14(b) displays the anticipated simulation waveform, also known as a timing diagram. It visually represents the execution of the 4-bit counter design, which counts from the hexadecimal value 4'h0 to 4'hf over a duration of 16 clock cycles. When the counter reaches its maximum value, it returns to 4'h0 and continues counting from the initial value. This cyclic behavior is clearly demonstrated by the repetitive pattern observed in the timing diagram.

To implement the counter design in simulation and synthesis, a 4-bit full adder is required as a fundamental building block. The 4-bit full adder,

TABLE 6.1
Counter IOs description.

Name	Direction	Bit Width	Description
clk	Input	1	Clock
rst	Input	1	Asynchronous reset, 0 valid
en	Input	1	Enable signal, 1 valid
cnt	Output	4	Counter output

depicted in Figure 6.14(c), is responsible for incrementing the counter by one within every clock cycle. Additionally, a 4-bit register is needed to cascade the output of the counter from one cycle to the next, enabling continuous counting. By combining the 4-bit full adder and the 4-bit register, the sequential counter design can achieve the desired functionality of counting from 4'h0 to 4'hf and then restarting the count. These components work in tandem to facilitate the sequential counting behavior exhibited in the timing diagram.

B. Verilog Design and Synthesis

As per the design specification, the Verilog code is provided below. Notably, the combinational circuit adder is implemented using a conditional *assign* block in line 4, while the sequential register is implemented in a separate *always* block spanning lines 6–12. This separation of the combinational and sequential circuits is considered a good coding practice as it promotes a concise and readable design structure.

```
1  // 0-f Counter Design
2  module cnt_0_f (input                 rst, clk, en,
3                  output reg [3:0] cnt          );
4
5  wire [3:0] nxt_cnt = en ? cnt+4'h1 : cnt;
6  always @ (negedge rst, posedge clk) begin
7    if (~rst) begin
8      cnt <= 4'h0     ;
9    end else begin
10     cnt <= nxt_cnt ;
11   end
12 end
13 endmodule
```

In Figure 6.15, the synthesis result aligns with the expected outcome. It can be segmented into a 4-bit adder and a 4-bit register, which perfectly matches the design code.

6.5.2 Timer 0–f

A. Design Specification

The counter design serves as the foundation for implementing a multi-level counter structure, commonly referred to as a timer. An example of a 2-level design structure of a timer is illustrated in Figure 6.16(a). The first level consists of a counter that counts 16 clock cycles, from hexadecimal 4'h0 to

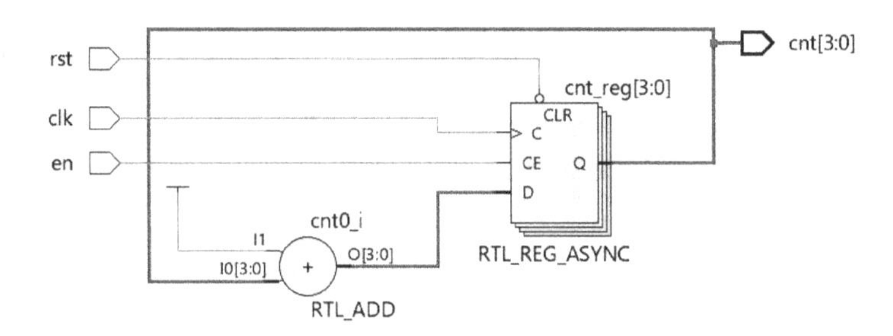

FIGURE 6.15
Synthesis result of counter from 0 to f.

4'hf. In the second level, the timer counts the number of units, with each unit comprising 16 clock cycles. Every 16 cycles, the timer increments by one, starting from hexadecimal 4'h0 and progressing up to the maximum value of hexadecimal 4'hf. Table 6.2 provides a summary of the IOs for the timer design.

The timing diagram in Figure 6.16(b) illustrates the behavior of the timer design. During the 16th clock cycle, the counter signal "cnt" reaches the maxi-

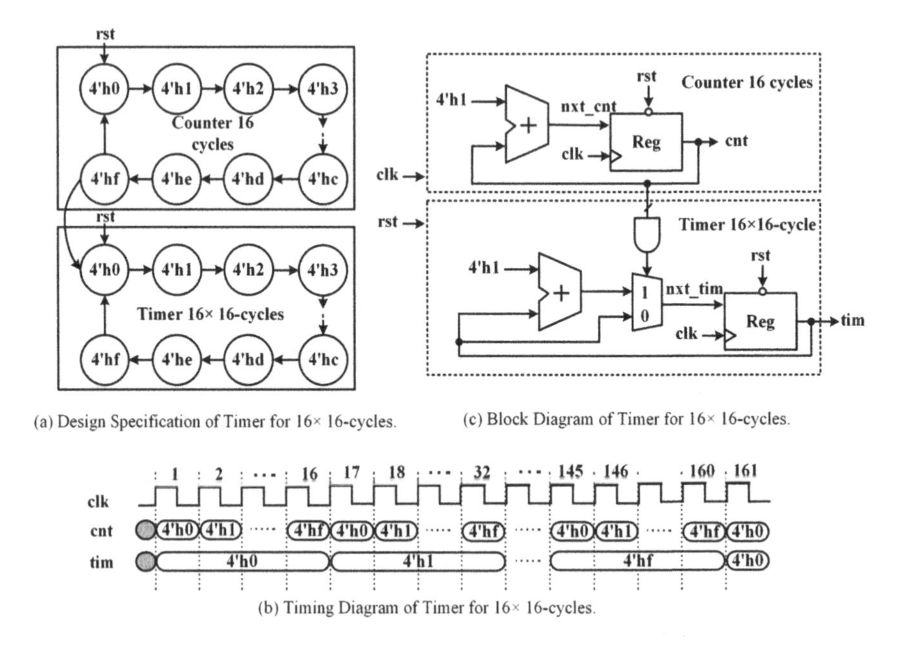

FIGURE 6.16
Design specification of timer for 16× 16-Cycles.

TABLE 6.2
Timer IOs description.

Name	Direction	Bit Width	Description
clk	Input	1	Clock
rst	Input	1	Asynchronous reset, 0 valid
en	Input	1	Enable signal, 1 valid
tim	Output	4	Timer output

mum value of hexadecimal 4'hf and resets to hexadecimal 4'h0 in the following 17th clock cycle. Simultaneously, the timer signal "tim" increments by one. Since both the counter and timer signals are 4-bit wide, they automatically restart from hexadecimal 4'h0 whenever they reach the maximum value of hexadecimal 4'hf, without requiring explicit initialization.

The block diagram in Figure 6.16(c) represents the design structure of the timer. As illustrated in the previous section, the counter utilizes a 4-bit adder and a 4-bit register. When the counter signal "cnt" reaches the value of hexadecimal 4'hf, indicating that all its signal bits are set to binary ones, the higher-level timer adds one to its value. To determine whether the incremented value should be propagated to the register in the second-level timer, a Reduction AND gate is employed. If the expression of the Reduction AND is TRUE, indicating that all bits of the "cnt" signal are binary ones, the timer output will reflect the incremented value of "tim+4'h1". Otherwise, the timer output will retain its previous value.

B. Verilog Design and Synthesis

The Verilog design on the timer is shown below. The first level, defined in lines 4–13, consists of a counter that increments by hexadecimal value 4'h1 in each clock cycle when the enable signal "en" is TRUE. The current count value is stored in the register and displayed by the "cnt" signal. Once the "cnt" signal reaches the maximum value of hexadecimal 4'hf, it automatically resets to 4'h0 in the next clock cycle.

The second level, as outlined in lines 15–23, pertains to the timer description, which advances by the hexadecimal value 4'h1 when the "cnt" signal reaches its maximum value of 4'hf. This condition signifies that all bits of "cnt" are set to binary ones. In line 16, a Reduction AND gate is employed to detect this condition. If such a scenario is detected, the "tim" output is incremented by the hexadecimal value 4'h1. Conversely, if the condition is not met, "tim" retains its previous value.

```
1  // Design on Timer 16x16-cycle
2  module timer_0_f_16_cycles (input    rst, clk, en,
```

```
3                                  output reg [3:0] tim);
4  // The first level counter
5  reg   [3:0] cnt;
6  wire [3:0] nxt_cnt = en ? cnt+4'h1 : cnt;
7  always @(posedge clk, negedge rst) begin
8    if(~rst) begin
9      cnt<=4'h0    ;
10   end else begin
11     cnt<=nxt_cnt;
12   end
13 end
14
15 // The second level timer
16 wire [3:0] nxt_tim = (en & &cnt) ? tim+4'h1 : tim;
17 always @(posedge clk, negedge rst) begin
18   if(~rst) begin
19     tim<=4'h0    ;
20   end else begin
21     tim<=nxt_tim;
22   end
23 end
24 endmodule
```

The design makes use of an *always* block that triggers on the positive clock edge ("clk"). Within this block, the counter and timer signals ("cnt" and "tim") are synchronously updated based on the current value of "cnt".

The synthesis result, depicted in Figure 6.17, confirms the anticipated 2-level design structure. It reveals the presence of a Reduction AND gate (labeled as "RTL_REDUCTION_AND") between the two levels. This gate is

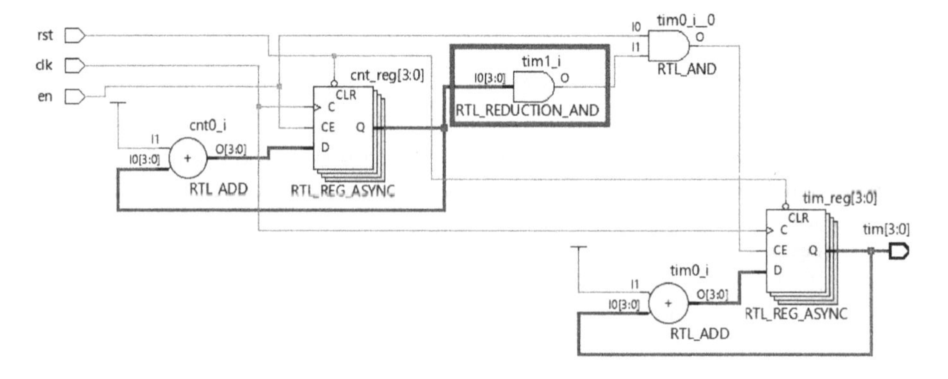

FIGURE 6.17
Synthesis result of timer for 16×16-Cycle.

responsible for determining the maximum value of the "cnt" signal (hexadecimal 4'hf) from the first level counter. When the maximum value is detected, it enables the register in the second level of the timer. The synthesis result aligns with the design specification, indicating the successful implementation of the counter and timer design.

Exercises

Problem 6.1. Given the Verilog designs "circuit1-circuit6" below, draw the hardware circuits.

```
module circuit1 (output f,
                 input  a, b, c, d, e, en);
assign f = en ? ~((a | b) | (c & d & e)) : 1'bz;
endmodule
```

```
module circuit2 (output reg c,
                 input      a, b, en);
wire nxt_c = a ^ b;
always @ (en, nxt_c) begin
  if (en) c <= nxt_c;
end
endmodule
```

```
module circuit3 (output reg e,
                 input      a, b, c, d,
                 input      clk, rst );
reg  r1, r2;
wire nxt_r1 = a & b  ;
wire nxt_r2 = c ^ d  ;
wire nxt_e  = r1 | r2;

always @(posedge clk, negedge rst) begin
  if (~rst) begin
```

```
11        r1  <= 1'b0    ;
12        r2  <= 1'b0    ;
13        e   <= 1'b0    ;
14      end else begin
15        r1  <= nxt_r1  ;
16        r2  <= nxt_r2  ;
17        e   <= nxt_e   ;
18      end
19    end
20    endmodule
```

```
1   module circuit4 (output         f,
2                    input          a, b, c, d, sel1,
3                    input   [1:0]  sel2             );
4   reg e;
5   assign f = sel1 ? e : 1'bz;
6   always @ (a, b, c, d, sel2) begin
7     case (sel2)
8       0:  e <= a;
9       1:  e <= b;
10      2:  e <= c;
11      3:  e <= d;
12      default: e <= a;
13    endcase
14  end
15  endmodule
```

```
1   module circuit5 (output reg [5:0] cnt,
2                    input            clk, rst, en);
3   wire [5:0] nxt_cnt = en ? cnt+6'h1 : cnt;
4   always @(posedge clk, negedge rst) begin
5     if (~rst) begin
6       cnt<=6'h0   ;
7     end else begin
8       cnt<=nxt_cnt;
9     end
10  end
11  endmodule
```

```
module circuit6 (input                  rst      ,
                 input                  clk      ,
                 input                  data_in  ,
                 output                 data_out ,
                 output reg [4:0] data_rev);

assign data_out = data_rev[4];
always @ (negedge rst, posedge clk) begin
  if (~rst) begin
    data_rev <= 5'h0                    ;
  end else begin
    data_rev <= {data_rev[3:0], data_in};
  end
end
endmodule
```

PBL 10: Four-Bit Full-Adder with Tri-State Buffer

Design a 4-bit adder that is connected to a tri-state buffer, as depicted in the block diagram shown in Figure 6.18. The input signals, "a" and "b", are both 4 bits wide, resulting in a 5-bit output signal, "c". To control the functionality of the tri-state buffer, an enable signal labeled as "en" is required. When the enable signal evaluates to FALSE (or binary zero), the output of the tri-state buffer will be in a high-impedance state, denoted as high-Z, indicating an open circuit. When the enable signal evaluates to TURE (or binary one), the output will be assigned as the adder's output.

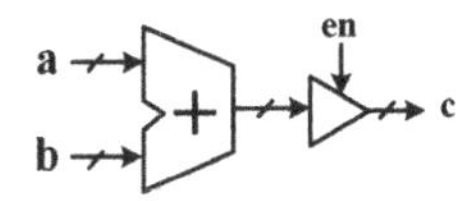

FIGURE 6.18
Block diagram of 4-bit adder with tri-state buffer.

PBL 11: Four-Bit Full-Adder with Latch

Design a 4-bit adder that is connected to a latch, as depicted in the block diagram shown in Figure 6.19. The input signals, "a" and "b", are both 4 bits wide, resulting in a 5-bit output signal, "c". The latch requires two control signals: an enable signal, denoted as "en" to activate the latch, and an asynchronous reset signal, denoted as "rst" to reset the latch output to hexadecimal 5'h0.

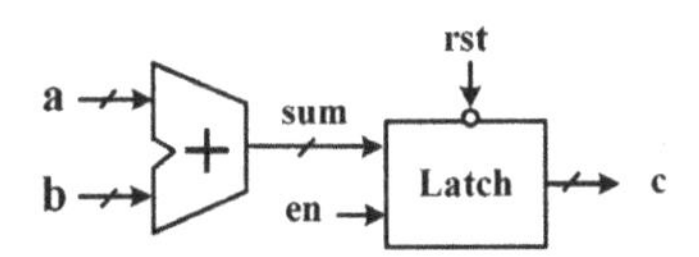

FIGURE 6.19
Block diagram of 4-bit adder with latch.

PBL 12: Four-Bit Full-Adder with Register

Design a 4-bit adder that is connected to a register, as depicted in the block diagram shown in Figure 6.20. The input signals, "a" and "b", are both 4 bits wide, resulting in a 5-bit output signal, "c". The register requires two control signals: a clock signal, denoted as "clk", and an asynchronous reset signal, denoted as "rst" to reset the register output to hexadecimal 5'h0.

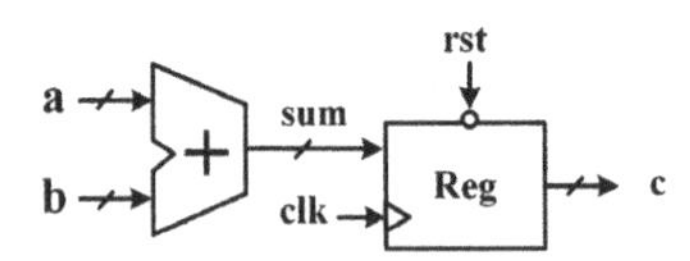

FIGURE 6.20
Block diagram of 4-bit adder with register.

PBL 13: Counter 0–9

Design a counter that counts clock cycles from the hexadecimal value 4'h0 to 4'h9, as specified in the design specification shown in Figure 6.21(a). The

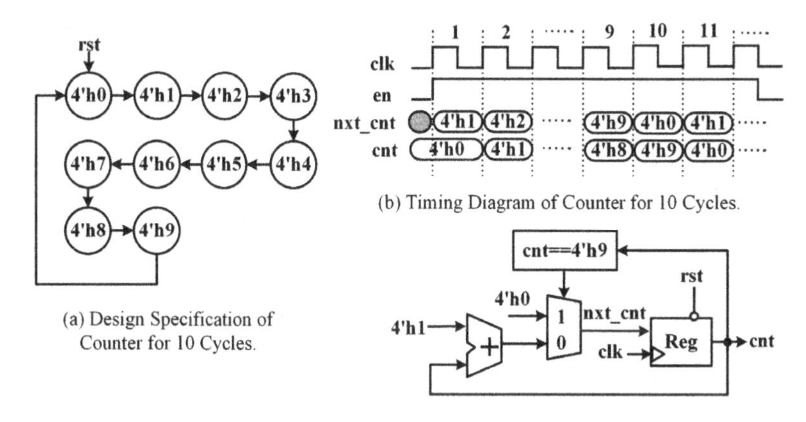

(a) Design Specification of Counter for 10 Cycles.

(b) Timing Diagram of Counter for 10 Cycles.

(c) Block Diagram of Counter for 10 Cycles.

FIGURE 6.21
Design specification of counter for 10 cycles.

counter has the following input/output connections, as summarized in Table 6.3. It operates based on an asynchronous reset and is triggered by the rising clock edge. When the counter reaches the maximum value of hexadecimal 4'h9, it restarts from 4'h0 in the next counting loop. The output of the counter is a 4-bit value, accommodating hexadecimal values ranging from 4'h0 to 4'h9.

Figure 6.21(b) illustrates the timing diagram of the counter, showcasing its behavior from hexadecimal 4'h0 to 4'h9. To handle the initialization of the counter when it reaches hexadecimal 4'h9, the utilization of a comparator and a multiplexer is required.

Figure 6.21(c) presents a block diagram depicting these components. The comparator compares the counter output with the hexadecimal value 4'h9. When the counter reaches 4'h9, the comparator output evaluates to TRUE. This triggers the multiplexer to switch its channels. If the comparator output is TRUE, the multiplexer selects the input signal representing 4'h0, causing the counter to reset back to hexadecimal 4'h0. On the other hand, if the comparator output is FALSE, the multiplexer selects the input signal representing

TABLE 6.3
Counter 0–9 IOs description.

Name	Direction	Bit Width	Description
clk	Input	1	Clock, rising edge trigger, 100 MHz
rst	Input	1	Asynchronous reset, 0 valid
en	Input	1	Enable signal, 1 valid
cnt	Output	4	Counter output

the current counter value plus hexadecimal 4'h1, incrementing the counter by one during each clock cycle.

PBL 14: Timer 0–9

Design a timer that counts from the hexadecimal value 4'h0 to 4'h9 with a unit of 0.1 microseconds (us). The design's input/output connections, summarized in Table 6.4, define the necessary interface for the timer design. The sequential circuit utilizes an asynchronous reset and is triggered by the rising clock edge.

Hint: It's essential to emphasize that with a clock frequency of 100 MHz, each clock cycle has a duration of 10 nanoseconds (ns). Therefore, to achieve a unit of 0.1 us, you would need to utilize ten times the number of clock cycles.

The design specification, presented in Figure 6.22(a), provides an overview of the design details for the timer. It illustrates a 2-level design structure for the timer implementation. The first-level counter is responsible for generating every 10-cycle unit, where each cycle corresponds to a duration of 0.01 us or 10 ns. The second-level timer counts the number of 10-cycle units and increments the "tim" output until it reaches the hexadecimal value 4'h9.

To gain a better understanding of the timer's behavior, a timing diagram in Figure 6.22(b) illustrates the output signal "tim" incrementing by hexadecimal value 4'h1 for every 10 clock cycles or when the counter reaches the value of 4'h9. After the 100th clock cycle, both the "cnt" and "tim" signals restart from the hexadecimal value 4'h0.

The block diagram in Figure 6.22(c) visually represents the components involved in the timer design. It showcases the counter and timer designs described earlier and emphasizes the logic necessary to reset both the counter and timer output. This reset condition is detected by an AND gate, which examines the outputs from the two comparators to ensure that both the counter

TABLE 6.4
Timer 0–9 IOs description.

Name	Direction	Bit Width	Description
clk	Input	1	Clock, rising edge trigger, 100 MHz
rst	Input	1	Asynchronous reset, 0 valid
en	Input	1	Enable the timer from 0 to 9
tim	Output	4	The timer output

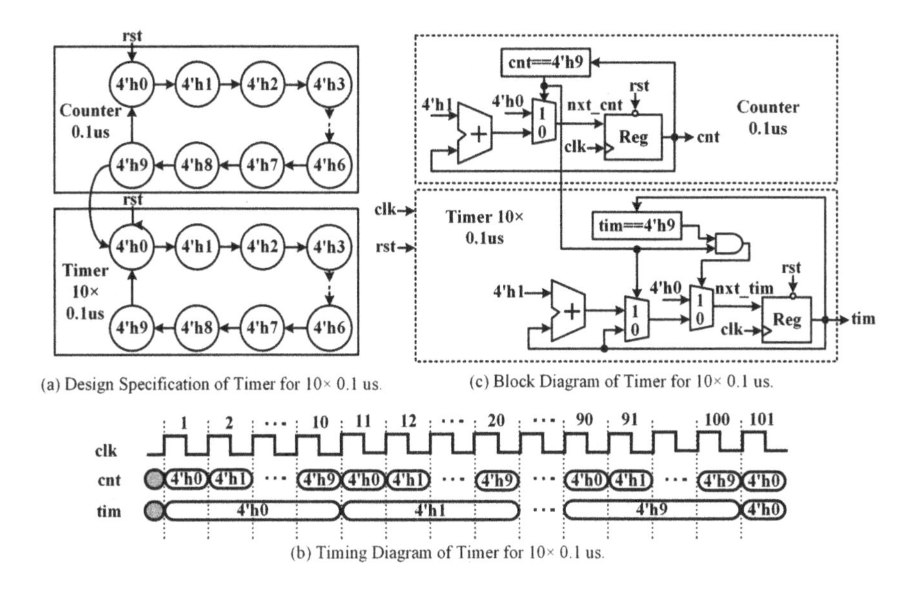

FIGURE 6.22
Design specification of timer for 10× 0.1 us.

and timer have reached the hexadecimal value 4'h9 before restarting the counting process.

PBL 15: Timing Interface

Design a timing interface as depicted in the timing diagram shown in Figure 6.23. The output signal, named "sig", is initially set to binary zero. It transitions to binary one within the time intervals of 0–0.3 us and 0.6–0.9 us. Between 0.4 us and 0.5 us, it resets back to binary zero. After 0.9 us, the signal is initialized to zero again. The IOs of the timing interface is summarized in Table 6.5.

According to the specification, the block diagram is depicted in Figure 6.24.

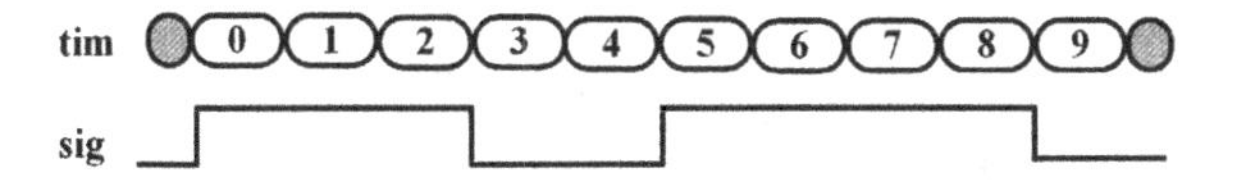

FIGURE 6.23
Timing diagram of timing interface.

TABLE 6.5
IOs description of timing interface.

Name	Direction	Bit Width	Description
clk	Input	1	Clock, 100 MHz
rst	Input	1	Asynchronous reset, 0 valid
en	Input	1	Enable signal
sig	Output	1	Signal output

The timer design remains the same as the design block shown in Figure 6.22. The output of the timer is compared with hexadecimal values 4'h3, 4'h4, and 4'h9. When the condition that the "tim" signal is "greater than 4'h4 AND less than 4'h9" is TRUE, the "sig" output is set to binary 1'b1. Another condition to set the "sig" output as binary 1'b1 is when the "tim" signal is less than hexadecimal 4'h3. Therefore, an AND gate and an OR gate are required to implement the circuit for the "sig" output.

PBL 16: Timing Controller

Design a circuit composed of a datapath and a timing controller, as specified in the block diagram shown in Figure 6.25(a). The datapath consists of three 8-bit binary design components: an adder followed by two multipliers in series. Assuming that each binary component takes one clock cycle from its input to output, the final output "g" of the entire datapath can be produced after three clock cycles once the inputs "a-d" are provided.

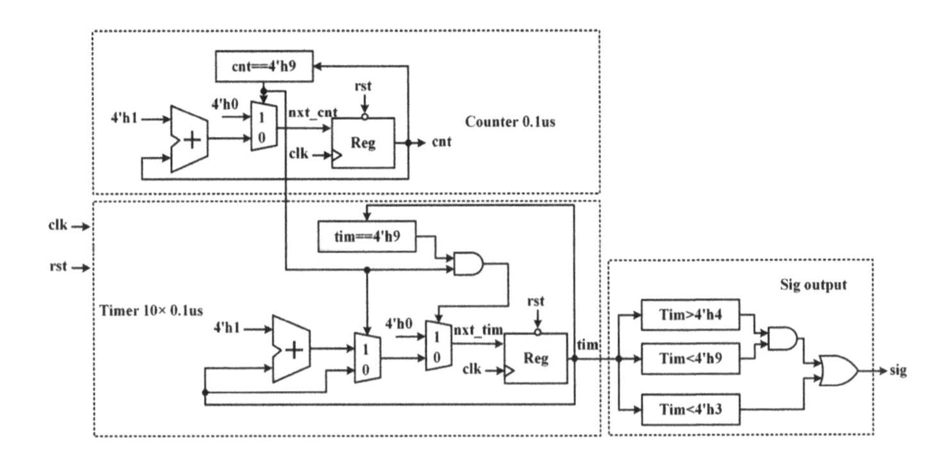

FIGURE 6.24
Block diagram of timing interface.

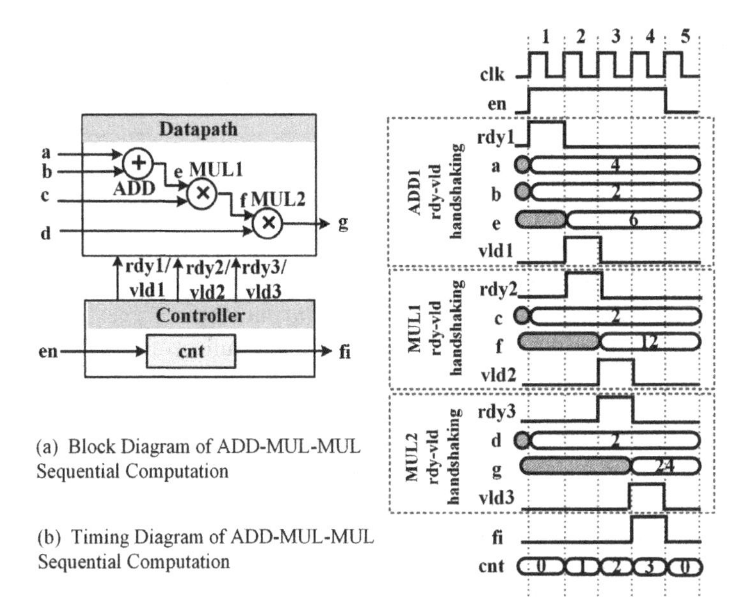

FIGURE 6.25
Design specification of timing controller.

To coordinate the binary operations within the datapath, a timing controller can be implemented using a counter. The enable input "en" activates the counter and the indicator output "fi" signals the completion of the data processing. All the inputs and outputs are summarized in Table 6.6.

Hint: The binary adder can be straightforwardly described in Verilog using the "+" operator, while the binary multiplier can be described using the "*" operator. In this case, we assume that there are no overflow situations for both the adder and multiplier designs. Between the ADD-MUL1-MUL2 components, three 8-bit registers are needed to isolate the datapath. These registers serve to store and propagate the intermediate values within the datapath, ensuring proper synchronization, timing, and speed for the data processing.

Between the datapath and the timing controller, the **ready-valid** mechanism is utilized to indicate the readiness of inputs and the validation of outputs for each component. Specifically, the signals "rdy1-vld1" are used for the adder (ADD), "rdy2-vld2" for the first multiplier (MUL1), and "rdy3-vld3" for the second multiplier (MUL2). These signals facilitate communication and synchronization between the components, ensuring that inputs are available and outputs are valid on the bus interfaces.

Specifically, the ready-valid protocol is primarily discussed in Figure 6.25(b). In the first clock cycle, the inputs "a-b" are ready on the buses, indicated by the asserted signal "rdy1". In the subsequent clock cycle, the

TABLE 6.6
IOs description of timing controller.

Name	Direction	Bit Width	Description
clk	Input	1	Clock, 100 MHz
rst	Input	1	Asynchronous reset, 0 valid
en	Input	1	Enable signal
a-d	Input	8	Input data
g	Output	8	Output data
fi	Output	1	Finish indicator

output "e" from the adder becomes valid on the bus, indicated by the asserted signal "vld1". Similarly, in the second clock cycle, the inputs "e" and "c" for the second multiplier are ready, indicated by "rdy2". Then, in the third clock cycle, the output "f" from the second multiplier becomes valid, indicated by "vld2". For the final data operation, the inputs "f" and "d" are ready in the third clock cycle, indicated by "rdy3". Finally, in the fourth clock cycle, the output "g" becomes valid, indicated by "vld3". It is important to note that the completion signal "fi" is asserted along with the "vld3" signal due to the last clock cycle for the entire data processing.

To follow the timing control, a 2-bit counter can be utilized. It progresses from decimal value 2'd0 to 2'd3 and then restarts to 2'b0. Once the data processing is enabled by the input "en", all ready-valid signals are generated in sequence, and the completion of the data processing is signaled by the output indicator "fi". The enable signal "en" should be set to a low state within the clock cycle when the "fi" output is asserted. The "enable-finish" handshaking guarantees the completion of data processing in the final clock cycle.

7

FSM Design

Timing controllers play a critical role in managing state transitions and coordinating operations within digital circuits. Various hardware components, such as counters/timers and finite state machines (FSMs), are commonly utilized for conducting timing control activities. Chapter 6 illustrates designs involving counters and timers, while Chapter 7 introduces designs involving various FSMs. Additionally, this chapter showcases direct applications of using FSMs, including the data package receiver, odd/even number of 1s checker, and sequence detectors. PBL 17 provides an FSM design practice using Verilog.

7.1 Modeling FSM

7.1.1 Introduction to FSM

An FSM is a design circuit that demonstrates behavior characterized by a finite set of states. FSMs are essential components in the design of controllers for digital circuits. The typical FSM model includes both combinational and sequential circuits, as depicted in Figure 7.1. The combinational circuit determines the next state and computes the outputs based on the current state and inputs, while the sequential circuit stores and reflects the machine's current state.

The sequential circuits are implemented using clock-driven registers, which act as memory elements capturing the calculated next state (referred to as "nxt_state") and displaying the current state (referred to as "cur_state") at their outputs over clock cycles. By considering the current state of the design machine and the input "x", transitions to the next state can be updated utilizing a combinational logic computation, creating a feedback loop from the state registers to the combinational circuit. Similarly, the output "y" can be calculated using the input and the current state through the combinational circuit.

The feedback mechanism enables the FSM to maintain its internal state and progress through different states based on the defined combinational logic. It is worth noting that in practical circuits, the state machine can consist of many inputs and outputs, making it highly versatile and capable of handling

DOI: 10.1201/9781003187080-7

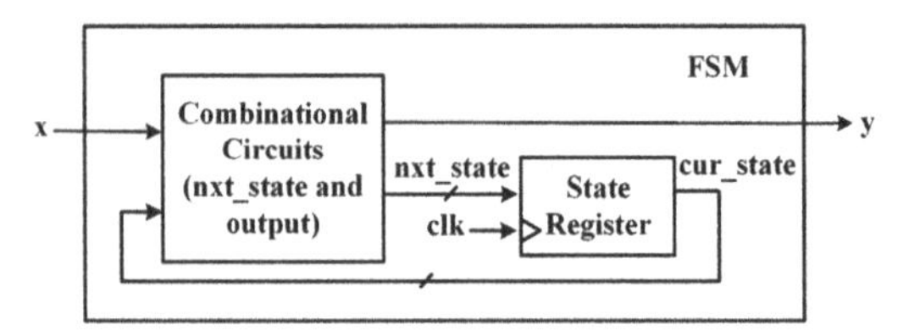

FIGURE 7.1
Finite state machine model.

complex tasks. The combination of sequential and combinational circuits in an FSM design allows it to function as a state controller, making decisions and executing actions based on its current state and inputs.

7.1.2 FSM design template with Verilog HDL

The Verilog template for the FSM model is presented below. This template demonstrates an FSM design with four states, where the current state and next state are declared as 2-bit *reg* data types. The four states are declared as *parameters* "S0", "S1", "S2", and "S4". In line 6, we encode the four states with binary 2'b00 to binary 2'b11.

```
1  module FSM_template (input       rst   ,
2                       input       clk   ,
3                       input       x     ,
4                       output reg  y     );
5  parameter SIZE = 2;
6  parameter S0 = 2'b00, S1 = 2'b01, S2 = 2'b10, S3=2'b11;
7
8  reg [SIZE-1:0] cur_state; // Sequential part of the FSM
9  reg [SIZE-1:0] nxt_state; // Combinational part of the FSM
10
11 //-------------------State Register -----------------
12 always @ (posedge clk, negedge rst) begin
13   if (~rst) begin
14     cur_state <= S0;
15   end else begin
16     cur_state <= nxt_state;
17   end
18 end
19
20 //---- Combinational Circuit for the Next State ---
21 always @ (cur_state, x, rst) begin
22   if (~rst) begin
23     nxt_state <= S0;
```

```
24      end else begin
25        case(cur_state)
26          S0      :
27          S1      :
28          S2      :
29          S3      :
30          default: nxt_state <= S0;
31        endcase
32      end
33    end
34
35    //----------Output Combinational Circuit--------------
36    always @ (cur_state, x, rst) begin
37      if (~rst) begin
38        y <= 1'b0;
39      end else begin
40        case(cur_state)
41          S0      :
42          S1      :
43          S2      :
44          S3      :
45          default: y <= 1'b0;
46        endcase
47      end
48    end
49    endmodule
```

The first *always* block in lines 11–18 describes the state register as shown in Figure 7.1. It can be triggered with rising clock edge and/or falling reset edge, showcasing a typical design on an asynchronous reset register. The second *always* block represents the combinational circuit for the next state transitions, and the third *always* block describes the combinational circuit to generate the output "y".

In order to enhance the code's conciseness and readability, it is recommended to separate the two *always* blocks for the computation of the next state and output. It's worth mentioning that the output design could be simplified using *assign* blocks. In such a case, the output signal "y" must be declared as a *wire* data type.

7.2 Mealy and Moore FSMs with Verilog HDL

7.2.1 Introduction to Mealy and Moore FSMs

FSMs can be divided into two types: Mealy machine and Moore machine. These two types differ in how they produce outputs. In a Mealy machine, the outputs are associated with both the current state and the inputs during the transition. On the other hand, in a Moore machine, the outputs are solely dependent on the current state and are not influenced by the inputs. The outputs of a Moore machine are associated with each state and remain constant throughout the state.

These distinctions in output behavior have implications for the design and implementation of FSMs, and the choice between Mealy and Moore machines depends on the specific requirements of the design circuits. In general, if the output should be simultaneously asserted with the input events, the Mealy machine is appropriate. If the output should be asserted after the input events, the Moore machine is preferable. Additionally, it's worth noting that Moore machines typically require more hardware resources due to their additional state transitions and memory utilization.

As illustrated in Figure 7.2, both Mealy and Moore machines use the same state register design and the "nxt_state" computation circuit. The main distinction lies in the output design circuit: Figure 7.2(a) represents the Mealy machine model, where the output circuit depends on both the "cur_state" from the feedback loop and the input "x". Conversely, Figure 7.2(b) demonstrates the Moore machine model, where the output circuit solely relies on the "cur_state".

7.2.2 Mealy machine

Designing and evaluating an FSM typically involve four main steps: 1) constructing a state graph, 2) creating the state table and encoding it into a transition table, 3) designing the FSM using Verilog HDL, and 4) analyzing the simulation and synthesis results using Karnaugh maps (commonly known as K-maps).

It's important to mention that synthesis tools in the IC design process can automate the K-map optimization, thus removing the necessity for manual K-map optimization by RTL designers. Nevertheless, having a grasp of the K-map simplification process can facilitate a deeper understanding of the

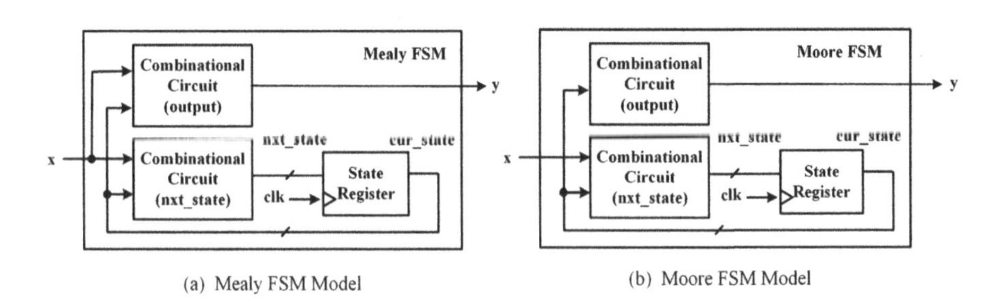

(a) Mealy FSM Model

(b) Moore FSM Model

FIGURE 7.2
Mealy and Moore FSM models.

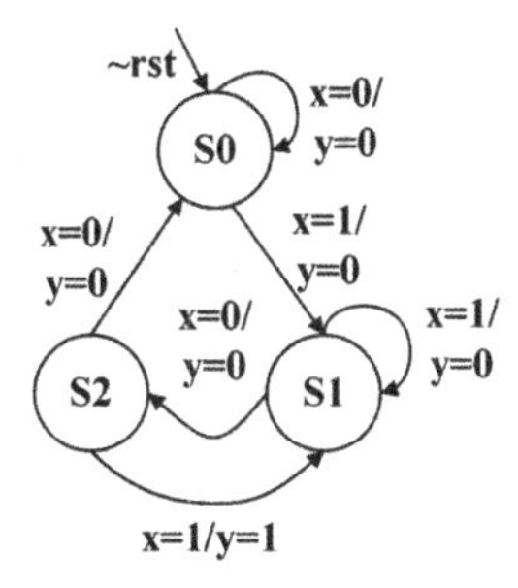

FIGURE 7.3
State graph of Mealy FSM.

underlying circuit transformation and optimization, transitioning from the register-transfer level design to gate-level implementation. Consequently, this section provides a comprehensive overview of the entire circuit design and analysis procedure.

7.2.2.1 State graph of Mealy machine

A state graph provides a visual representation of the state transitions and outputs in an FSM, serving as a foundational tool for the FSM design and implementation. As an example, a state graph of a Mealy machine is depicted in Figure 7.3. The graph illustrates three current states, represented by circles labeled as "S0", "S1", and "S2". When the circuit is reset, the state machine initializes to the initial state, "S0".

The arcs connecting the current states to the next states represent the state transitions that occur over clock cycles. In every clock cycle, one of the arcs will be executed to transition the machine state into the next state. For instance, in the current state "S0", the next state will be "S0" if the input "x" is binary zero, and it will be "S1" if the input "x" is binary one. Utilizing registers to memorize the behavior, the current state will demonstrate the state transition in the following clock cycle.

Additionally, each arc also includes the input "x" and its corresponding output "y". The output "y" is determined based on the current states (in the circle) and the input "x" (on the arc). In both transition arcs starting from "S0", for instance, the output "y" is binary zero. Following the same pattern, the transitions from the current states "S1" and "S2" are depicted through their respective arcs. In this example, only when the current state is "S2" and the input is binary one, the output "y" can be asserted.

7.2.2.2 State and transition tables of Mealy machine

To clarify the next state transitions and the output computation, the state graph can be translated into a state table, as shown in Figure 7.4(a). In the

cur_state	nxt_state		y	
	x=0	x=1	x=0	x=1
S0	S0	S1	0	0
S1	S2	S1	0	0
S2	S0	S1	0	1

(a) State Table of the Mealy FSM

q1q0	q1+q0+		y	
	x=0	x=1	x=0	x=1
00	00	01	0	0
01	10	01	0	0
10	00	01	0	1

S0=2'b00, S1=2'b01, S2=2'b10

(b) Transition Table of the Mealy FSM

FIGURE 7.4
State and transition tables of Mealy FSM.

second and third columns, it shows the next state transitions as the current states ("S0-S2" in the first column) and the input "x" ("x=0" and "x=1" in the second row).

Remember that in a Mealy machine, the output "y" is influenced not just by the current state but also by the input "x". As a result, in the last two columns, the input "x" is listed to determine different results for the output "y". In the table provided, for instance, the last row of the state table demonstrates this behavior: if the current state is "S2" and the input "x" is binary zero, the output "y" will be binary zero; however, if the input "x" is binary one, the output "y" will be binary one.

In Figure 7.4(b), the transition table is presented by encoding the state table. As there are three states being represented, a 2-bit register is required. In the encoding, the binary value 2'b00 corresponds to state "S0", 2'b01 represents state "S1", and 2'b10 represents state "S2". The binary value 2'b11 is not utilized in the encoding. The "q1" and "q0" bits refer to the most significant bit (MSB) and least significant bit (LSB) of the current state, respectively. On the other hand, "q1+" and "q0+" represent the MSB and LSB of the next state. The next state ("q1+q0+") serves as the input to the state register, while the current state ("q1q0") serves as the output of the state register. Therefore, the next state transitions can be memorized by the register to show the current state within each clock cycle.

It is crucial to note that state encoding is not unique, and there may be slight variations in the hardware results based on the specific encoding chosen. Different encodings can result in different assignments of state values to the binary representations.

7.2.2.3 Verilog HDL design of Mealy machine

Based on the state graph or state table, below is the the Verilog description of the Mealy machine using the FSM template. In this code, the three states are declared as *parameters* in line 7 to assign meaningful names to their binary representations.

```
module mealy_fsm(input       rst    ,
                 input       clk    ,
                 input       x      ,
                 output      y      );

parameter SIZE = 2;
parameter S0 = 2'b00, S1 = 2'b01, S2 = 2'b10;

reg [SIZE-1:0] cur_state; // Sequential part of the FSM
reg [SIZE-1:0] nxt_state; // Combinational part of the FSM

//-------------------State Register ----------------
always @ (posedge clk, negedge rst) begin
  if (~rst) begin
    cur_state <= S0;
  end else begin
    cur_state <= nxt_state;
  end
end

//-------Next State Combinational Circuit-----------
always @ (cur_state, x, rst) begin
  if (~rst) begin
    nxt_state <= S0;
  end else begin
    case(cur_state)
      S0     : if(x)  nxt_state <= S1;
      S1     : if(~x) nxt_state <= S2;
      S2     : if(x)  nxt_state <= S1; else nxt_state <= S0;
      default: nxt_state <= S0;
    endcase
  end
end

//----------Output Combinational Circuit--------------
assign y = (cur_state==S2) & x;
endmodule
```

In lines 12–19, the register for the current state is described with a clock-edge-triggered *always* block. The next state transition is described in lines 21–33 using a *case – endcase* statement in a level-triggered *always* block. It is important to note that the missing *else* in each *case* statement represents the scenario where the next state stays the same as the current state. For instance, in line 27 it shows that the next state will remain as "S0" if the current state is "S0" and the input "x" is binary zero, where describes the same transition in the state graph (Figure 7.3) and state table (in Figure 7.4).

The computation of the output "y" is described using a concurrent *assign* block in lines 35–36. Therefore, the output "y" is declared in line 4 in a default *wire* data type.

7.2.2.4 Simulation and synthesis analysis of Mealy machine

A. Synthesized Circuit

Following the Verilog code, the synthesis tool can be used to automate the conversion from the register-transfer level description into the gate-level netlist. However, to gain a better understanding of the hardware implementation, we manually convert the behavioral model-level design into logic gates by analyzing the transition table, as illustrated in Figure 7.5. This process involves performing K-map optimization to minimize the gate logic required for the implementation.

Figure 7.5(a) depicts the three K-map simplifications for output "y", and the MSB "q1+" and LSB "q0+" of the next state. It is important to note that the binary value 2'b11 is not utilized by any machine states, resulting in the K-map table being filled with 2'bxx to indicate unknown data when the current state is 2'b11.

Based on the Boolean expressions of the K-map result, the implementation for "q1+" can be achieved using one AND gate and one inverter, as depicted in Figure 7.5(b). On the other hand, the "q0+" computation can be directly connected to the input "x" since the K-map result indicates a direct mapping. Similarly, the output "y" can be implemented using one AND gate with inputs "x" and "q1". This gate-level implementation is based on the simplified expressions obtained from the K-map optimization process. The sequential circuit,

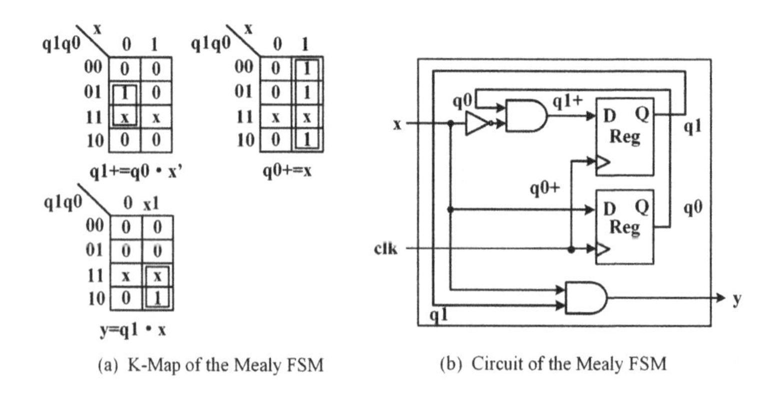

(a) K-Map of the Mealy FSM (b) Circuit of the Mealy FSM

FIGURE 7.5
K-Map optimization and circuit of Mealy FSM.

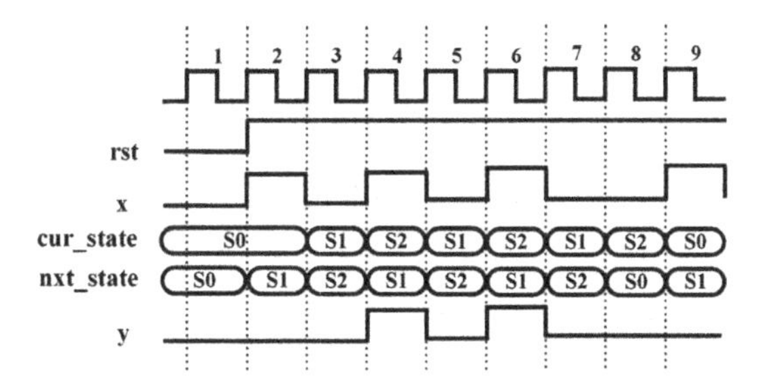

FIGURE 7.6
Timing diagram of Mealy FSM.

which consists of two registers, is responsible for storing the current state and updating it based on the next state at each rising clock edge.

B. Timing Diagram

To illustrate the state transitions in conjunction with the RTL design, a timing diagram for the Mealy machine is presented in Figure 7.6. The state register is reset to the initial state "S0" in the first clock cycle. During the second clock cycle, the reset process is finalized, transitioning the state machine into its normal operating mode. In this specific clock cycle, the state machine is in the "S0" state, and the input "x" is set to binary one. According to line 27 in the Verilog code, the next state transitions to "S1", which is demonstrated in the timing diagram. Likewise, the output "y" is dependent on both the current state and the input "x". As specified in line 36 of the Verilog description, the output "y" is binary zero in this clock cycle, as depicted in the figure.

In the third clock cycle, the current state follows the next state ("S1") at the rising edge and remains stable throughout the entire clock cycle. As a result, the next state computation yields "S2" (as indicated in Verilog code line 28), and the output "y" remains at zero because the input is set to binary zero during this cycle. Moving to the fourth clock cycle, the current state progresses to the next state ("S2") at the rising clock edge and maintains this state throughout the cycle. Consequently, the next state computation results in "S1" (as shown in Verilog code line 29). During this cycle, the output is asserted because the current state is "S2", and the input is binary one (as depicted in Verilog code line 36).

Following the same manner, the update of the current state occurs at every rising clock edge, based on the next state, resembling the behavior of a sequential register. Conversely, the update of the next state and the output is determined by the current state and input values, processed by the combinational circuit.

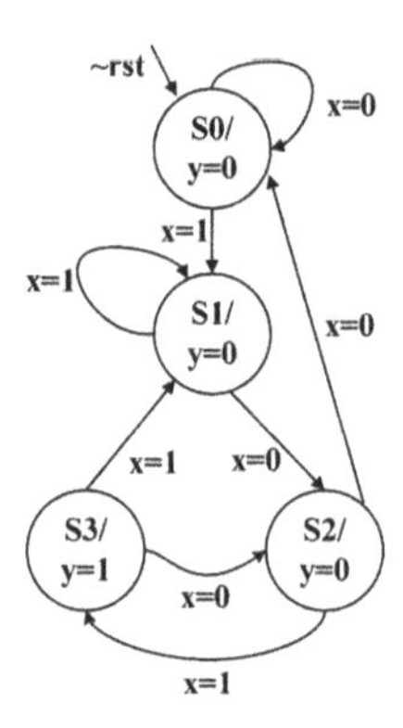

FIGURE 7.7
State graph of Moore FSM.

7.2.3 Moore machine

7.2.3.1 State graph of Moore machine

Figure 7.7 illustrates an example of the state graph for a Moore machine. In this graph, the current states ("S0", "S1", "S2", and "S3") and the output "y" are all represented within the four circles. Because in a Moore machine the output "y" depends solely on the current state and remains constant within each state. Consequently, it takes an additional state, "S3", to represent the output "y=1" compared to a Mealy machine.

The arcs connecting the current states to the next states represent the transitions that happen over clock cycles. For instance, when the current state is "S0", the next state will be "S0" if the input "x" is binary zero, and the next state will be "S1" if the input "x" is binary one. The output "y" is zero in this current state. In the current state "S3", the output "y" is asserted. The arcs illustrates that the next state will be "S2" if the input "x" is binary zero, and the next state will be "S1" if the input "x" is binary one.

7.2.3.2 State and transition tables of Moore machine

Based on the state graph, the state table is presented in Figure 7.8(a). The second and third columns represent the next state transitions, taking account of the current states ("S0-S3" in the first column) and the corresponding input "x" ("x=0" and "x=1" in the second row). In the Moore machine, the output "y" is solely determined by the current state. Therefore, the last column displays the output "y" without considering the input "x". Specifically, it is evident from the table that the output "y" is binary one only when the current state is "S3". For the other states, "S0" to "S2", the output remains binary zero.

In Figure 7.8(b), the transition table is depicted as the encoding of the state table. In this encoding scheme, binary 2'b00 represents state "S0", binary

cur_state	nxt_state		y
	x=0	x=1	
S0	S0	S1	0
S1	S2	S1	0
S2	S0	S3	0
S3	S2	S1	1

(a) State Table of the Moore FSM

q1q0	q1+q0+		y
	x=0	x=1	
00	00	01	0
01	10	01	0
10	00	11	0
11	10	01	1

S0=2'b00, S1=2'b01,
S2=2'b10, S3=2'b11

(b) Transition Table of the Moore FSM

FIGURE 7.8
State and transition tables of Moore FSM.

2'b01 represents state "S1", binary 2'b10 represents state "S2", and binary 2'b10 represents state "S3". The transition of the next state ("q1+q0+") is determined by the combination of the current state ("q1q0") and the input "x".

7.2.3.3 Verilog HDL design of Moore machine

The Verilog description of the Moore machine is provided below. The state register and the next state transition designs (in lines 12–34) are similar to that of the Mealy machine. However, the key distinction lies in the output assignment. In line 37, the output "y" is set to binary one only when the current state is "S3". The output computation does not depend on the input "x". Another way to express this assignment is "y = cur_state[1] & cur_state[0]", which demonstrates the utilization of an AND gate for the output calculation.

```
1  module moore_fsm(input        rst    ,
2                   input        clk    ,
3                   input        x      ,
4                   output       y      );
5
6  parameter SIZE = 2;
7  parameter S0 = 2'b00, S1 = 2'b01, S2 = 2'b10, S3 = 2'b11;
8
9  reg [SIZE-1:0] cur_state; // Sequential part of the FSM
10 reg [SIZE-1:0] nxt_state; // Combinational part of the FSM
11
12 //------------------State Register -----------------
13 always @ (posedge clk, negedge rst) begin
14   if (~rst) begin
15     cur_state <= S0;
16   end else begin
17     cur_state <= nxt_state;
```

```
18      end
19  end
20
21  //-------Next State Combinational Circuit-----------
22  always @ (cur_state, x, rst) begin
23    if (~rst) begin
24      nxt_state <= S0;
25    end else begin
26      case(cur_state)
27        S0     : if(x)  nxt_state <= S1;
28        S1     : if(~x) nxt_state <= S2;
29        S2     : if(x)  nxt_state <= S3; else nxt_state <= S0;
30        S3     : if(x)  nxt_state <= S1; else nxt_state <= S2;
31        default: nxt_state <= S0;
32      endcase
33    end
34  end
35
36  //----------Output Combinational Circuit-------------
37  assign y = (cur_state==S3);
38  endmodule
```

7.2.3.4 Simulation and synthesis analysis of Moore machine

A. Synthesized Circuit

The RTL analysis for the combinational circuits is depicted in Figure 7.9(a). The K-map result illustrates the utilization of AND, NOT, and OR gates for the "q1+" computation. On the other hand, the "q0+" calculation simply

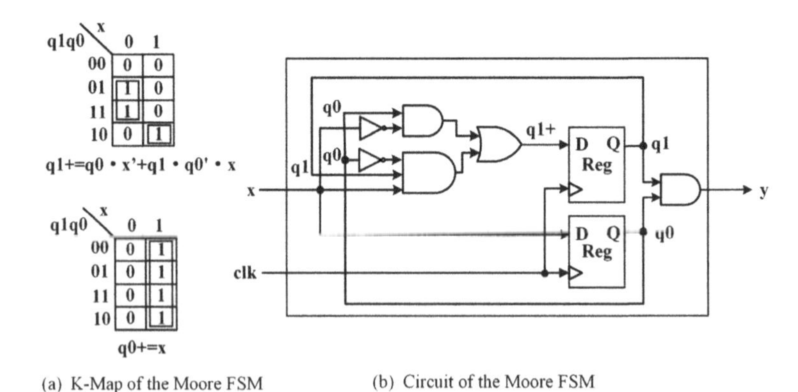

(a) K-Map of the Moore FSM (b) Circuit of the Moore FSM

FIGURE 7.9
K-Map optimization and circuit of Moore FSM.

involves a direct wire connection between the input "x" and "q0+". For the "y" output, an AND gate is employed, taking "q1" and "q0" as inputs. This can be represented by the concurrent *assign* "y = cur_state[1] & cur_state[0]", or alternatively, "y = (cur_state == S3)".

Figure 7.9(b) provides an overview of the hardware implementation, which consists of a 2-bit state register and the combinational circuits responsible for the "y" output and next state computations.

B. Timing Diagram

As an example, the timing diagram in Figure 7.10 illustrates the behavior of the Moore machine. The state transitions in the Moore machine are similar to those in the Mealy machine, where the current state, being the output of state registers, follows the next state at every rising clock edge. Additionally, the next state can be immediately computed by the combinational circuit, based on the current state and the input "x".

It is important to note that the output "y" of the Moore machine is solely determined by the current state, regardless of the input "x". In the case of the Moore machine, when the current state is "S3", the output "y" is set to binary one, in the clock cycles 5 and 7. Therefore, the Moore machine introduces a one clock cycle delay for the output "y", compared to the Mealy machine. This delay is attributed to the introduction of an extra state, "S3", which serves as a clock-triggered register output specifically designed to represent the state when "y=1".

7.3 Design Example: Sequence Detector

A widely employed application of an FSM is the sequence detector. It analyzes a stream of digit 0s and 1s supplied to the input "x" and produces an output "y=1" exclusively when a particular input sequence is detected. In this section,

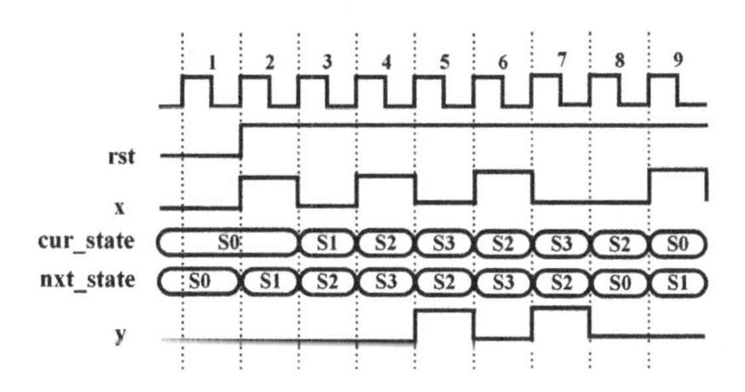

FIGURE 7.10
Timing diagram of Moore FSM.

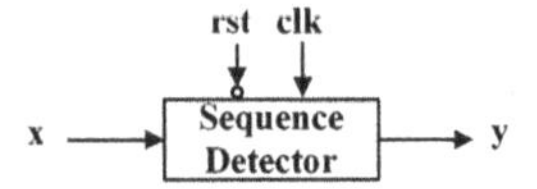

FIGURE 7.11
A Sequence detector model.

two examples are presented using Verilog HDL: one for a Mealy machine and one for a Moore machine.

7.3.1 Introduction to sequence detector

Figure 7.11 illustrates a sequence detector model that asserts an output signal "y=1" when a specified pattern of consecutive bits is received in the serial input stream "x". The input stream is fed into the detector from left to right, with only one bit digit per clock cycle.

As a case study, Figure 7.12 illustrates the details about a 0101 sequence detection. In this example, the output "y" is set to binary one whenever the digit string 0101 occurs. For example, when the first four digits form the string 0101, the first occurrence of "y=1" is generated to indicate the successful receipt of the string.

It is important to note that the digits in the input can be overlapping between two consecutive occurrences of the desired 0101 string. For instance, consider the scenario where the last two digits "01" from the previous desired "0101" string serve as the first two digits of the following desired string. In the case of the input sequence "010101", as illustrated in the figure, the output "y=1" occurs twice. This indicates the successful detection of the first four digits "0101" and the successful detection of the last four digits "0101", with two intermediate digits "01" overlapping between them. Likewise, the input sequence 01010101 results in three output "y=1" as shown in the figure.

To create the sequence detector, we first identify the essential inputs and outputs, as summarized in Table 7.1. The asynchronous reset is utilized for initializing the register, while the clock is employed to sample each digit input. A state register is a crucial component in this design, responsible for storing

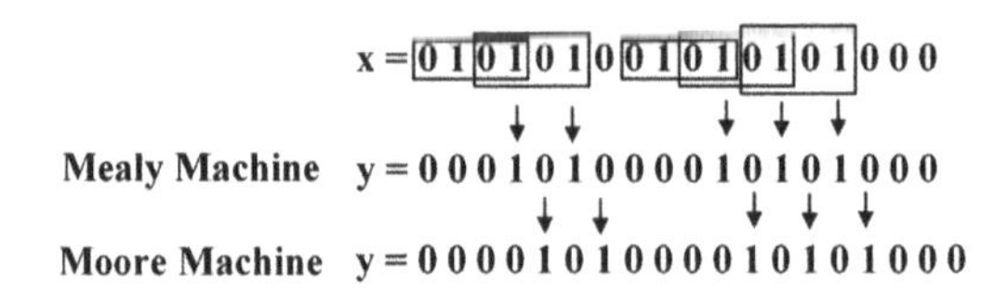

FIGURE 7.12
0101 sequence detection utilizing Mealy and Moore FSM.

TABLE 7.1
Sequence detector IOs description.

Name	Direction	Bit Width	Description
clk	Input	1	Clock, rising edge trigger
rst	Input	1	Asynchronous reset, 0 valid
x	Input	1	Digit string input
y	Output	1	Output 1 when 0101 string detected

and updating the status of the digit string input. It ensures that the state can be updated on every active clock cycle, enabling the sequence detector to effectively track and detect the desired digit pattern.

7.3.2 Mealy FSM: 0101 sequence detector

7.3.2.1 Mealy state graph of 0101 sequence detector

This section focuses on designing a Mealy machine to implement the detection of the digit string 0101. Figure 7.13 depicts the state graph including four states: the initial state ("S0"), the state when the first desired digit 0 is detected ("S1"), the state when the first two desired digits 01 are detected ("S2"), and the state when the first three desired digits 010 are detected ("S3"). On the arcs, the first digit represents the "x" input, and the following digit represents the output "y".

The state graph starts from the initial state "S0". If the first desired digit zero is received, the next state transitions to "S1", representing that the first desired zero digit is detected. Otherwise, if a digit one is received as the first digit, the next state returns to "S0", because the first desired digit must be

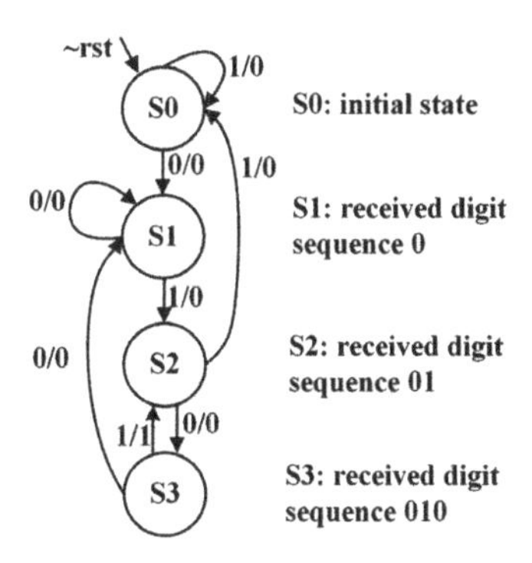

FIGURE 7.13
State graph of 0101 sequence detector design with Mealy FSM.

zero to form the desired string 0101. In both cases, the output "y" is zero since only one digit is received at this moment.

From the current state "S1", if the second desired digit one is received, the next state transitions to "S2", indicating that the two desired 01 digits are received. Otherwise, if a digit zero is received, the next state remains in "S1" to indicate the ongoing reception of the first desired zero digit.

In the current state "S2", meaning that the first two desired digits 01 are received, the next state can be "S3" if the third desired digit zero is received. However, if a digit one is received as the third digit, the next state returns to the initial state "S0" since the received string 011 is not the desired string, and any digit segment, whether the last digit 1 or the last two digits 11, does not constitute the desired digit string.

In the current state "S3", the first three desired digits 010 are received. It is important to note that in the Mealy FSM design, there is no separate state dedicated to receiving the entire desired string 0101. Instead, when the current state "S3" is reached and the following input digit is one, the output "y" is asserted to indicate the successful reception of the string 0101. Simultaneously, the next state transitions back to state "S2" to ensure that the last two digits 01 can be memorized and reutilized as the first two desired digits for the following digit string detection. This mechanism allows for the detection of multiple occurrences of the desired string with partial digits overlapping between them.

Conversely, when the input digit is zero while in the current state "S3", forming a digit string of "0100", only the last digit "0" needs to be remembered since it can serve as the first digit for the subsequent detection of the desired "0101" string. Consequently, the next state transitions to "S1", and the output remains "y=0".

7.3.2.2 Mealy State and transition tables of 0101 sequence detector

As shown in Figure 7.14(a), the state graph is translated into the state table. In the state table, the second and third columns show the next state transitions based on the current states and the input "x". It is worth noting that the output "y" of the Mealy machine depends on both the current state and the input "x". Therefore, the last two columns indicate the available options for input "x" that determine the output "y".

In Figure 7.14(b), the state table is encoded into a transition table. The state register requires two bits, denoted as "q1" and "q0", to represent the four states from "S0" to "S3". The next state can be represented as "q1+q0+" for the MSB and the LSB of the register input.

7.3.2.3 Verilog HDL design of 0101 Mealy sequence detector

The state graph or state table can be described using Verilog as shown below. The state register is described in lines 12–19, and the next state transitions is designed in lines 21–34. In line 37, the output signal "y" is asserted when the current state is "S3", indicating that the first three digits 010 have been

cur_state	nxt_state		y	
	x=0	x=1	x=0	x=1
S0	S1	S0	0	0
S1	S1	S2	0	0
S2	S3	S0	0	0
S3	S1	S2	0	1

(a) State Table of the 0101 Sequence Detector Design with Mealy FSM

q1q0	q1+q0+		y	
	x=0	x=1	x=0	x=1
00	01	00	0	0
01	01	10	0	0
10	11	00	0	0
11	01	10	0	1

S0=2'b00, S1=2'b01, S2=2'b10, S3=2'b11

(b) Transition Table of the 0101 Sequence Detector Design with Mealy FSM

FIGURE 7.14
State and transition tables of 0101 sequence detector design with Mealy FSM.

received, and the subsequent input "x" is one, completing the desired digit string 0101.

```
1  module seq_det_0101_mealy (input  rst,
2                             input  clk,
3                             input  x   ,
4                             output y   );
5
6  parameter SIZE = 2;
7  parameter S0 = 2'b00, S1 = 2'b01, S2 = 2'b10, S3 = 2'b11;
8
9  reg [SIZE-1:0] cur_state; // Sequential part of the FSM
10 reg [SIZE-1:0] nxt_state; // Combinational part of the FSM
11
12 //-----------------State Register ----------------
13 always @ (posedge clk, negedge rst) begin
14   if (~rst) begin
15     cur_state <= S0;
16   end else begin
17     cur_state <= nxt_state;
18   end
19 end
20
21 //-------Next state combinational circuit----------
22 always @ (cur_state, x, rst) begin
23   if (~rst) begin
24     nxt_state <= 2'b00;
25   end else begin
26     case(cur_state)
27       S0:       if(~x) nxt_state <= S1;
28       S1:       if(x)  nxt_state <= S2;
29       S2:       if(~x) nxt_state <= S3; else nxt_state <= S0;
30       S3:       if(~x) nxt_state <= S1; else nxt_state <= S2;
```

```
31          default: nxt_state <= S0;
32        endcase
33    end
34  end
35
36  //---------Output combinational circuit-------------
37  assign y = (cur_state==S3) & x;
38  endmodule
```

7.3.2.4 Simulation and synthesis analysis of 0101 Mealy sequence detector

A. Synthesized Circuit

By analyzing the transition table and performing the K-map simplification, the Boolean expressions for the next state can be derived, as shown in Figure.7.15(a). The expression for "q1+" involves an inverter, AND gates, and an OR gate, while the expression for "q0+" can be implemented using a single inverter.

In Figure.7.15(b), the circuit is depicted, which includes the combinational circuit for computing "q1+q0+" as well as the output "y", along with the sequential circuit for the state register design.

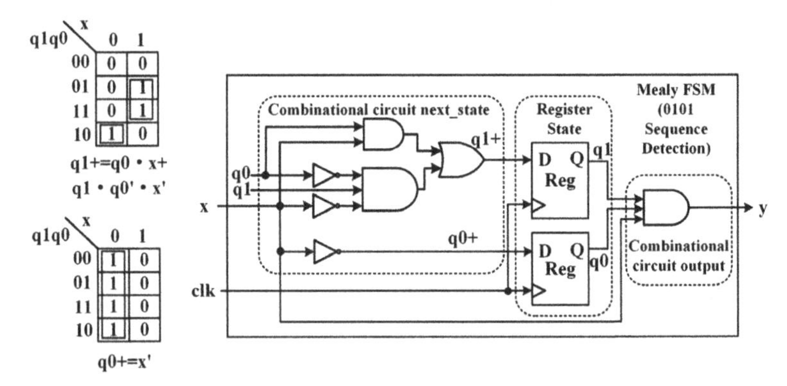

(a) K-Map of the 0101 Sequence Detector Design with Mealy FSM

(b) Circuit of the 0101 Sequence Detector Design with Mealy FSM

FIGURE 7.15
K-Map optimization and circuit of 0101 sequence detector design with Mealy FSM.

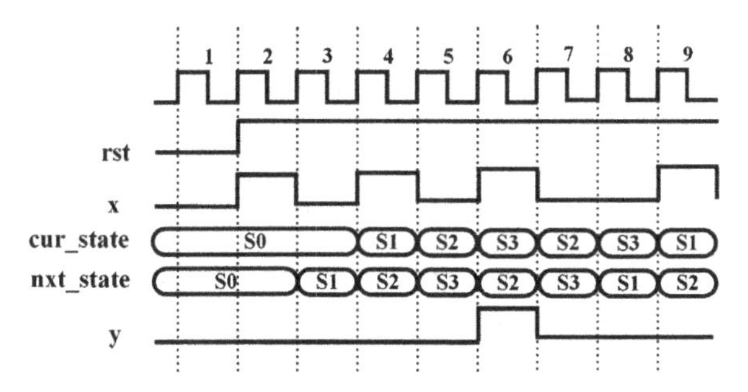

FIGURE 7.16
Timing diagram of 0101 sequence detector design with Mealy FSM.

B. Timing Diagram

Figure 7.16 presents a timing diagram for detecting the digit string 0101. The next state computation in this Mealy machine is based on a combinational circuit, which means that the next state updates immediately when the current state and/or the input "x" change.

During the second clock cycle, following the reset phase's completion, the current state is set to "S0". As the input "x" is binary one, which does not match the first desired digit for detecting the string "0101", the next state remains "S0". With the arrival of the next rising clock edge, this "S0" state is stored in the register during the subsequent third clock cycle. In this cycle, the detection of the first desired digit zero occurs, leading to a state transition to "S1" to signify the reception of the initial desired digit.

In clock cycle 4, the current state reflects this transition and shows as "S1". Since the input is one in this cycle, forming the first two desired string 01, the next state is computed into "S2". In the clock cycle 5, the current state "S2" and the input zero drives the next state into "S3". Continuing the state transitions, in clock cycle 6, the current state transitions to "S3", indicating that three desired digits 010 are received. In the same clock cycle, the "y" output is asserted because the input "x" is detected as one within the current "S3" state, forming the entire string 0101.

7.3.3 Moore FSM: 0101 sequence detector

7.3.3.1 Moore state graph of 0101 sequence detector

To compare the difference between Mealy and Moore machines, Figure 7.17 illustrates the state graph of a Moore machine for detecting the digit string 0101. The states defined in the graph are similar to those in the Mealy machine design. "S0" represents the initial state, "S1" indicates the detection of the first desired digit zero, "S2" denotes the detection of the first two desired digits 01, and "S3" represents the detection of the first three desired digits 010.

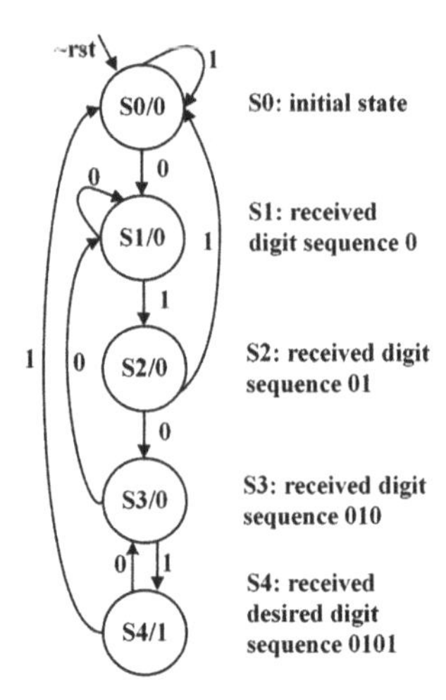

FIGURE 7.17
State graph of 0101 sequence detector design with Moore FSM.

The main difference between the Moore machine and the Mealy machine is that the Moore machine has a separate state, "S4", to indicate the successful detection of the entire digit string 0101. In the state graph, the circle labeled "S4" represents this final state, and the output "y" is asserted as one within this state. In all other cases, the output "y" remains zero, indicating that the entire desired 4-digit sequence has not been received yet.

It is important to note that in the "S4" state, if the current input "x" is a digit zero, the next state will transition back to "S3". This transition represents the formation of a new desired string 010 by including the last two digits 01 of the previous string and the current input digit zero. The transition to state "S3" indicates that another desired digit string 010 is formed, and the machine is ready to detect the next occurrence of the desired 0101 sequence.

7.3.3.2 Moore state and transition tables of 0101 sequence detector

Figure 7.18 presents the state table and the transition table for the Moore machine. In this design, a 3-bit register is required to encode the four states of the Moore machine. Furthermore, the last column of both tables indicates that the output "y" is solely determined by the current state, regardless of the input "x".

7.3.3.3 Verilog HDL design of 0101 Moore sequence detector.

According to the state graph or state table, the Verilog code is as follows. The state register is defined in lines 13–20, while the next state transitions are specified in lines 22–36. In line 39, the output "y" is activated exclusively when

cur_state	nxt_state		y
	x=0	x=1	
S0	S1	S0	0
S1	S1	S2	0
S2	S3	S0	0
S3	S1	S4	0
S4	S3	S0	1

(a) State Table of the 0101 Sequence Detector Design with Moore FSM

q2q1q0	q2+q1+q0+		y
	x=0	x=1	
000	001	000	0
001	001	010	0
010	011	000	0
011	001	100	0
100	011	000	1

S0=3'b000, S1=3'b001, S2=3'b010, S3=3'b011, S4=3'b100

(b) Transition Table of the 0101 Sequence Detector Design with Moore FSM

FIGURE 7.18
State and transition tables of 0101 sequence detector design with Moore FSM.

the current state is "S4". This illustrates the input independence characteristic of a Moore machine.

```
1  module seq_det_0101_moore (input  rst,
2                             input  clk,
3                             input  x  ,
4                             output y  );
5
6  parameter SIZE = 3;
7  parameter S0 = 3'b000, S1 = 3'b001, S2 = 3'b010,
8            S3 = 3'b011, S4 = 3'b100;
9
10 reg [SIZE-1:0] cur_state; // Sequential part of the FSM
11 reg [SIZE-1:0] nxt_state; // Combinational part of the FSM
12
13 //-----------------State Register ----------------
14 always @ (posedge clk, negedge rst) begin
15   if (~rst) begin
16     cur_state <= S0;
17   end else begin
18     cur_state <= nxt_state;
19   end
20 end
21
22 //-------Next state combinational circuit----------
23 always @ (cur_state, x, rst) begin
24   if (~rst) begin
25     nxt_state <= S0;
26   end else begin
27     case(cur_state)
28       S0:      if(~x) nxt_state <= S1;
```

```
29          S1:       if(x)  nxt_state <= S2;
30          S2:       if(x)  nxt_state <= S0; else nxt_state <= S3;
31          S3:       if(x)  nxt_state <= S4; else nxt_state <= S1;
32          S4:       if(x)  nxt_state <= S0; else nxt_state <= S3;
33          default: nxt_state <= S0;
34        endcase
35      end
36    end
37
38    //----------Output combinational circuit-------------
39    assign y = (cur_state==S4) ;
40    endmodule
```

7.3.3.4 Simulation and synthesis analysis of 0101 Moore sequence detector

A. Synthesized Circuit

The design circuit can be obtained by analyzing the transition table and using K-maps. In Figure 7.19(a), the combinational circuits for computing the output "y" and the next state, represented as "q2+q1+q0+", can be constructed using a combination of AND gates, OR gates, and inverters. The circuit also includes a 3-bit register to reflect the current states.

Figure 7.19(b) provides a visual representation of the combinational circuit responsible for calculating the next state and the output. It also depicts the sequential circuit consisting of the state register, which retains the current state within each clock cycle.

B. Timing Diagram

Figure 7.20 presents the timing diagram illustrating the detection of the digit string "0101" using the Moore machine. The sequence detection commences in the second clock cycle, following the reset phase. From clock cycles 3 to 5, the input digits "010" are sequentially received, leading to the machine transitioning to the "S3" state by clock cycle 6. During clock cycle 6, upon receiving the final digit one, the computation of the next state results in "S4". In clock cycle 7, the current state reflects the transition to "S4", indicating the successful acquisition of the entire digit string "0101".

It's important to highlight that the output "y" can only be asserted when the machine is in the current state "S4". As a result, an additional clock cycle, clock cycle 7, is required to update the current state to "S4" and simultaneously activate the output "y".

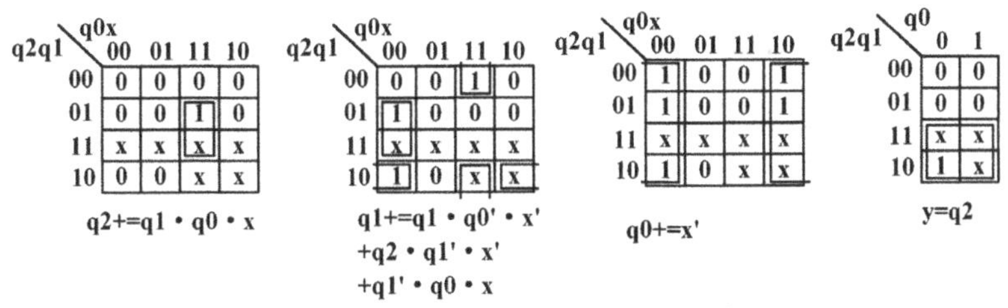

(a) K-Map of the 0101 Sequence Detector Design with Moore FSM

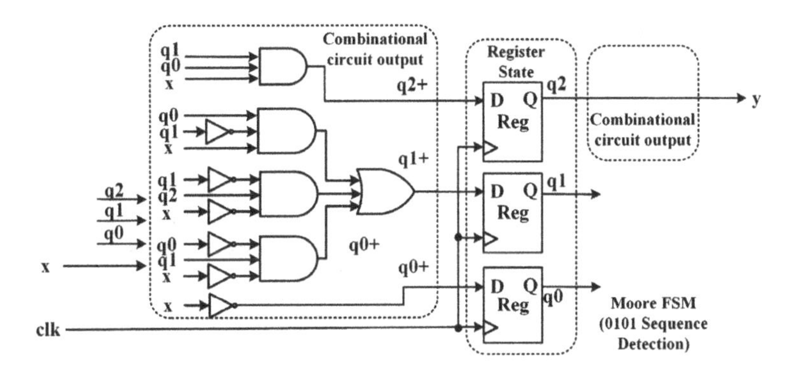

(b) Circuit of the 0101 Sequence Detector Design with Moore FSM

FIGURE 7.19
K-Map optimization and circuit of 0101 sequence detector design with Moore FSM.

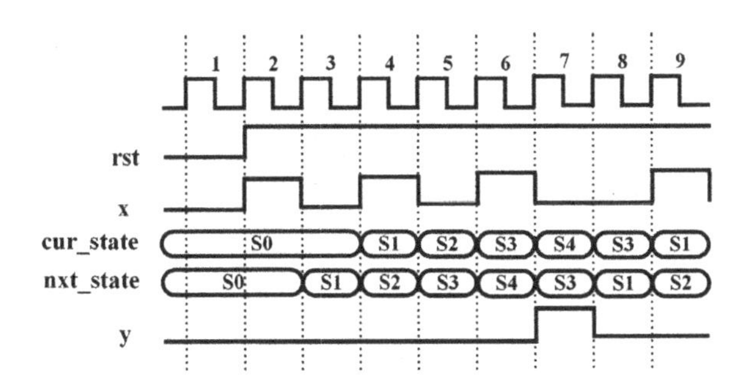

FIGURE 7.20
Timing diagram of 0101 sequence detector design with Moore FSM.

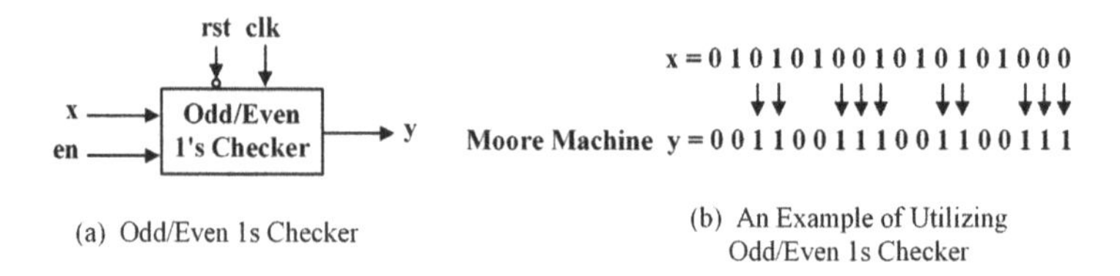

(a) Odd/Even 1s Checker

(b) An Example of Utilizing Odd/Even 1s Checker

FIGURE 7.21
Odd/Even number of 1s checker.

7.4 Design Example: Odd/Even Number of 1s Checker

7.4.1 Introduction to odd/even number of 1s checker

An odd/even number of 1s checker is a digital circuit that determines whether a given binary sequence contains an odd or even number of 1s. This type of circuit is commonly used in various applications, such as error detection, parity checking, and data transmission. It is a fundamental components in digital circuit that provides important parity information and is essential for ensuring data integrity in various circuits and systems.

As shown in Figure 7.21(a), the design model operates by examining the binary sequence on the input "x" and counting the number of 1s present in the sequence. It then determines whether the count is odd or even and produces a corresponding output signal "y" to indicate the result. For example, in Figure 7.21(b), the output "y" is asserted whenever an odd number of 1s is received, while it remains zero for an even number of 1s.

The odd/even number of 1s checker can be implemented using a simple Moore machine. The sequential circuit keeps track of the received 1s using a state register, while the combinatorial circuit typically handles the state transitions based on the current state and the input data. By analyzing the transition table, the circuit can be implemented to accurately determine whether the number of 1s is odd or even.

7.4.2 State graph of odd/even 1s checker

Figure 7.22 shows the state graph for the checker design. It starts in the initial state "INI", where the output "y" remains zero (represented as "$\sim y$" in the circle). If the input "x" is zero (represented as "$\sim x$" on the arc), the next state stays in the initial state, maintaining the output at zero. However, if the input is one (represented as "x" on the arc), the circuit transitions to the state "ODD", indicating that an odd number of 1s is received, and the output "y" is asserted (represented as "y" in the circle).

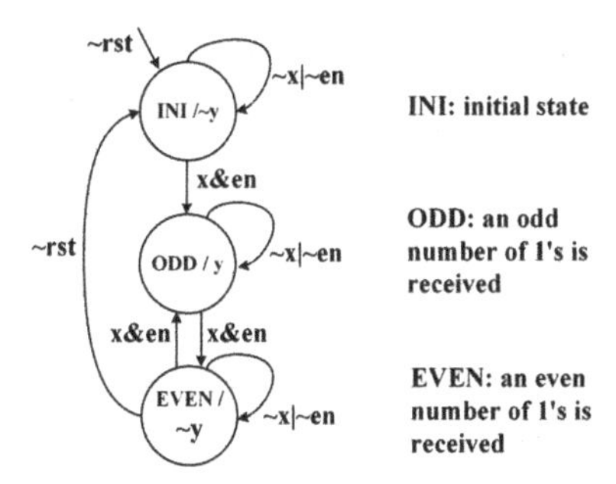

FIGURE 7.22
State graph of odd/even 1s checker.

In the "ODD" state, the next state remains in "ODD" if the input is zero, as the odd count of 1s continues. However, if the input is one, the graph transitions to the state "EVEN", indicating that an even number of 1s is received. In the "EVEN" state, the next state stays in "EVEN" if the input is zero, while it transitions back to the "ODD" state if the input is one, indicating the count of 1s becomes odd again.

Additionally, the state remains unchanged if the enable signal "en" is OFF (represented as "$\sim en$" on the arc), meaning that the circuit is temporarily inactive. If a reset occurs (represented as "$\sim rst$" on the arc), the state returns to the initial state "INI", because the entire state machine should be initiated.

7.4.3 State and transition tables of odd/even 1s checker.

Figure 7.23 depicts the state table and the following transition table to encode the current and next states. In this example, binary 2'b00 indicates the "INI" state, 2'b01 represents the "ODD" state, and 2'b10 denotes the "EVEN" state. Since the FSM is a Moore machine, the output depends on only the current state as shown in the last column of the two tables.

For the sake of simplicity, it's essential to emphasize that the enable input is factored into the state and transition tables, while the reset is intentionally

cur_state	nxt_state (en, x)			y
	~en	~x	en&x	
INI	INI	INI	ODD	0
ODD	ODD	ODD	EVEN	1
EVEN	EVEN	EVEN	ODD	0

(a) State Table of Odd/Even 1s Checker

q1q0	q1+q0+ (en, x)			y
	~en	~x	en&x	
00	00	00	01	0
01	01	01	10	1
10	10	10	01	0

INI=2'b00, ODD=2'b01, EVEN=2'b10

(b) Transition Table of Odd/Even 1s Checker

FIGURE 7.23
State and transition tables of odd/even 1s checker.

excluded. If the circuit is disabled, the state remains unchanged in its current state, as illustrated in the second column. As the input zero doesn't alter the ODD or EVEN state, the state also remains identical to the current state, which is shown in the third column. The state only updates when both the enable "en" and "x" inputs are set to ones, as depicted in the fourth column.

It's crucial to verify that all input combinations are included in the state table, particularly in the second row in Figure 7.23(a). The Boolean sum of all the inputs in this row should evaluate to one, as represented by "$(\sim en) + (\sim x) + (en\&x) = 1$". This comprehensive coverage of input combinations is vital for ensuring a comprehensive and precise state transition description.

7.4.4 Verilog HDL design of odd/even 1s checker.

The state graph or state table can be represented in Verilog as follows. The state register is defined in lines 12–19, and the next state transition is established in lines 21–33. In line 36, the output "y" can be activated when the current state is "ODD", signifying the reception of an odd number of 1s.

```
1  module odd_even_1s_checker(input  rst ,
2                             input  clk ,
3                             input  en  ,
4                             input  x   ,
5                             output y   );
6  parameter SIZE = 2;
7  parameter INI = 2'b00, ODD = 2'b01, EVEN = 2'b10;
8
9  reg [SIZE-1:0] cur_state; // Sequential part of the FSM
10 reg [SIZE-1:0] nxt_state; // Combinational part of the FSM
11
12 //------------------State Register -----------------
13 always @ (posedge clk or negedge rst) begin
14   if (~rst) begin
15     cur_state <= INI      ;
16   end else begin
17     cur_state <= nxt_state;
18   end
19 end
20
21 //-------Next state combinational circuit-----------
22 always @ (cur_state, en, x, rst) begin
23   if(~rst) begin
24     nxt_state <= INI;
25   end else begin
26     case(cur_state)
27       INI     : if(x&en) nxt_state <= ODD ;
```

```
28          ODD     : if(x&en) nxt_state <= EVEN;
29          EVEN    : if(x&en) nxt_state <= ODD ;
30          default : nxt_state <= INI;
31        endcase
32      end
33  end
34
35  //---------Output combinational circuit------------
36  assign y = (cur_state==ODD);
37  endmodule
```

7.4.5 Simulation and synthesis analysis of odd/even 1s checker.

A. Synthesized Circuit

As per the transition table, the K-map simplification can be executed as shown in Figure 7.24(a). It results in the gate-level circuit demonstrated in Figure 7.24(b), containing the combinational circuit for computing the next state and the output "y", and the sequential circuit for implementing the state register.

B. Timing Diagram

Figure 7.25 illustrates a timing diagram for the odd/even number of 1s checker. It showcases the state transitions and output behavior of the circuit over nine

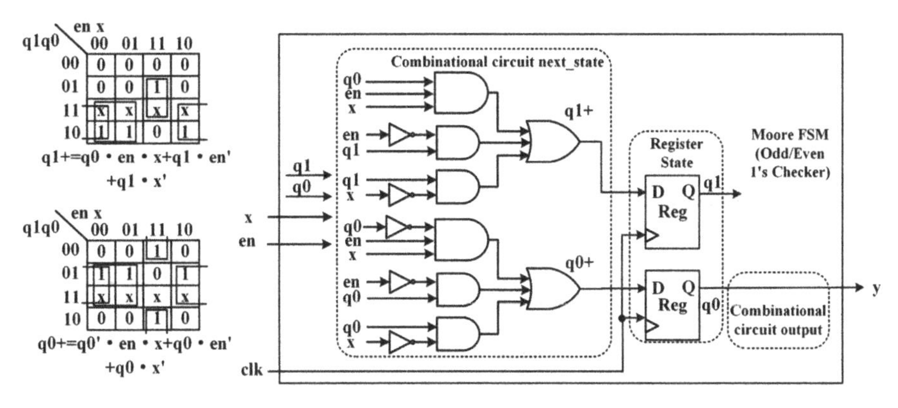

(a) K-Map of Odd/Even 1s Checker (b) Circuit of Odd/Even 1s Checker

FIGURE 7.24
K-Map optimization and circuit of odd/even 1s checker.

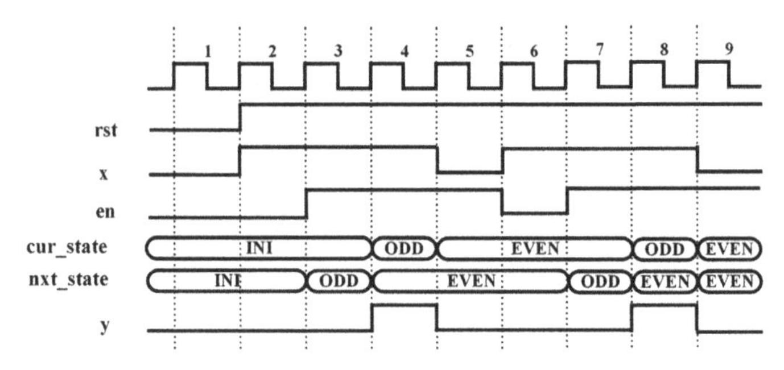

FIGURE 7.25
Timing diagram of odd/even 1s checker.

clock cycles. The entire circuit is reset to the initial states in the first clock cycle. In the second clock cycle, the circuit remains in the initial state "INI" because the enable signal "en" is deasserted, indicating that no input is being processed.

Starting from the third clock cycle, the circuit receives the first digit one as the enable signal is asserted. This triggers a next-state transition from the initial state to the "ODD" state, indicating that an odd number of 1s is being detected. Due to a Moore machine behavior, consequently, in the fourth clock cycle, the current state is updated into the "ODD" state and the output "y" is asserted. In clock cycle 4, a continuous digit one is received, leading to the next state transitioning to the "EVEN" state. Consequently, in clock cycle 5, the current state reflects the transition to the "EVEN" state, causing the output "y" to be deasserted.

The same pattern repeats in subsequent clock cycles, with the circuit transitioning between the "ODD" and "EVEN" states based on the input sequence of ones and zeros. In clock cycle 8, the output "y" is asserted again within the current state of "ODD".

7.5 Design Example: Data Package Receiver

7.5.1 Introduction to data package receiver

In the majority of serial bus communications, data is transmitted in the form of packets, which consist of multiple data segments. Consequently, a data packet receiver plays a pivotal role by handling the reception and extraction of data from incoming packets. As an example, Figure 7.26(a) presents a data package receiver, which enables data reception through the serial input signal "data_in". It indicates the successful receipt of the data package header, command, and stop tail through the output indicators "header", "cmd", and

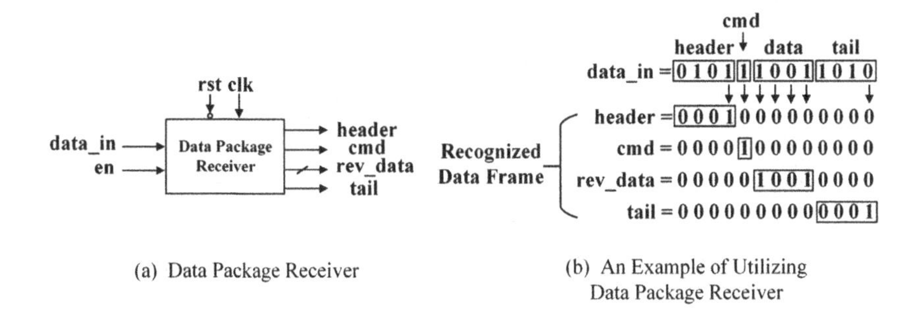

(a) Data Package Receiver

(b) An Example of Utilizing Data Package Receiver

FIGURE 7.26
Data package receiver.

"tail", respectively. Furthermore, the extracted data from the data package is shown through the output "rev_data".

Figure 7.26(b) illustrates the details of the data reception and extraction process. Upon the arrival of a data packet, the receiver initiates synchronization by first identifying the data frame header. In this example, the data header is represented as the binary string 0101. Once the frame header is detected, the receiver triggers the "header" output, signifying the successful reception of the data package header. This alignment ensures that the receiver is synchronized with the sender, enabling it to accurately interpret the subsequent commands and data.

Once the receiver is synchronized, it proceeds to extract the command and data components from the data package. The extracted command is available on the "cmd" output as a single-bit value (for instance, zero for a control data transaction and one for a status data transaction), while the extracted data is presented on the "rev_data" bus as a 4-bit value. These extracted components can be further processed or utilized by the receiving circuit. Moreover, the data package receiver actively identifies the tail within the data package and activates the "tail" indicator to signal the completion of data reception. In this particular instance, the data package tail is required to match the binary string 1010, confirming the successful reception of the entire data package.

7.5.2 State graph of data package receiver

As the design requirements, a Mealy FSM is presented in this section to receive and extract data packages illustrated in Figure 7.26. The FSM governs the state transitions and controls the reception and extraction of the data packages. Figure 7.27 depicts the state graph that defines ten states to receive and extract the data package. The states are as follows: the four states, "H0", "H1", "H2", and "H3", represent the reception of the four digits 0101 in the header; the "CM" state represents the reception of the command bit (either zero or one); the "RD" state represents the reception of the 4-bit data; and

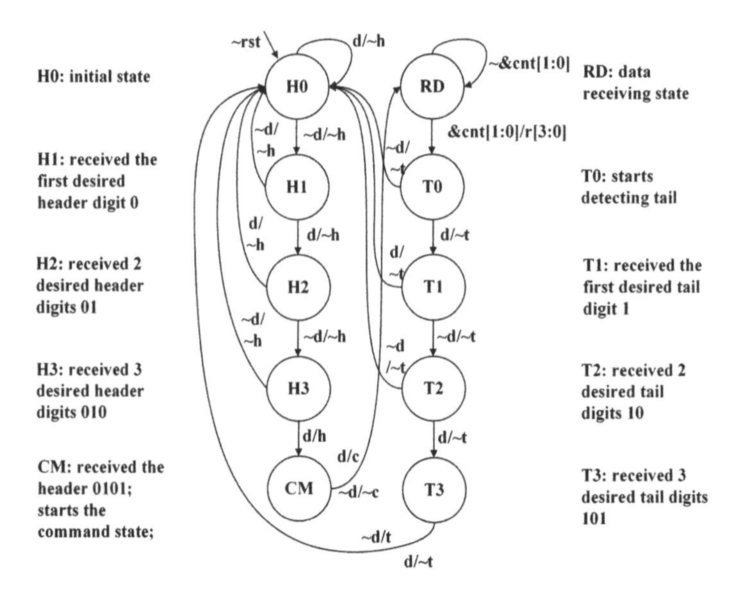

FIGURE 7.27
State graph of data package receiver.

the four states, “T0”, “T1”, “T2”, and “T3”, represent the reception of the last digits 1010 in the tail.

The Mealy machine outputs, represented as “h”, “c”, and “t”, indicate the successful reception of the header, command, and tail, respectively. To track the progress of receiving the 4-bit data, denoted as output “r[3:0]”, a 2-bit counter (“cnt[1:0]”) is utilized. The counter starts counting when the FSM is in the “RD” state and increments until it reaches the maximum value of 2’h3, indicating that all four data bits have been received. For simplicity, the input “data_in” corresponds to the “d” input in the state graph.

7.5.3 State and transition tables of data package receiver

The state table, representing the state graph, is summarized in Figure 7.28(a). Most state transitions are determined by the current state and the input “d”. However, in the data receiving state “RD”, the next state transition depends on the current state and the counter value (represented as $\&cnt$ or $\sim\&cnt$ in the second row). When the counter reaches its maximum value of 2’h3 (expressed as $\&cnt$), the next state transitions to the first tail detection state “T0”. Otherwise (expressed as $\sim\&cnt$), the next state remains in the current data receiving state “RD” to continue collecting the input data. In the output columns, the output data “r[3:0]” accumulates the received digits over four clock cycles, and the output indicators, “h”, “c”, and “t”, update with the current state and the input “d”.

cur_state	nxt_state				h		c		t		r[3:0]	
	~d	d	&cnt	~&cnt	~d	d	~d	d	~d	d	&cnt	~&cnt
H0	H1	H0	-	-	0	0	0	0	0	0	-	-
H1	H0	H2	-	-	0	0	0	0	0	0	-	-
H2	H3	H0	-	-	0	0	0	0	0	0	-	-
H3	H0	CM	-	-	0	1	0	0	0	0	-	-
CM	RD	RD	-	-	0	0	0	1	0	0	-	-
RD	-	-	T0	RD	0	0	0	0	0	0	r[3:0]	d
T0	H0	T1	-	-	0	0	0	0	0	0	-	-
T1	T2	H0	-	-	0	0	0	0	0	0	-	-
T2	H0	T3	-	-	0	0	0	0	0	0	-	-
T3	H0	H0	-	-	0	0	0	0	1	0	-	-

(a) State Table of Data Package Receiver

cur_state	nxt_state				h		c		t		r[3:0]	
	~d	d	&cnt	~&cnt	~d	d	~d	d	~d	d	&cnt	~&cnt
0	1	0	-	-	0	0	0	0	0	0	-	-
1	0	2	-	-	0	0	0	0	0	0	-	-
2	3	0	-	-	0	0	0	0	0	0	-	-
3	0	4	-	-	0	1	0	0	0	0	-	-
4	5	5	-	-	0	0	0	1	0	0	-	-
5	-	-	6	5	0	0	0	0	0	0	r[3:0]	d
6	0	7	-	-	0	0	0	0	0	0	-	-
7	8	0	-	-	0	0	0	0	0	0	-	-
8	0	9	-	-	0	0	0	0	0	0	-	-
9	0	0	-	-	0	0	0	0	1	0	-	-

H0=4'd0, H1=4'd1, H2=4'd2, H3=4'd3; CM=4'd4, RD=4'd5;
T0=4'd6, T1=4'd7, T2=4'd8, T3=4'd9;

(b) Transition Table of Data Package Receiver

FIGURE 7.28
State and transition tables of data package receiver.

Figure 7.28(b) displays the transition table, which assigns encodings to the states as follows: 4'd0-4'd3 for "H0-H3", 4'd4 for "CM", 4'd5 for "RD", and 4'd6-4'd9 for "T0-T3".

7.5.4 Verilog HDL design of data package receiver

The Verilog description for the receiver design is provided below. All states are defined as *parameters* in lines 10–19. The state register is described in lines 24–31, and the transitions to the next state are designed in lines 33–68. The counter, used in the next state description, is defined in lines 34–42. In the "RD" state, the counter increments by one until it reaches the maximum value of 2'h3. This counter can trigger the next state transition within the current state "RD" as shown in line 57.

```
1  module data_pack_rec(input                rst      ,
2                       input                clk      ,
```

```
3                        input              en        ,
4                        input              data_in   ,
5                        output             header    ,
6                        output             cmd       ,
7                        output reg [3:0]   rev_data  ,
8                        output             tail      );
9   parameter SIZE = 4;
10  parameter H0 = 4'b0000;
11  parameter H1 = 4'b0001;
12  parameter H2 = 4'b0010;
13  parameter H3 = 4'b0011;
14  parameter CM = 4'b0100;
15  parameter RD = 4'b0101;
16  parameter T0 = 4'b0110;
17  parameter T1 = 4'b0111;
18  parameter T2 = 4'b1000;
19  parameter T3 = 4'b1001;
20
21  reg [SIZE-1:0] cur_state; // Sequential part of the FSM
22  reg [SIZE-1:0] nxt_state; // Combinational part of the FSM
23
24  //------------------State Register -----------------
25  always @ (posedge clk, negedge rst) begin
26    if(~rst) begin
27      cur_state <= H0       ;
28    end else begin
29      cur_state <= nxt_state;
30    end
31  end
32
33  //-------Next state combinational circuit----------
34  reg  [1:0] cnt;
35  wire [1:0] nxt_cnt = (cur_state==RD) ? cnt+2'h1 : 2'h0;
36  always @ (posedge clk, negedge rst) begin
37    if(~rst) begin
38      cnt <= 2'h0    ;
39    end else begin
40      cnt <= nxt_cnt;
41    end
42  end
43
44  always @ (cur_state, en, data_in, rst, cnt) begin
45    if(~rst) begin
46      nxt_state <= H0;
47    end else begin
48      case(cur_state)
49        H0    : if(~data_in & en) nxt_state <= H1;
50        H1    : if(data_in )      nxt_state <= H2;
51                else              nxt_state <= H0;
```

```
52        H2     : if(~data_in)      nxt_state <= H3;
53                 else              nxt_state <= H0;
54        H3     : if(data_in )      nxt_state <= CM;
55                 else              nxt_state <= H0;
56        CM     :                   nxt_state <= RD;
57        RD     : if(&cnt)          nxt_state <= T0;
58        T0     : if(data_in)       nxt_state <= T1;
59                 else              nxt_state <= H0;
60        T1     : if(~data_in)      nxt_state <= T2;
61                 else              nxt_state <= H0;
62        T2     : if(data_in)       nxt_state <= T3;
63                 else              nxt_state <= H0;
64        T3     :                   nxt_state <= H0;
65        default : nxt_state <= H0;
66       endcase
67    end
68  end
69
70  //---------Output combinational circuit-------------
71  assign header = (cur_state==H3) & data_in ;
72  assign cmd    = (cur_state==CM) & data_in ;
73  assign tail   = (cur_state==T3) & ~data_in;
74
75  wire [3:0] nxt_rev_data;
76  assign nxt_rev_data=(cur_state==RD)?{rev_data[2:0],data_in}
77                                         : rev_data;
78  always @ (posedge clk, negedge rst) begin
79    if(~rst) begin
80      rev_data <= 4'h0         ;
81    end else begin
82      rev_data <= nxt_rev_data;
83    end
84  end
85  endmodule
```

The output computations are described in lines 70–84. Lines 71–73 define the indicators "header", "cmd", and "tail" when the current states correspond to their respective states and the expected inputs are received. Lines 75–84 demonstrate the data collection over four clock cycles within the "RD" state. The conditional *assign* block captures the single-bit "data_in" into the 4-bit signal "nxt_rev_data". In what follows, the subsequent *always* block updates the received data in the "rev_data" register over clock cycles.

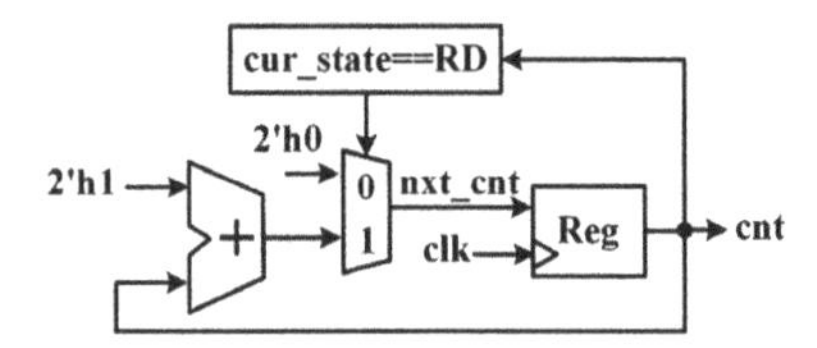

FIGURE 7.29
Data receiver counter.

7.5.5 Simulation and synthesis analysis of data package receiver

A. Synthesized Circuit

Similar to other FSMs, the design circuit primarily consists of a sequential state register and a combinational circuit responsible for computing the next state and outputs. Additionally, the counter design, depicted in Figure 7.29, includes a trigger signal generated by a comparator. When the current state transitions into the "RD" state, the counter is activated to count the clock cycles, beginning from hexadecimal 4'h0 and incrementing up to 4'h3.

While in the "RD" state, the serial data on the "data_in" input can be directed into the shift registers for gathering the 4-bit data package, as illustrated in Figure 7.30. Each clock cycle results in a one-bit rightward shift in the input, from the MSB to the LSB of the 4-bit data package, following a big-endian data collection approach. After four clock cycles, the accumulated data is stored in the "rec_data[3:0]" register, effectively representing the final received data.

B. Timing Diagram

Figure 7.31 presents a timing diagram illustrating the behavior of receiving and extracting data packages using the FSM. In the first clock cycle, the entire state machine is reset. From clock cycles 2 to 5, the current state transitions from "H0" to "H3". Within the clock cycle 5, the data input becomes logic one,

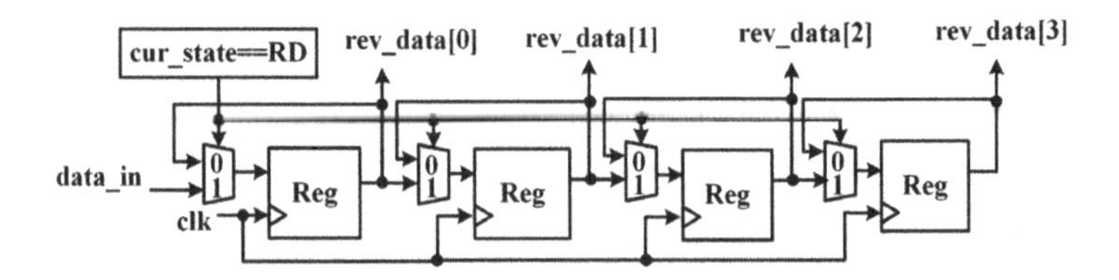

FIGURE 7.30
Data receiver shifter (big-endian input).

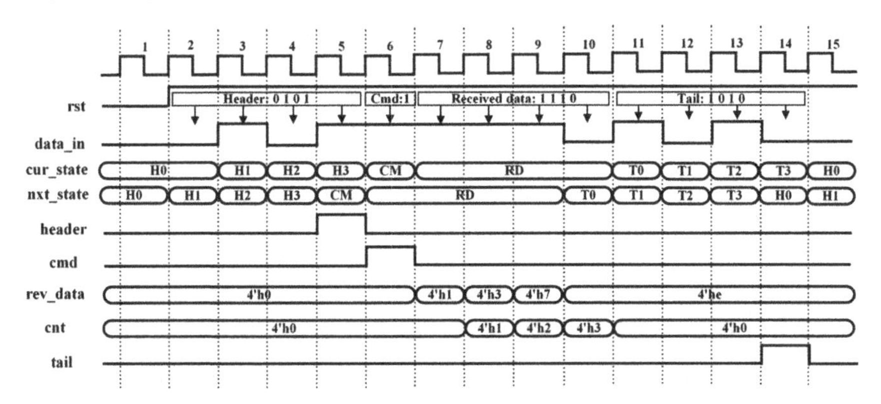

FIGURE 7.31
Timing diagram of data package receiver.

indicating the successful reception of the header string 0101. Consequently, the "header" indicator activates in the same clock cycle.

Similarly, the current state transitions to "CM" in clock cycle 6. Since the current input is logic one, the "cmd" indicator reflects the command as a status data transaction. In the subsequent four clock cycles, 7 to 10, the current state transitions into the "RD" state, and the digit string 1110 is received and shown on the "rev_data[3:0]" output.

In the final four clock cycles, 11 to 14, the current state transitions through the "T0-T3" states to detect the tail in the data package. Once the tail digit string 1010 is successfully extracted, the final "tail" indicator is asserted, indicating the completion of the data package reception.

Exercises

Problem 7.1. Design a 101 sequence detector utilizing both Mealy and Moore FSMs. As depicted in Figure 7.11, the sequence detector will examine a string of 0's and 1's provided to the input "x" and produce an output "y=1" only when the input sequence concludes with the pattern 101.

For instance, Figure 7.32 visually demonstrates the operation of the sequence detector. It processes serial data on the "x" input in a left-to-right fashion, handling each digit per clock cycle. Notably, there can be an overlap between digits in two consecutive instances of the desired 101 string. For example, in the second and third occurrences of the 101 string, the last digit "1" from the second occurrence serves as the first digit "1" of the third instance.

Perform the following four steps for both Mealy FSM and Moore FSM designs:

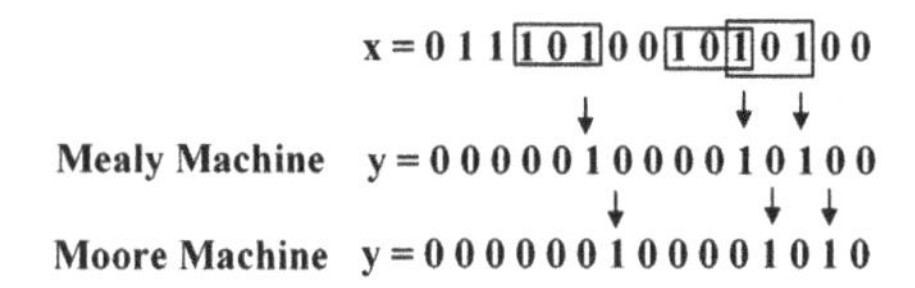

FIGURE 7.32
101 sequence detection utilizing Mealy and Moore FSM.

1) Build a state graph.

2) Generate the state table and encode it into a transition table.

3) Create block diagrams illustrating the expected hardware results, depicting the sequential registers and combinational circuits for both output and next state. Evaluate the combinational circuits using K-maps to optimize the logic.

4) Analyze the simulation results using the input stimulus shown in Figure 7.33. The input signals "x" and the asynchronous reset "rst" are provided. Draw the current state "cur_state" and the next state "nxt_state", as well as the output "y".

PBL 17: Sequence Detector

1) Design a 101 sequence detector with Verilog HDL, utilizing both Mealy and Moore FSMs. Table 7.2 provides an overview of the design's IOs. In addition to the input "x" and output "y", the state machine necessitates an asynchronous reset for initializing the state register and a positive edge-triggered clock for sampling each state.

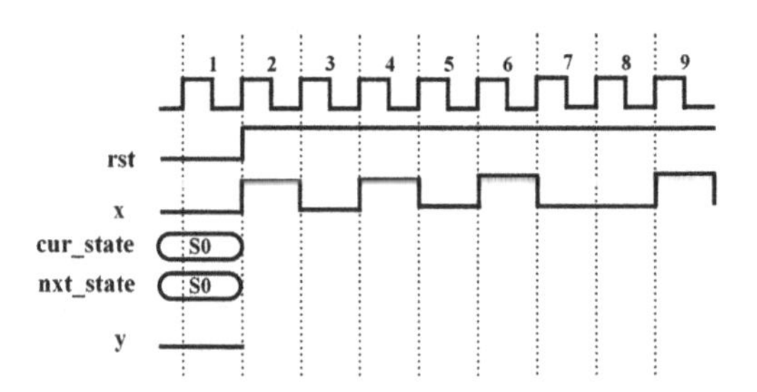

FIGURE 7.33
Simulation results analysis.

TABLE 7.2
Sequence detector IOs description.

Name	Direction	Bit Width	Description
clk	Input	1	Clock, rising edge trigger
rst	Input	1	Asynchronous reset, 0 valid
x	Input	1	Digit string input
y	Output	1	Output 1 when 101 string detected

2) Create a testbench to simulate the design, following the input stimulus depicted in Figure 7.33. Compare the simulation outcomes from ModelSim with the analysis results from Problem 7.1.
3) Execute synthesis and implementation in Vivado. Compare the RTL analysis results obtained in Vivado with the hardware diagrams generated in Problem 7.1.

8

FSM-Datapath Design and Bus Communication

The integration of timing controllers and datapaths constitutes a prevalent design paradigm in digital circuitry. The datapath primarily involves various data operations and processing, while the timing controller mainly coordinates the execution of operations over clock cycles or transition states. It typically employs FSMs for timing control, resulting in the controller-datapath design being named FSM-Datapath, or FSMD. Nevertheless, other design components containing information on execution latency/states can be utilized to form the timing controller, including counters and timers.

Chapter 8 provides insights into the typical constructions of FSMD for bus communication applications. It introduces an industry-standard protocol – I2C (inter-integrated circuit), and a self created master-slave bus. Additionally, the chapter explores widely-used interconnection bus mechanisms, including the ready-valid protocol, enable-finish handshaking, and request-grant arbitration. These bus protocols and standards play a pivotal role in IC design, facilitating command and data movement among diverse design components and devices.

PBL 18–21 showcase the implementation of FSMD designs tailored specifically for floating-point (FP) data operations. These advanced designs involve utilizing high-speed FP operators, creating pipeline design structures, and implementing multi-core processing with parallel computing. Through these advanced techniques, the projects aim to enhance computational efficiency and optimize performance in handling FP data-intensive tasks.

8.1 FSM-Datapath and Bus Communication

8.1.1 FSMD construction

Figure 8.1 presents a block diagram illustrating the interaction between the timing controller (represented as the FSM) and the datapath. The state controller is tasked with managing state transitions and coordinating operations within the circuit, while the datapath consists of functional components responsible for executing computations and data processing tasks. This includes

DOI: 10.1201/9781003187080-8

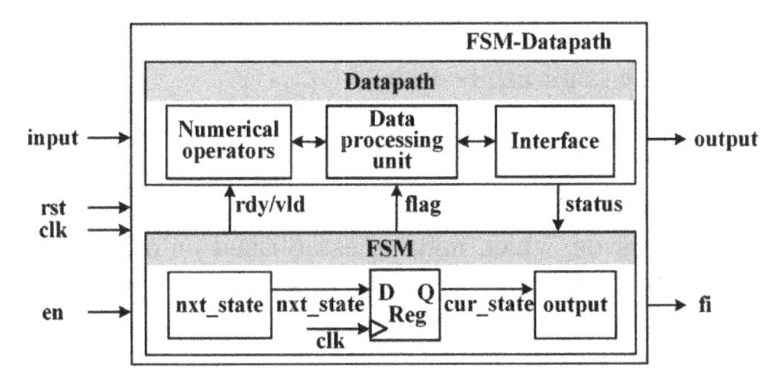

FIGURE 8.1
FSMD design structure.

numerical operators, memory blocks, and custom-designed elements tailored to the circuit's specific functional requirements. Together, the state controller and datapath form the foundation of many digital circuit designs.

Control signals and status signals are essential for facilitating communication between the controller and the datapath. Control signals, which originate from the controller, are responsible for initiating or activating specific components within the datapath. On the other hand, status signals serve as a means of providing feedback from the datapath to the controller. This figure provides examples, such as the ready-valid mechanism (referred to as "rdy-vld"), various flag signals, and a range of status indicators. Furthermore, it showcases the enable-finish handshaking (referred to as "en-fi") mechanism, which serves as the external bus communication to enable the FSMD design and signal the completion of data processing.

8.1.2 Bus communications

Before we explore bus communication mechanisms, it's crucial to define our terminology. The term "design components" encompasses smaller-scale elements and interconnected modules, such as the state controller and the datapath. In contrast, "hardware devices" refer to hardware designs operating at the system level. These devices can include master or slave devices engaged in communication through industry-standard bus protocols such as I2C and UART.

A. Bus Protocols for Design Components

In RTL design, bus protocols play a vital role in enabling control and data communication among design components. These protocols establish standardized procedures for exchanging commands and data between design blocks within a digital circuit. In this chapter, we will introduce three fundamental bus mechanisms commonly employed for connecting and facilitating communication between design components: the ready-valid mechanism, enable-finish handshaking, and request-grant arbitration.

B. Bus Protocols for Hardware Devices

Bus communications for hardware devices are typically defined in accordance with industry standards, which may be established by organizations such as IEEE, as well as by self-design teams. In this chapter, we will introduce two design examples that utilize the FSMD structure for facilitating bus communication among hardware devices.

Firstly, we will illustrate the design of an I2C bus, an industry-standard serial bus protocol commonly chosen for low-speed device communication due to its simplicity and adaptability. It proves particularly suitable when a lower data transfer rate suffices. Conversely, applications necessitating high-speed data transfer typically require a system-on-chip (SoC) bus protocol with greater bandwidth to accommodate increased data throughput. Therefore, our second example will feature a high-speed bus protocol, namely master-slave bus (MSBUS), to demonstrate data communication between typical master and slave components. This illustration offers valuable insights into how a master device communicates with a slave device using bus transactions.

8.2 Bus Communication Mechanisms

8.2.1 Ready-valid protocol

A. Bus Protocol

The ready-valid mechanism is widely adopted in design scenarios where the data processing unit receives ready-to-use data input and specifies valid data output in specific clock cycles. As depicted in Figure 8.2(a), the input data can be fed into the data processing engine through the "a" input, and the output data can be pushed out via the output "b". The ready-valid mechanism is employed in this case to specify the time slots for valid data input and output. Specifically, an asserted ready signal indicates that the data on the "a" input are ready in the current clock cycles; while an asserted valid signal indicates that the data on the "b" output are valid in the current clock cycles.

Figure 8.2(b) presents a timing diagram to illustrate the mechanism. In the first clock cycle, the ready signal is asserted, indicating that the input data on the "a" bus is ready to be processed. Assume that the data processing unit takes three clock cycles. As a result, the valid output data on the "b" bus can be signaled by the valid indicator "vld" in clock cycle 4, with a three-clock-cycle delay. This timing diagram visually demonstrates the synchronization

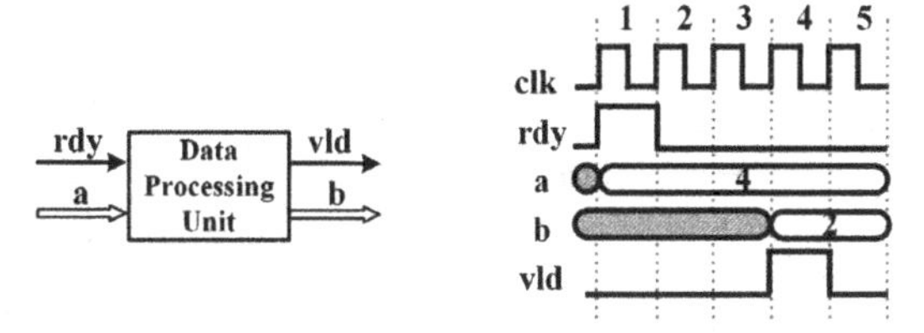

(a) Block Diagram and Connection (b) Timing Diagram and Bus Protocol

FIGURE 8.2
Ready-valid protocol.

and coordination achieved through the ready-valid mechanism, ensuring that data is only processed when it is ready or valid.

B. Design Examples

Signaling ready/valid data on the bus is a crucial aspect of RTL design, as data processing and movement are dependent on the availability of data within specific time slots. As an example depicted in Figure 8.3, consider the scenario where input data is read from a memory block, and the output data must be written back into another memory block. In this case, let's assume that the memory read/write takes one clock cycle, and memory access is always allowed.

To maintain uninterrupted data flow, it's essential to issue the memory read command in advance, guaranteeing the presence of input data on the "a" bus during the data processing. This timing optimization ensures that the data processing unit receives input data ready for immediate processing as soon as it becomes available. Likewise, for write operations, the command should be synchronized with the "vld" signal, ensuring that valid data on the "b" bus is used for accurate write-back into the memory block.

Another design example will be presented in Section 12.1, demonstrating how AMBA AXI bus protocol is carried out using the valid-ready mechanism. While the ready-valid processing in the AXI bus protocol may have slightly different interpretations compared to those in this section, the fundamental

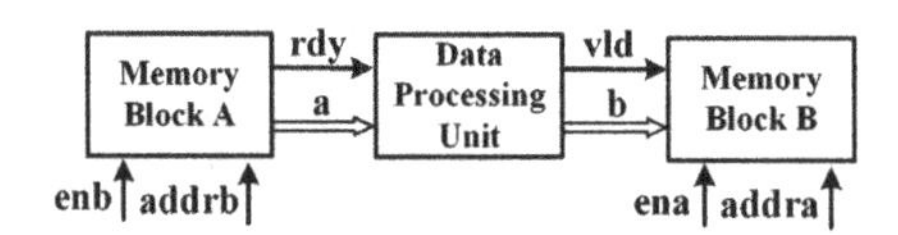

FIGURE 8.3
An example of ready-valid protocol (data movement).

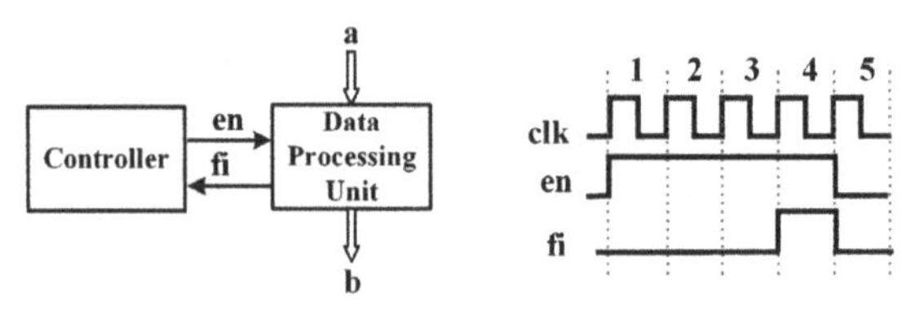

(a) Block Diagram and Connection (b) Timing Diagram and Bus Protocol

FIGURE 8.4
Enable-finish handshaking.

concept remain consistent, ensuring efficient and coordinated data transfer between the source and destination design components.

8.2.2 Enable-finish handshaking

A. Bus Protocol

Enable-finish handshaking involves the use of an enable signal to initiate a data processing operation and a finish signal to indicate the completion of the process. As shown in Figure 8.4(a), the enable signal "en" triggers the start of the data processing, and the finish signal "fi" asserts once the operation is finished. This handshaking helps managing the timing of data processing operations and ensures proper synchronization between different components.

Figure 8.4(b) presents a timing diagram illustrating the enable-finish handshaking mechanism. The enable signal "en" activates in the first clock cycle to initiate the data processing unit and remains asserted during subsequent clock cycles, waiting for the finish signal "fi" to be asserted to complete the handshaking process. Assuming that the data processing takes three clock cycles, the finish signal "fi" is asserted in the fourth clock cycle, deasserting the enable signal and finalizing the handshaking process. This coordinated enable-finish handshaking ensures that the data processing unit operates only when explicitly enabled and signals its completion once it finishes processing the data.

B. Design Examples

Bus handshaking is a fundamental idea of hardware design, serving to streamline the coordination, control, and status monitoring among interconnected design components. As an example depicted in Figure 8.5, a multi-core scheduler is responsible for enabling different cores by asserting their enable signals ("en1" and "en2" in this figure). It then monitors the status of each core by observing the completion status of the final handshaking ("en1-fi1" and "en2-fi2" in this figure). By using the handshaking protocol, the scheduler can

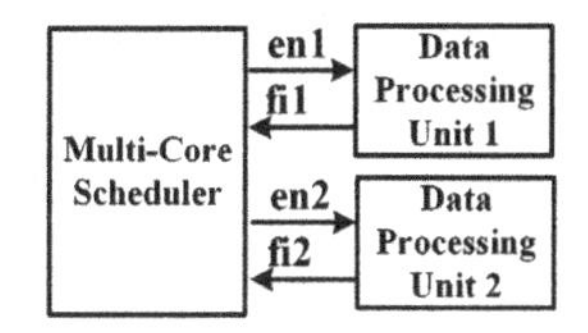

FIGURE 8.5
An example of enable-finish handshaking (multi-core scheduling).

effectively manage and control the execution between interfaced cores, ensuring that they operate in a synchronized and orderly manner.

8.2.3 Request-grant arbitration.

Request-grant arbitration is a commonly used method for bus access control in multi-master systems, as depicted in Figure 8.6(a). When a master needs to access a shared bus, it sends a request signal labeled as "req". In situations where multiple masters request the bus simultaneously (as shown with "req1" and "req2" in the figure), a bus arbiter makes the decision regarding which request to grant based on their priorities and the arbitration design in place. Once a master is granted access to the bus, the slave sends a grant signal labeled as "gnt" to acknowledge the request. This indicates that the corresponding master has gained control of the bus, and its commands and data can now be processed by the slave.

In addition to the arbitration process, the timing of the request-grant communication follows a similar pattern to the enable-finish handshaking. As depicted in Figure 8.6(b), the request is chosen by the arbiter in the first clock cycle. It takes until the fourth clock cycle for the slave to respond with a grant signal, pulling down the request and thereby completing the request-grant handshaking process.

Request-grant arbitration plays a crucial role in ensuring fair and efficient bus access in multi-master systems. It helps prevent conflicts and maintains

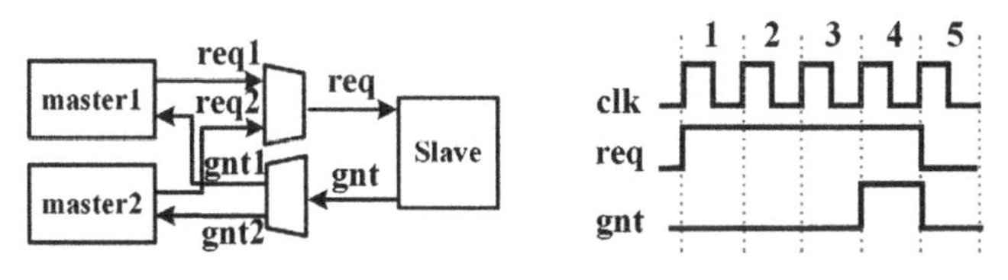

(a) Block Diagram and Topology (b) Timing Diagram and Bus Protocol

FIGURE 8.6
Request-grant arbitration.

a well-coordinated operation of the design system. A practical application of arbitration is demonstrated in Section 12.2, where DMA (Direct Memory Access) arbitration is showcased.

8.3 Design Example: I2C Write

Serial buses have become ubiquitous in modern applications, ranging from general-purpose micro-controllers and processors to various devices like EEPROMs and Flash Controllers. There are several serial bus families available, including I2C, SPI, SDIO, GPIO, UART, and many more. From an IC design perspective, all of these serial buses can be implemented using FSMD designs. To illustrate this, this section demonstrates the design of an I2C master using the FSMD approach.

8.3.1 I2C bus protocol

The I2C bus, invented by Philips Semiconductor in 1982, has become a widely adopted de facto serial bus standard for connecting low-speed peripheral ICs to processors. Its popularity stems from its hardware efficiency and circuit simplicity. The I2C bus possesses several key features:

- Only two single-bit lines are required: a serial data line (SDA) and a serial clock line (SCL).
- There is no strict baud rate requirement as the I2C master provides the bus clock.
- Serial, 8-bit oriented, bidirectional data transfers can be performed at different speeds including standard-mode (up to 100 kbit/s), fast-mode (up to 400 kbit/s), fast-mode plus (up to 1 Mbit/s), and high-speed mode (up to 3.4 Mbit/s).
- Simple master-slave relationships exist among all I2C devices. Each device connected to the bus can be addressed using a unique device address. The controller-target relationships are straightforward, allowing controllers to operate as either controller-transmitters or controller-receivers.
- The I2C bus supports a simple multi-master configuration with collision detection and arbitration mechanisms to prevent data corruption when multiple controllers initiate data transfers simultaneously.

For the sake of simplicity, we will focus on the design of I2C write operations with a single-master and single-slave bus structure.

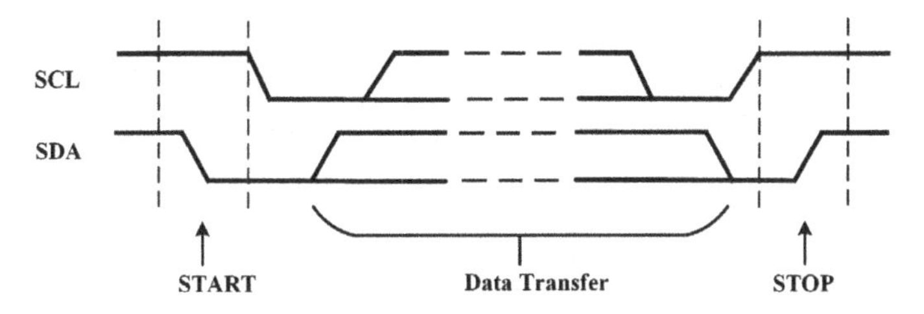

FIGURE 8.7
START and STOP conditions.

8.3.1.1 START and STOP Conditions

Figure 8.7 illustrates the I2C bus protocol, where all transactions commence with a START (S) signal and conclude with a STOP (P) signal. A transition from HIGH to LOW on the SDA line while SCL is HIGH indicates the occurrence of a START condition. Conversely, a transition from LOW to HIGH on the SDA line while SCL is HIGH signifies a STOP condition. It is important to note that the generation of START and STOP conditions is exclusively the responsibility of the master controller. Once the START condition is established, the bus is considered busy. After the STOP condition, the bus is considered free again after a certain period of time.

A repeated START condition (Sr) is similar to a regular START condition, but it occurs before a STOP condition, even when the bus is not idle. While a regular START condition signifies the beginning of a new communication and is followed by a STOP condition, a repeated START condition allows the master to initiate a new communication without letting the bus go idle. In terms of functionality, both the START and repeated START conditions keep the bus busy. They are essentially equivalent in this regard.

8.3.1.2 Data validity and byte format

Figure 8.8 demonstrates the data transfer over SCL cycles. Each clock pulse of the SCL bus corresponds to the transfer of one data bit on the SDA line. A byte consists of eight bits on the SDA line and can represent various information such as a device address, register address, or data to be written into or read from a slave device. The data is transmitted most significant bit (MSB) first and least significant bit (LSB) last. In the example provided, the data byte being transferred is binary 8'b10101010 or hexadecimal 8'haa.

It is important to note that any number of data bytes can be transferred from the master to slave between the START and STOP conditions. During the high phase of the clock period, the data on the SDA line must remain stable, as any changes in the data line when the SCL bus is high will be interpreted as control commands such as START or STOP.

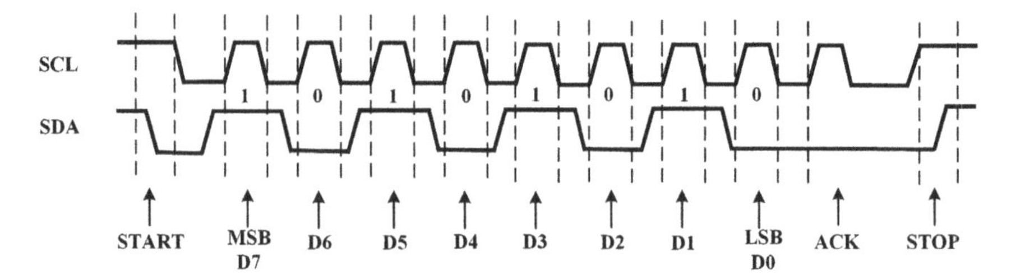

FIGURE 8.8
Single-byte data transfer with ACK.

8.3.1.3 Acknowledge (ACK) and Not Acknowledge (NACK)

After each data byte is transferred, the slave must send an ACK (Acknowledge) bit to the master. This ACK bit serves as confirmation to the master that the data byte was successfully received, and it signals that another data byte can be sent. To respond with an ACK bit, as depicted in Figure 8.8, the master must release the SDA line, allowing the slave to control the line. The slave will pull down the SDA line during the low phase of the ACK/NACK-related clock period (after the last cycle of the data bit 0), ensuring that the SDA line remains stably low during the high phase of the ACK/NACK-related clock period.

If the SDA line remains high during the ACK/NACK-related clock period, as shown in Figure 8.9, this is interpreted as a NACK (Not Acknowledge). There are several conditions that can lead to the generation of a NACK:

- The slave device is not ready to communicate with the master and is unable to receive or transmit data.
- The slave device receives data or commands that it does not understand during the transfer.

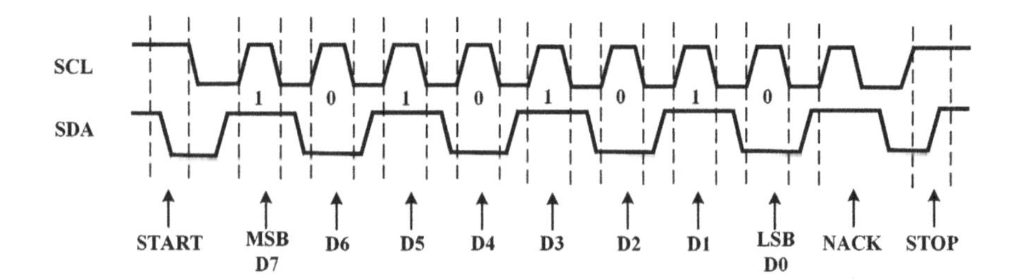

FIGURE 8.9
Single-byte data transfer with NACK.

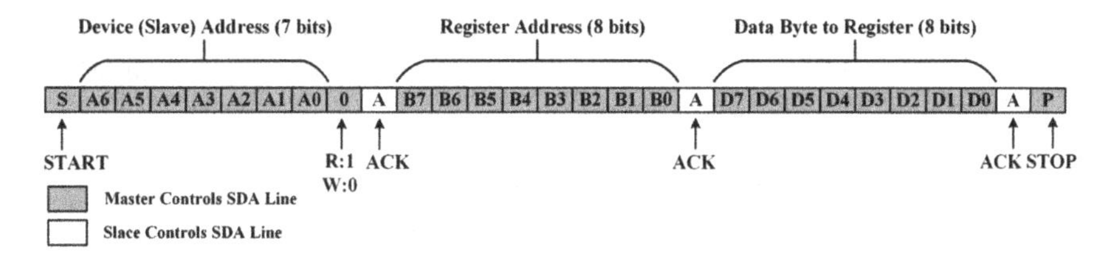

FIGURE 8.10
I2C Write to device's register.

- The slave device reaches a point where it cannot receive any more data bytes.
- The master device indicates that it has finished reading data and communicates this to the slave through a NACK signal.

The first three conditions are considered abnormal cases, while the last condition is the normal case to conclude the data reading transactions on the I2C bus.

8.3.1.4 Write operations on I2C bus

Building upon the introduction of I2C timing for START/STOP, DATA Transmission, and ACK/NACK processing, this section delves into write and read operations on the I2C bus. The process of writing on the I2C bus is demonstrated in the example shown in Figure 8.10. In the diagram, the gray blocks represent the commands, addresses, and data sent by the master, while the white blocks represent the acknowledgments sent by the slave. Here's a step-by-step explanation of the protocol:

- The master initiates communication by generating a START condition on the bus.
- Subsequently, the master transmits a 7-bit device address (A6-A0) with the final bit (R/W bit) set to zero, indicating a write operation. Upon receiving the device address, the slave confirms the transmission by sending an ACK bit.
- After the acknowledgment, the master proceeds to send the 8-bit register address (B7-B0) of the specific register it intends to write to. The slave acknowledges the register address by transmitting an ACK bit, signifying its readiness to receive the data.
- The master then begins transmitting the actual data (D7-D0) to be written into the register. This data can consist of one or more bytes, depending on the requirements. The slave acknowledges the data transmission, affirming successful receipt.

- Upon transmitting all the necessary data, the master concludes the operation by issuing a STOP condition on the bus. This STOP condition signifies the completion of the I2C write operation.

8.3.1.5 Read operations on I2C bus.

Reading from a slave in the I2C protocol follows a similar process to writing, as shown in Figure 8.11. Here is a step-by-step explanation of the read operation:

- The master initiates communication by generating a START condition on the bus.
- It is important to note that the master should initiate the transmission by sending the device address with the R/W bit set to zero, indicating a write operation firstly. It also includes the register address it wishes to read from. The slave acknowledges both the device address and the register address.
- The master sends a repeated START condition on the bus, followed by the device address with the R/W bit set to one, indicating a read operation. Here, it starts the I2C read operation. The slave acknowledges the read request.
- The master releases the SDA line, allowing the slave to transmit data. The master continues to provide clock pulses to synchronize the transmission. The slave, acting as the slave-transmitter, sends data on the SDA line during each SCL pulse. After receiving each byte of data, the master sends an ACK signal to the slave, indicating that it is ready for more data. Once the master has received the desired number of bytes, it sends a NACK signal to the slave, signaling the end of the communication and instructing the slave to release the bus.
- The master concludes the transaction with a STOP condition.

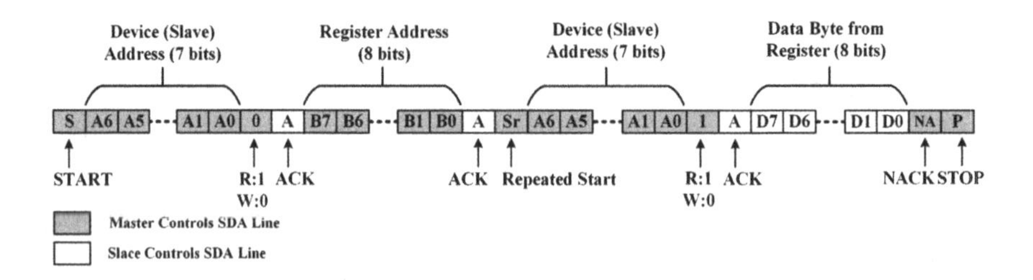

FIGURE 8.11
I2C read from device's register.

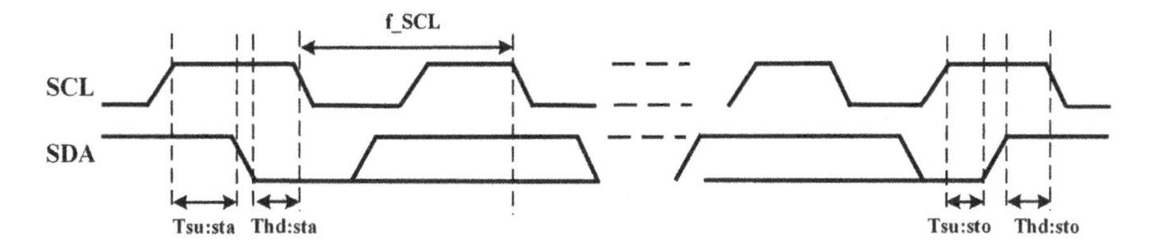

FIGURE 8.12
I2C write and read timing constraints.

8.3.2 An FSMD design example: I2C write operations

A. Design Requirements

To demonstrate the practical application using the I2C bus, this section introduces an I2C master design specifically developed for writing the functional registers of an I2C-enabled device. The timing requirements for the I2C write transactions are outlined in Figure 8.12, which provides details on the various timing parameters.

The detailed timing constraints are summarized in Table 8.1. Specifically, the SCL frequency should not exceed 400 KHz as a fast-mode I2C-enabled device. The setup time (rising edge of SCL to falling edge of SDA) and hold time (falling edge of SDA to falling edge of SCL) of the START condition are specified as a minimum of 5 μs. Similarly, the setup time and hold time of the STOP condition are specified as a minimum of 5 μs.

B. Design Specifications

To design the I2C master, two main aspects need to be considered. Firstly, the design should adhere to the I2C protocol, as depicted in Figure 8.10, to initiate the I2C START and STOP conditions and transmit the required data including the device address, register address, and register data. As a result, Figure 8.13 depicts an FSM controller to manage the state transitions including seven states: the initial state (INI), I2C START state (START),

TABLE 8.1
Timing requirements of I2C device.

Symbols	**Timing Constraints**	**Min**	**Max**	**Unit**
f_SCL	Clock Frequency	–	400	KHz
Tsu:sta	START Condition Setup Time	5	–	μs
Thd:sta	START Condition Hold Time	5	–	μs
Tsu:sto	STOP Condition Setup Time	5	–	μs
Thd:sto	STOP Condition Hold Time	5	–	μs

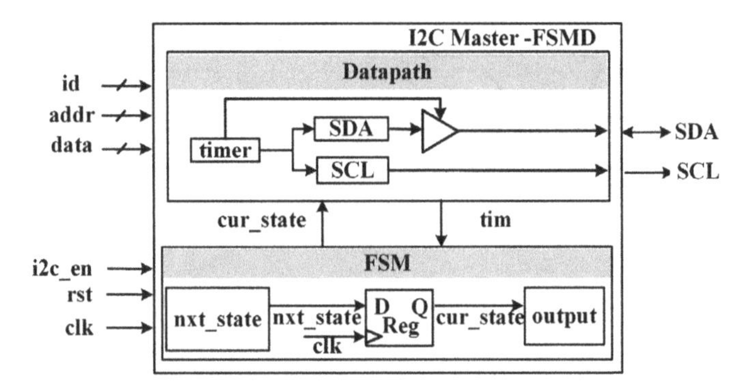

FIGURE 8.13
I2C Master design with FSMD.

device address sending (ID), register address sending (ADDR), data sending (DATA), and the final STOP state (STOP).

Secondly, it is crucial to adhere to the timing requirements specified in Table 8.1 to ensure proper communication with the I2C device. For instance, the serial clock (SCL) generated by the I2C master must not exceed the specified limit of 400 KHz. Assuming that a 50 MHz clock is available, each clock period is 20 ns. To maximize an I2C SCL frequency close to 400 KHz, the SCL bus should alternate between zero and one every 64 clock cycles, as a timer unit of 64×20 ns $= 1.28$ μs, which produces an SCL period of 1.28 μs $\times 2 = 2.56$ μs. This way, the SCL frequency will be around $1/2.56$ μs ≈ 400 KHz.

As per the design specification, a timer is essential for counting half of SCL cycles, which equates to $64 \times$ 50-MHz cycles, or 1.28 μs. This timer serves the purpose of generating START, STOP, ACK/NACK signals, and facilitating data transmission on the I2C bus. For instance, each set of four timer units takes approximately 5 μs (4×1.28 μs ≈ 5 μs), meeting the setup time and hold time requirements for the START and STOP conditions, as outlined in Table 8.1.

8.3.3 Datapath design

As mentioned earlier, Figure 8.14 illustrates the standard procedure for writing to a register through the I2C bus. To maintain precise timing during data transactions on the I2C bus and adhere to the specified timing requirements for communication with the I2C device, a timer with a unit of 1.28 μs is employed. This timer manages the timing of the SCL and SDA buses, guaranteeing their compliance with the timing constraints detailed in Table 8.1.

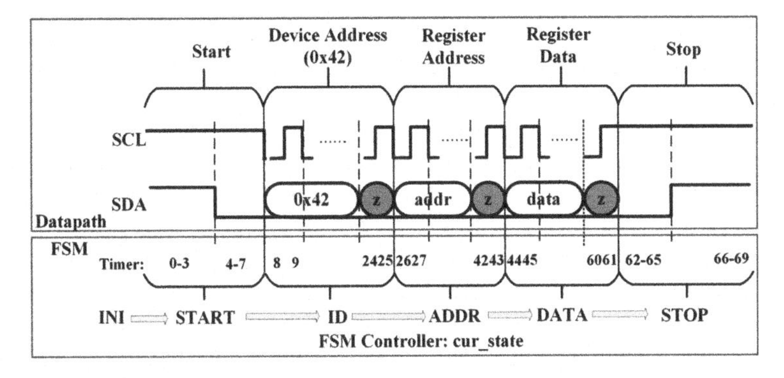

FIGURE 8.14
I2C Write timing diagram with FSMD.

A. START and STOP Stages

To meet the timing requirements for the setup and hold phases (each requiring at least 5 μs), the setup stage employs a duration equivalent to 4× the timer units, resulting in 5.12 μs, while the hold stage also necessitates 4× the timer units, totaling 5.12 μs. Specifically, during the START state, the SCL bus can transition to a high state when the timer's decimal values range from 0 to 7. Simultaneously, the SDA bus can transition to a high state during timer values 0–3 and to a low state during timer values 4–7, thus establishing the I2C START condition.

Conversely, during the STOP state, the SCL bus can transition to a high state when the timer's decimal values are within the range of 62 to 69. Simultaneously, the SDA bus can shift from a low state during timer values 62–65 to a high state during timer values 66–69, thereby forming the I2C STOP condition.

B. ID, ADDR, and DATA Stages

Assuming that the I2C device features a 7-bit unique device ID, represented as 7'b0010001 in binary, the device address can be expressed as 8'b00100010 by appending the last digit zero, which represents write operations. Alternatively, in hexadecimal notation, this address is represented as 8'h42. As illustrated in the figure, between the START and STOP conditions, the I2C master transmits the unique device address (8'h42), followed by the register address, and the corresponding data in a sequential manner.

In each of the three FSM states, ID, ADDR, and DATA, the SCL bus transitions low when the timer reaches decimal values 8, 10, 12, and so forth, up to 60. Conversely, the SCL bus transitions high when the timer reaches decimal values 9, 11, 13, and so on, up to 61, effectively forming a 400 KHz clock signal. The SDA bus is synchronized with this clock and transmits specific bits during each SCL cycle. In the ID state, the 8-bit data on the SDA bus should represent the device address (8'h42). In the ADDR state, the register address is transmitted bit by bit on the SDA bus, and in the DATA state, the register data is sent via the SDA bus.

Importantly, after the transmission of the 8-bit device address, register address, and register data, a high-impedance (high-Z) status is employed to release the bus from the I2C master's control. Subsequently, the responsibility for driving the bus with either an ACK or NACK status, indicating the slave's readiness to receive the byte or not, falls upon the slave device. To facilitate the transition between actively driving the bus and releasing it, the RTL design requires a tri-state buffer in the datapath. When the timer reaches decimal values 24–25 within the ID state, 42–43 within the ADDR state, and 60–61 within the DATA state, the I2C master releases the bus by deasserting the tri-state buffer, allowing it to receive the ACK/NACK responses from the slave device.

8.3.4 FSM design

Figure 8.15 illustrates the state graph for the FSM design. The input signals are as follows: the hardware reset (denoted as "rst"), enable ("i2c_en"), timer indicator ("tim"), and the counter ("cnt").

Initially, the FSM is reset to the initial state, denoted as INI. Upon ac-

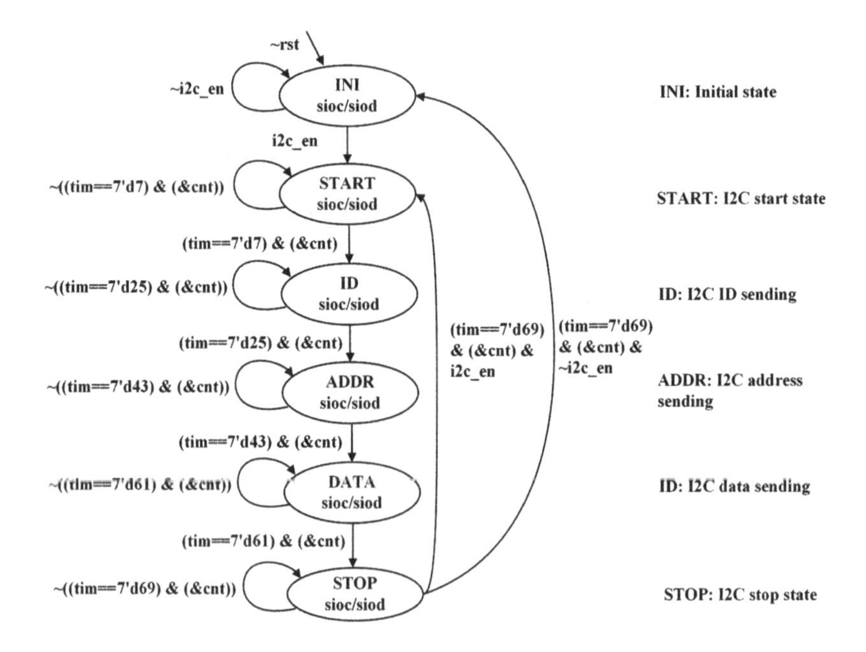

FIGURE 8.15
I2C write state graph.

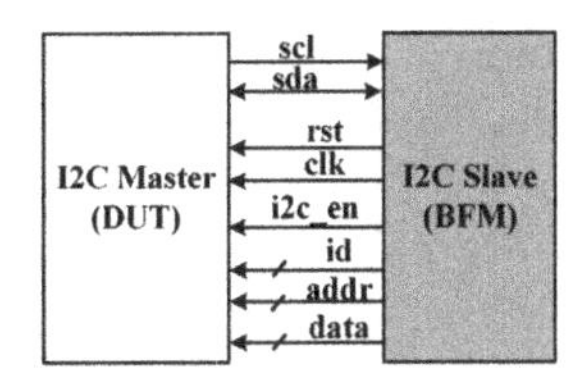

FIGURE 8.16
I2C Write simulation testbench.

tivation of the "i2c_en" input, the state machine becomes enabled, and the transition initiates from the INI state to the START state. The state remains in the START state as long as the timer values range from decimal 7'd0 to 7'd7. In the final clock cycle, when the timer reaches decimal value 7'd7 and the counter approaches hexadecimal 6'h3f (expressed as $\&cnt[5:0]$ in Verilog), the state progresses to the subsequent ID state. Between timer values of 7'd24 and 7'd25 within the ID state, the I2C master deasserts the SDA bus and awaits a response from the I2C device.

This pattern continues as the state machine transitions to the ADDR state, followed by the DATA state, sequentially transmitting the register address and register data. In the last clock cycle within the DATA state, when the counter approaches hexadecimal 6'h3f and the timer reaches decimal 7'd61, the state transitions to the STOP state, signaling the STOP condition to conclude the I2C write operation.

8.3.5 I2C Master design with verilog HDL

Figure 8.16 illustrates the simulation testbench between the I2C master and the I2C slave. The I2C master serves as the Verilog design-under-test, while the I2C slave functions as the bus functional model within the testbench. In addition to the SCL and SDA interfaces, the circuit also necessitates inputs for the reset signal ("rst"), clock signal ("clk"), and enable signal ("i2c_en"). Within the design testbench, the device ID, register address, and register data are provided through 8-bit interfaces labeled as "id", "addr", and "data".

Provided below is the Verilog description of the I2C master design. Lines 14–35 outline the descriptions of the counter and timer, ensuring compliance with the timing requirements of the I2C device. Starting from line 37, the FSM is described, including the sequential circuit for the current state registers (lines 47–53) and the combinational circuit for the next state transitions (lines 55–74).

```
1 module i2c_master  (input       clk     ,
2                     input       rst     ,
```

```
3                          input        i2c_en    ,
4                          input  [7:0] id        ,
5                          input  [7:0] addr      ,
6                          input  [7:0] data      ,
7                          inout        sda       ,
8                          output reg   scl       );
9   reg [5:0] cnt       ;
10  reg [6:0] tim       ;
11  reg [2:0] nxt_state ;
12  reg [2:0] cur_state ;
13
14  //-------------------------------------------------------//
15  // cnt: 0~64, timer unit - 64x50MHz clock is 1.28 us //
16  // tim: timing control for scl and sda                   //
17  //-------------------------------------------------------//
18  wire [5:0]  nxt_cnt = i2c_en ? cnt+6'd1 : cnt;
19  always @(posedge clk, negedge rst) begin
20    if(~rst) begin
21      cnt <= 6'd0    ;
22    end else begin
23      cnt <= nxt_cnt;
24    end
25  end
26
27  wire [6:0] nxt_tim = tim==7'd69 & &cnt ? 7'd0 :
28                                 i2c_en & &cnt ? tim+7'd1 : tim;
29  always @(posedge clk, negedge rst) begin
30    if(~rst) begin
31      tim <= 7'd0    ;
32    end else begin
33      tim <= nxt_tim;
34    end
35  end
36
37  //-------------------------------------------------------//
38  //-----------    finite state machine      -----------//
39  //-------------------------------------------------------//
40  parameter INI  =3'h0;
41  parameter START=3'h1;
42  parameter ID   =3'h2;
43  parameter ADDR =3'h3;
44  parameter DATA =3'h4;
45  parameter STOP =3'h5;
46
47  always @(posedge clk, negedge rst) begin
48    if(~rst) begin
49      cur_state <= INI       ;
50    end else begin
51      cur_state <= nxt_state;
```

```
52    end
53  end
54
55  always @(rst, cur_state, i2c_en, cnt, tim) begin
56    if (~rst) begin
57      nxt_state <= INI;
58    end else begin
59      case (cur_state)
60        INI  : if(i2c_en)                nxt_state <= START;
61        START: if(tim==7'd7  & (&cnt)) nxt_state <= ID   ;
62        ID   : if(tim==7'd25 & (&cnt)) nxt_state <= ADDR ;
63        ADDR : if(tim==7'd43 & (&cnt)) nxt_state <= DATA ;
64        DATA : if(tim==7'd61 & (&cnt)) nxt_state <= STOP ;
65        STOP : if(tim==7'd69 & (&cnt)) begin
66                 if(i2c_en) begin
67                   nxt_state <= START ;
68                 end else begin
69                   nxt_state <= INI   ;
70                 end
71               end
72      endcase
73    end
74  end
75
76  //---------------------------------------------------//
77  //-------------------   scl   ----------------------//
78  //---------------------------------------------------//
79  always @(rst, cur_state, tim) begin
80    if (~rst) begin
81      scl <= 1'b1;
82    end else begin
83      case (cur_state)
84        INI            : scl<=1'b1      ;
85        START          : scl<=1'b1      ;
86        ID, ADDR, DATA: scl<=tim[0]    ;
87        STOP           : scl<=1'b1      ;
88        default        : scl<=1'b1      ;
89      endcase
90    end
91  end
92
93  //---------------------------------------------------//
94  //-------------------   sda   ----------------------//
95  //---------------------------------------------------//
96  reg sda_reg;
97  assign sda=(tim>=7'd24 & tim<=7'd25) |
98             (tim>=7'd42 & tim<=7'd43) |
99             (tim>=7'd60 & tim<=7'd61) ? 1'bz : sda_reg;
100
```

```
101 always @(rst, cur_state, tim, id, addr, data) begin
102   if (~rst) begin
103     sda_reg <= 1'b1;
104   end else begin
105     case (cur_state)
106       INI  :   sda_reg <= 1'b1;
107       START:   sda_reg <= (tim>=7'd0 & tim<=7'd3);
108       ID   : case (tim)
109                8, 9  : sda_reg<=id[7];
110                10,11 : sda_reg<=id[6];
111                12,13 : sda_reg<=id[5];
112                14,15 : sda_reg<=id[4];
113                16,17 : sda_reg<=id[3];
114                18,19 : sda_reg<=id[2];
115                20,21 : sda_reg<=id[1];
116                22,23 : sda_reg<=id[0];
117              endcase
118       ADDR : case (tim)
119                26,27 : sda_reg<=addr[7];
120                28,29 : sda_reg<=addr[6];
121                30,31 : sda_reg<=addr[5];
122                32,33 : sda_reg<=addr[4];
123                34,35 : sda_reg<=addr[3];
124                36,37 : sda_reg<=addr[2];
125                38,39 : sda_reg<=addr[1];
126                40,41 : sda_reg<=addr[0];
127              endcase
128       DATA : case (tim)
129                44,45 : sda_reg<=data[7];
130                46,47 : sda_reg<=data[6];
131                48,49 : sda_reg<=data[5];
132                50,51 : sda_reg<=data[4];
133                52,53 : sda_reg<=data[3];
134                54,55 : sda_reg<=data[2];
135                56,57 : sda_reg<=data[1];
136                58,59 : sda_reg<=data[0];
137              endcase
138       STOP :   sda_reg <= (tim>=7'd66 & tim<=7'd69);
139       default: sda_reg <= 1'b1;
140     endcase
141   end
142 end
143 endmodule
```

The datapath design generates the timing diagram for the SCL and SDA buses based on the current state and timer results. Lines 76–91 detail the I2C SCL bus design, while lines 93–142 illustrate the I2C SDA bus design.

Specifically, in the START state, the I2C SCL bus is activated (line 85), and the SDA bus is activated during timer intervals ranging from 0 to 3 (line 107). This timing aligns with the timing diagram presented in Figure 8.14, illustrating the START condition in the I2C protocol.

In the ID/ADDR/DATA states, the SCL clock toggles between zero and one with each timer increment, as demonstrated in line 86. The SDA bus is populated from the MSB to the LSB of the input signals "id", "addr", and "data", respectively, as illustrated in lines 108–137. This configuration facilitates data transactions on the I2C bus, encompassing the device address, register address, and register data.

Within the STOP state, the SCL clock remains active during the timer intervals ranging from decimal 62 to 69 (line 87). In all other instances, the I2C clock is deactivated. Regarding the SDA data bus, it is deactivated when the timer falls between decimal 62 and 65 and is subsequently activated when the timer ranges from decimal 66 to 69, as indicated in line 138.

As illustrated in the timing diagram shown in Figure 8.14, it is essential for the SDA bus to be relinquished when the timers reach decimal values 24–25, 42–43, and 60–61. This release is crucial to allow the I2C slave to assume control of the bus. This behavior can be observed in lines 97–99.

8.4 Design Example: MSBUS Communication

In SoC designs, high-speed bus protocols are developed to efficiently transmit substantial amounts of data within clock cycles. This section presents a case study that serves as a practical example of a straightforward implementation of an SoC bus involving a single-master and multi-slave for interfacing purposes.

8.4.1 MSBUS Protocol

8.4.1.1 MSBUS architecture

Figure 8.17 depicts a simple bus architecture where the CPU functions as the master of the SoC Bus. The other devices connected to the bus, such as audio/video processing units, flash controllers, DMAs linked to the memory controller, and peripherals like I2C and UART, are all designated as bus slaves. Therefore, the single-master and multi-slave bus is named MSBUS, which primarily enables the CPU master to exert control over these slave devices. This control encompasses issuing commands and facilitating data transfer according to the control and data processing/movement requirements.

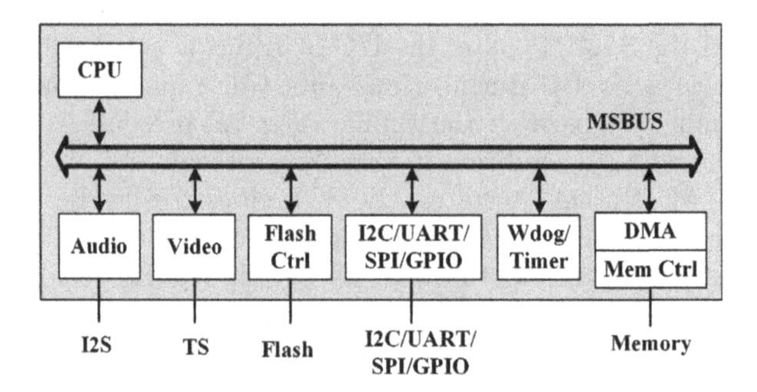

FIGURE 8.17
MSBUS SoC architecture.

8.4.1.2 Data transaction protocol

The MSBUS operates as a half-duplex bus, which means it doesn't allow simultaneous execution of write and read operations. Its primary purpose is to serve as a control bus for configuring functional registers using the SINGLE transfer mode. This mode comprises two stages for each data transaction: the command stage and the data stage, each requiring at least one cycle. Specifically, Figure 8.18 illustrates a typical data transaction on MSBUS, which involves a write operation followed by a read operation. In the timing diagram, signals prefixed with "m_" represent the IOs driven by the MSBUS master, while signals prefixed with "s_" represent the IOs driven by the slave.

In the initial clock cycle, the master initiates a write command by activating the enable signal "m_ce", setting the write indicator "m_wr" to binary one (indicating a write operation), and providing the address on the "m_addr_wdata" bus. During the subsequent clock cycle, the master transmits the write data over the "m_addr_wdata" bus. Assuming that the slave is

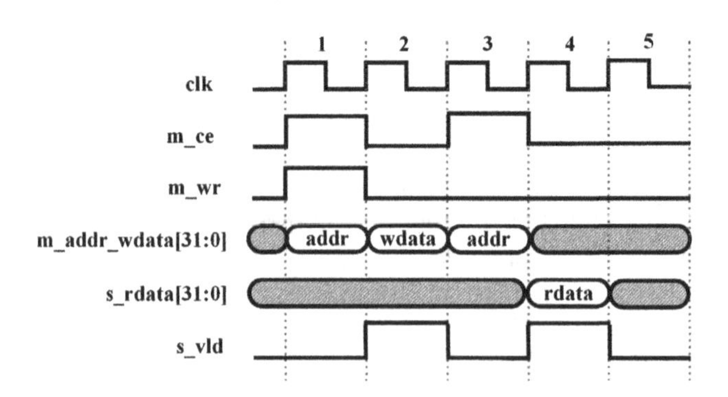

FIGURE 8.18
MSBUS timing diagram.

consistently prepared to receive the data, this results in the slave responding to the write command and data by asserting the "s_vld" signal. This signal serves as an indicator that the corresponding data has been successfully received by the slave.

Similarly, in the third cycle, the master initiates a read command by asserting the enable signal "m_ce" and de-asserting the "m_wr" signal (indicating a read operation). The data is read out within the following clock cycle, at which point the slave provides the data with a valid "s_vld" signal.

It is important to note that the bus "m_addr_wdata" serves as a shared bus, accommodating the write address and write data. While this shared bus requires one more clock cycle to send both register address and register data, it enhances wire usage efficiency and simplifies hardware interconnections.

8.4.1.3 Functional register access with MSBUS

A. Byte-Size MSBUS

Depending on the size of the MSBUS, the commands and data transmitted on the bus will vary. As an example, Figure 8.19(a) demonstrates a bus transfer scenario where byte-sized commands and data are utilized. This particular example involves four write commands, each followed by their corresponding read operations. The write/read addresses for these commands are 8'h0, 8'h1, 8'h2, and 8'h3, while the data written into and read from the memory correspond to 8'ha0, 8'ha1, 8'ha2, and 8'ha3, respectively.

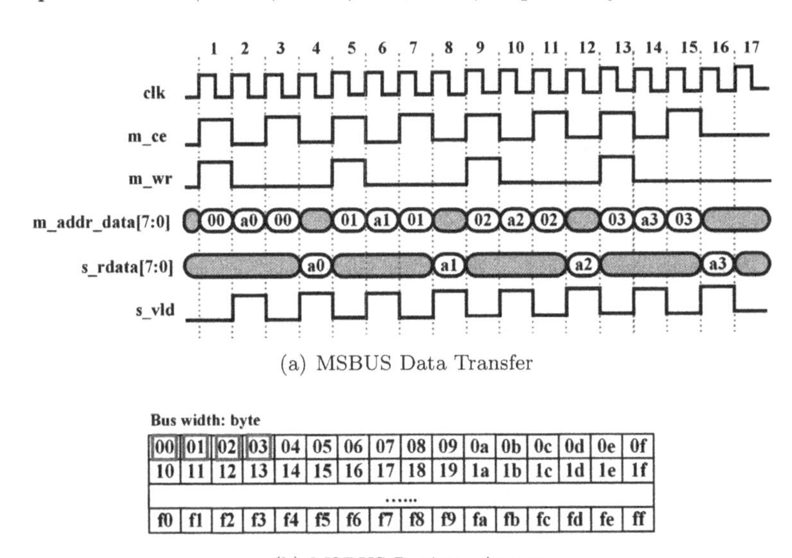

(a) MSBUS Data Transfer

(b) MSBUS Register Access

FIGURE 8.19
Register file access with byte-sized MSBUS.

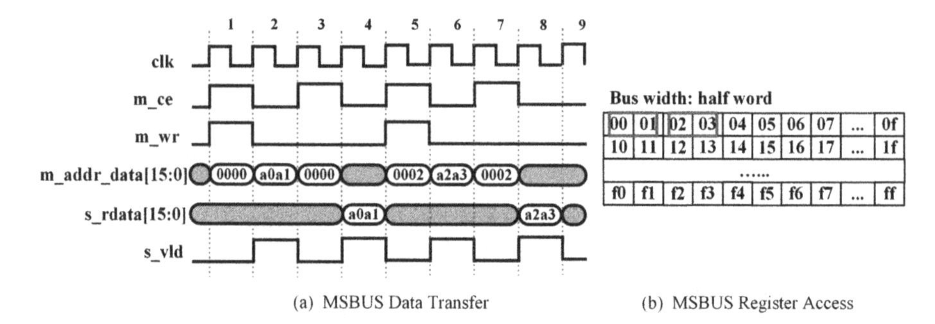

(a) MSBUS Data Transfer (b) MSBUS Register Access

FIGURE 8.20
Register file access with half-word-sized MSBUS.

The memory accesses depicted in Figure 8.19(b) correspond to the same byte-sized commands as shown in the MSBUS transfer diagram. Data in each of the four memory cells can be written in or read out using a single memory write or read command.

B. Half-Word-Size MSBUS

In the case of a half-word-size MSBUS, the addresses transmitted on the bus should align with half-word boundaries. This means that the LSB of the address will be disregarded or set to binary zero during MSBUS transactions. For instance, Figure 8.20(a) presents an example involving two write commands followed by their corresponding read operations. The write/read addresses in this case are 16'h0 and 16'h2, respectively. The data written into and read from memory correspond to 16'ha0a1 and 16'ha2a3, respectively. The memory access illustrated in Figure 8.20(b) demonstrates the two operations based on the half-word-size MSBUS. It can be seen that each bus access can write into or read out data by two bytes.

To illustrate the impact of half-word alignment, let's consider an unusual command scenario in which the CPU sends a memory address of $0x1$. At the hardware level, the bus master will discard the LSB of the bus address and set it to binary zero because the bus size is half-word. Consequently, the address transmitted on the bus will be adjusted to 16'h0. Similarly, in a similar abnormal command scenario with an address of $0x3$, the hardware will modify it to 16'h2 to ensure half-word alignment. This adjustment ensures that data is correctly accessed according to the half-word bus protocol.

C. Word-Size MSBUS

In the case of a word-size MSBUS, the addresses transmitted on the bus should align with word boundaries. This means that the least-significant two bits of the memory address will be disregarded and set to zeros during MSBUS

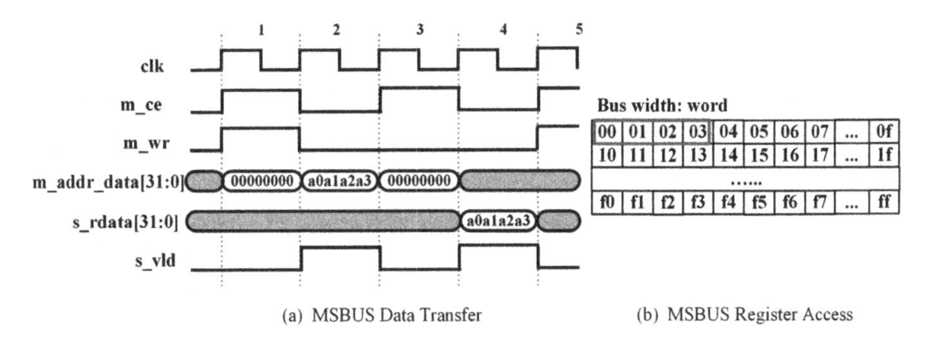

FIGURE 8.21
Register file access with word-sized MSBUS.

transactions. In Figure 8.21(a), we can observe an example of a single write command followed by its corresponding read operation. The write/read address in this case is 32'h0, while the data written into and read from memory correspond to 32'ha0a1a2a3. The memory access depicted in Figure 8.21(b) demonstrates the operation based on the word size.

Let's consider an abnormal command where the CPU sends a memory address of $0x03$. At the hardware level, the bus master will disregard the least-significant two bits of the bus address and set them to binary 2'b00. As a result, the address transmitted on the bus will be corrected to 32'h0. Similarly, abnormal commands with memory addresses of $0x1$ and $0x2$ will be adjusted to 32'h0 on the hardware side to ensure word-size alignment.

8.4.2 An FSMD design example: MSBUS master and slave

A. Design Structure

In this section, we delve into a case study focused on master-slave communication utilizing the MSBUS protocol. Figure 8.22 provides an illustration of the design structure, featuring an MSBUS master responsible for transmitting commands and an MSBUS slave responsible for responding with data.

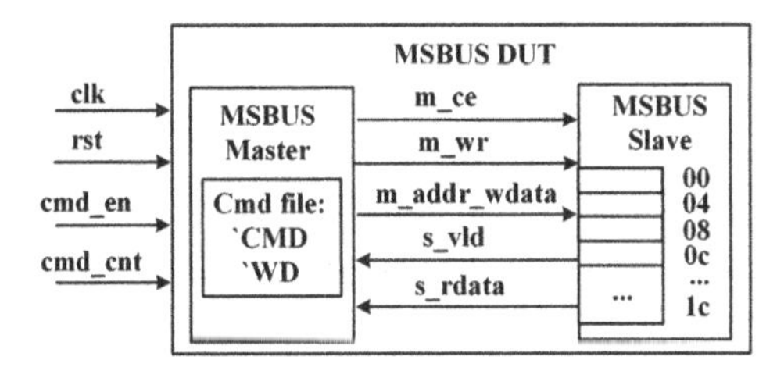

FIGURE 8.22
MSBUS master-slave design structure.

The input "cmd_en" facilitates the transmission of both commands and data between the master and slave, while "cmd_cnt" specifies the register number accessed by the master. The "cmd_cnt" value spans from decimal zero to seven, incrementing by one between neighboring registers. The corresponding bus addresses range from hexadecimal 32'h00 to 32'h1c, incrementing by 32'h4 due to the word-sized design with MSBUS.

To facilitate bus communication between the master and slave devices, the MSBUS interface is employed to interconnect the two design modules. All essential commands and data are housed in a command file within the MSBUS master. When register access is enabled, the master transmits commands and write data to the slave. After each register write, a read operation is executed to confirm the successful write operation. The MSBUS slave's primary function is to receive the data sent by the master and write it into the register files based on the provided command instructions.

B. Design Block Diagram

Figure 8.23 illustrates the FSMD block diagram, comprising primarily of two FSMs: one for the master and another for the slave. The master FSM is responsible for generating control signals, including the flag signal to enable the master write command ("mwc_f"), the flag signal to enable the master read command ("mrc_f"), and the write data flag to enable the master write data ("mwd_f").

The datapath acknowledges the master FSM by issuing a valid signal ("s_vld"). In the context of a write operation, when the "s_vld" signal is

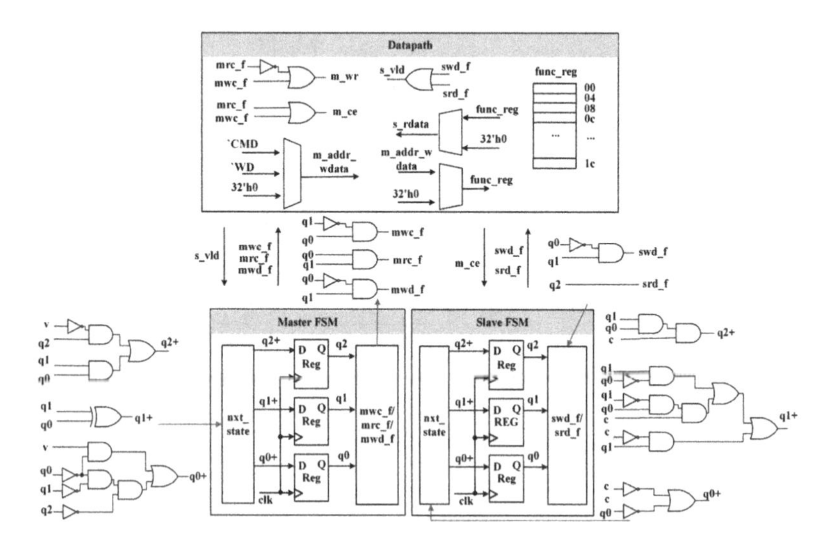

FIGURE 8.23
MSBUS master-slave FSMD structure.

asserted, it signifies the successful reception of the write data by the slave. Likewise, in the case of a read operation, the assertion of the "s_vld" signal indicates the validity of the data on the read data bus.

Similarly, the slave FSM generates control signals to signify the write data stage ("swd_f") and the read data stage ("srd_f"). The datapath, in response, provides the master enable signal ("m_ce") to the slave FSM, enabling it to carry out the required operations.

8.4.3 MSBUS master design with Verilog HDL

A. FSM Design of MSBUS Master

Figure 8.24 presents the state graph of the FSM in the MSBUS master design. Upon reset, the state machine enters the initial state (INI). When the "cmd_en" signal is activated, the MSBUS write-read operation begins. The state machine initiates in the write command (WC) state, where the write command flag ("mwc_f") is asserted as a Moore machine's output.

The subsequent state is the write data (WD) state, where the master flag indicating the write data ("mwd_f") is asserted. In the WD state, the state machine awaits the data valid signal "s_vld" from the slave. If the valid signal is received, signifying that the slave successfully receives the write data, the state machine transitions to the read command (RC) state and asserts the read command flag ("mrc_f").

The final state is the read data (RD) state. Once the slave asserts the "s_vld" signal, indicating that the data on the read data bus is valid, the state machine transitions to the subsequent state. Within the RD state, if the "cmd_en" signal remains asserted, indicating the presence of another command, the state machine transitions back to the WC state to continue with

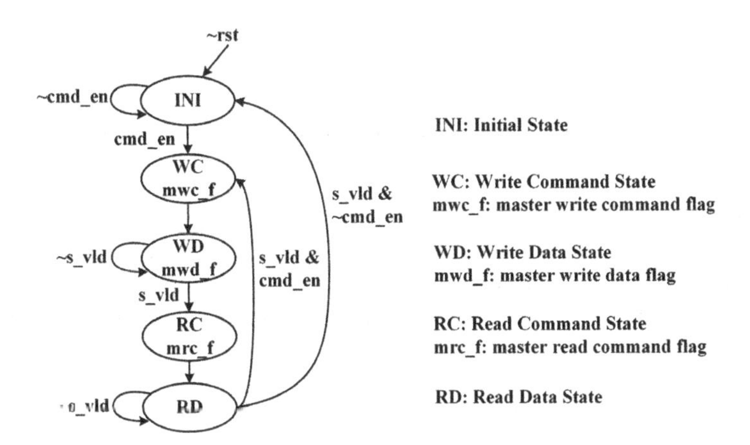

FIGURE 8.24
MSBUS master state graph.

the next round of the write-read operation. On the other hand, if the "cmd_en" signal is de-asserted, indicating the absence of further commands, the state machine resets to the INI state to await the next command sequence.

B. MSBUS Master Design with Verilog HDL

Prior to discussing the master design, let's review the command file "command.v" that defines all the commands and data for the register configuration. The command file includes various entries and values, such as CMD_1C (specifying that the MSBUS address as hexadecimal 32'h1c) and WD_1C (defining that the MSBUS data as hexadecimal 32'hf7). These entries and their corresponding values in the command file determine the specific register configurations for the master design.

```
1  `define CMD_1C  32'h1c
2  `define CMD_18  32'h18
3  `define CMD_14  32'h14
4  `define CMD_10  32'h10
5  `define CMD_0C  32'hc
6  `define CMD_08  32'h8
7  `define CMD_04  32'h4
8  `define CMD_00  32'h0
9
10 `define WD_1C  32'hf7
11 `define WD_18  32'hf6
12 `define WD_14  32'hf5
13 `define WD_10  32'hf4
14 `define WD_0C  32'hf3
15 `define WD_08  32'hf2
16 `define WD_04  32'hf1
17 `define WD_00  32'hf0
```

Here is the Verilog description of the MSBUS master responsible for loading each command and data and transmitting them to the slave via the MSBUS. The five machine states are declared as *parameters* in lines 11–15. In what follows, lines 17–27 represent the sequential circuitry for the state machine design, while lines 29–48 describe the combinational circuit responsible for the next state transitions. Additionally, lines 50–55 outline the output signals of the Moore machine, which include the flag signals for the master write command ("mwc_f"), master read command ("mrc_f"), and master write data ("mwd_f").

```
1  `include "command.v"
2  module msbus_master (input                clk          ,
3                       input                rst          ,
4                       input                cmd_en       ,
5                       input         [2:0]  cmd_cnt      ,
6                       output               m_ce         ,
7                       output               m_wr         ,
8                       output reg [31:0]    m_addr_wdata ,
9                       input      [31:0]    s_rdata      ,
10                      input                s_vld        );
11 parameter INI = 3'h0  ;
12 parameter WC  = 3'h1  ;
13 parameter WD  = 3'h2  ;
14 parameter RC  = 3'h3  ;
15 parameter RD  = 3'h4  ;
16
17 //------------------------------------------
18 //------- current state--------------
19 //------------------------------------------
20 reg [2:0] cur_state, nxt_state;
21 always @(posedge clk, negedge rst) begin
22   if(~rst) begin
23     cur_state <= INI      ;
24   end else begin
25     cur_state <= nxt_state;
26   end
27 end
28
29 //------------------------------------------
30 //------- next state----------------
31 //------------------------------------------
32 always @(rst, cur_state, cmd_en, s_vld) begin
33   if(~rst) begin
34     nxt_state <= INI;
35   end else begin
36     case (cur_state)
37       INI: if(cmd_en) nxt_state <= WC;
38       WC:             nxt_state <= WD;
39       WD:  if (s_vld) nxt_state <= RC;
40       RC:             nxt_state <= RD;
41       RD:  if (s_vld) begin
42              if (cmd_en) nxt_state <= WC ;
43              else        nxt_state <= INI;
44            end
45       default:        nxt_state <= INI;
46     endcase
47   end
```

```
48 end
49
50 //---------------------------------
51 //------- output logic------------
52 //---------------------------------
53 wire mwc_f = cur_state==WC;
54 wire mrc_f = cur_state==RC;
55 wire mwd_f = cur_state==WD;
56
57 //-----------------------------------------
58 //------- master data path---------------
59 //-----------------------------------------
60 always @(rst, mwc_f, mrc_f, cmd_cnt, mwd_f) begin
61   if(~rst) begin
62     m_addr_wdata <= 32'h0;
63   end if (mwc_f | mrc_f)  begin
64     case (cmd_cnt)
65       3'd7 : m_addr_wdata <= `CMD_1C;
66       3'd6 : m_addr_wdata <= `CMD_18;
67       3'd5 : m_addr_wdata <= `CMD_14;
68       3'd4 : m_addr_wdata <= `CMD_10;
69       3'd3 : m_addr_wdata <= `CMD_0C;
70       3'd2 : m_addr_wdata <= `CMD_08;
71       3'd1 : m_addr_wdata <= `CMD_04;
72       3'd0 : m_addr_wdata <= `CMD_00;
73       default: m_addr_wdata <= `CMD_00;
74     endcase
75   end else if (mwd_f) begin
76     case (cmd_cnt)
77       3'd7 : m_addr_wdata <= `WD_1C;
78       3'd6 : m_addr_wdata <= `WD_18;
79       3'd5 : m_addr_wdata <= `WD_14;
80       3'd4 : m_addr_wdata <= `WD_10;
81       3'd3 : m_addr_wdata <= `WD_0C;
82       3'd2 : m_addr_wdata <= `WD_08;
83       3'd1 : m_addr_wdata <= `WD_04;
84       3'd0 : m_addr_wdata <= `WD_00;
85       default: m_addr_wdata <= `WD_00;
86     endcase
87   end
88 end
89
90 assign m_ce = mwc_f |  mrc_f;
91 assign m_wr = mwc_f | ~mrc_f;
92 endmodule
```

These output signals from the Moore machine are utilized for coordinating and controlling the master datapath in accordance with the MSBUS protocol. The description of the datapath can be found in lines 57–88. Within the specified "cmd_cnt" value range, the corresponding command and data can be asserted on the write command bus and the write data bus ("m_addr_wdata"). It's essential to emphasize that these write command and write data buses are shared resources, and assignments occur in separate stages. These stages are delineated by the flag signals, namely the command stage (indicated by "mwc_f" or "mrc_f") or the data stage ("mwd_f").

The master outputs include the "m_ce" signal, which enables the MSBUS transaction, and the "m_wr" signal, which indicates whether a write or read operation is being performed on the MSBUS. As demonstrated in lines 90–91, the enable signal is asserted whenever the write or read command flag becomes active, while the write/read indicator is asserted when the associated command signifies a write access.

8.4.4 MSBUS slave design with Verilog HDL

A. FSM Design of MSBUS Slave

Figure 8.25 displays the state graph of the FSM in the MSBUS slave design. Similar to the master state graph, the state machine begins in the initial state (INI). Upon activation of the "cmd_en" signal, the register write-read operation commences. The data transfer process starts in the write command (WC) state. Upon receiving the "m_en" signal, the state machine transitions to the write data (WD) state, where the write data flag ("swd_f") is asserted. Subsequently, the state machine progresses to the read command (RC) state.

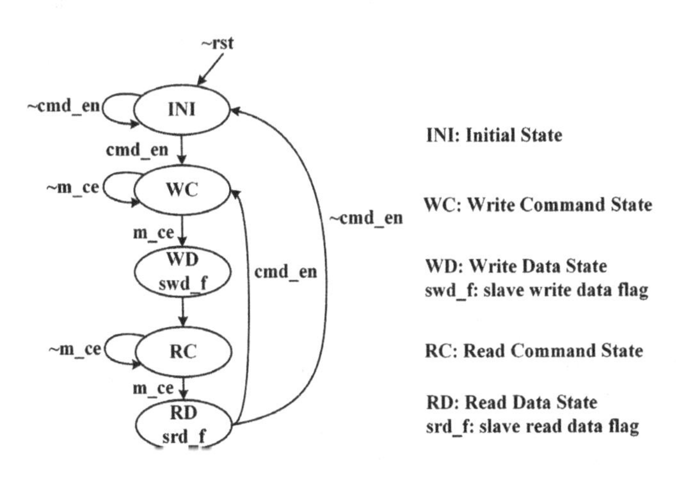

FIGURE 8.25
MSBUS slave state graph.

Upon receiving the "m_en" signal, it transitions to the read data (RD) state, where the read data flag ("srd_f") is asserted.

During the RD state, if the "cmd_en" signal remains active, indicating the existence of another command, the state machine reverts to the WC state to proceed with the subsequent phase of the write-read operation. Conversely, if the "cmd_en" signal is deactivated, indicating no more commands, the state machine resets to the INI state, ready to await the next command sequence.

Big-Endian vs. Little-Endian

In SoC bus architecture, big-endian and little-endian are two byte-ordering schemes that determine how multi-byte data is stored in memory and transmitted over data buses.

Big-endian format refers to the scheme where the most significant byte of a multi-byte data is stored at the lowest memory address or transmitted first over the bus. The subsequent bytes are then stored or transmitted in decreasing order of significance. For example, in a 32-bit word (32'haab-bccdd) with address 32'h0, the most significant byte (8'haa) should be stored at address 32'h0, followed by the second byte (8'hbb) at address 32'h1, the third byte (8'hcc) at address 32'h2, and the least significant byte (8'hdd) at address 32'h3.

In contrast, little-endian format refers to the scheme where the least significant byte of a multi-byte data is stored at the lowest memory address or transmitted first over the bus. The subsequent bytes are then stored or transmitted in increasing order of significance. For example, in a 32-bit word (32'haabbccdd) with address 32'h0, the least significant byte (8'hdd) would be stored at address 32'h0, followed by the second byte (8'hcc) at address 32'h1, the third byte (8'hbb) at address 32'h2, and the most significant byte (8'haa) at address 32'h3.

B. MSBUS Slave Design with Verilog HDL

Here is the Verilog description of the MSBUS slave design. In line 11, the functional register array is declared, consisting of 32 bytes or 8 words. Given that the MSBUS size is word-based, each bus access can write or read four bytes.

Similar to the MSBUS master design, lines 19–29 describe the sequential circuitry for the state machine design, while lines 31–51 present the combinational circuit responsible for the next state computation. The Moore machine outputs are described in lines 53–57. During the write data state (WD), the write data flag ("swd_f") is asserted, indicating that data is being written into the registers. Conversely, in the read data state, the read data flag ("srd_f") is asserted, signifying that data is being read from the registers. These flag signals are utilized for coordinating and controlling the slave datapath actions in accordance with the MSBUS protocol.

```
1  `include "command.v"
2  module msbus_slave (input                clk           ,
3                      input                rst           ,
4                      input                cmd_en        ,
5                      input   [7:0]        cmd_cnt       ,
6                      input                m_ce          ,
7                      input                m_wr          ,
8                      input   [31:0]       m_addr_wdata  ,
9                      output  [31:0]       s_rdata       ,
10                     output               s_vld         );
11 reg [7:0] func_reg[0:31] ;
12
13 parameter INI = 3'h0;
14 parameter WC  = 3'h1;
15 parameter WD  = 3'h2;
16 parameter RC  = 3'h3;
17 parameter RD  = 3'h4;
18
19 //-----------------------------------
20 //------- current state--------------
21 //-----------------------------------
22 reg [2:0] cur_state, nxt_state;
23 always @(posedge clk, negedge rst) begin
24   if(~rst) begin
25     cur_state <= INI      ;
26   end else begin
27     cur_state <= nxt_state;
28   end
29 end
30
31 //--------------------------------
32 //------- next state--------------
33 //--------------------------------
34 always @(rst, cur_state, cmd_en, m_ce) begin
35   if(~rst) begin
36     nxt_state <= INI;
37   end else begin
38     case (cur_state)
39       INI: if (cmd_en) begin
40              nxt_state <= WC;
41            end else begin
42              nxt_state <= INI;
43            end
44       WC:  if(m_ce)   nxt_state <= WD ;
```

```
45        WD:                nxt_state <= RC ;
46        RC:  if(m_ce)     nxt_state <= RD ;
47        RD:  if(cmd_en) nxt_state <= WC ;
48             else        nxt_state <= INI;
49      endcase
50    end
51  end
52
53  //-----------------------------------------
54  //------- output logic---------------
55  //-----------------------------------------
56  wire swd_f = cur_state==WD;
57  wire srd_f = cur_state==RD;
58
59  //-----------------------------------------
60  //------- slave data path- -------------
61  //-----------------------------------------
62  integer i;
63  always @(rst, swd_f, m_addr_wdata) begin
64    if(~rst) begin
65      for (i=0; i<32; i=i+1) begin
66        func_reg[i] <= 8'h0;
67      end
68    end else if (swd_f) begin
69  `ifdef BIG_ENDIAN
70      func_reg[4*cmd_cnt+3]  <= m_addr_wdata[7 : 0];
71      func_reg[4*cmd_cnt+2]  <= m_addr_wdata[15: 8];
72      func_reg[4*cmd_cnt+1]  <= m_addr_wdata[23:16];
73      func_reg[4*cmd_cnt  ]  <= m_addr_wdata[31:24];
74  `elsif LITTLE_ENDIAN
75      func_reg[4*cmd_cnt+3]  <= m_addr_wdata[31:24];
76      func_reg[4*cmd_cnt+2]  <= m_addr_wdata[23:16];
77      func_reg[4*cmd_cnt+1]  <= m_addr_wdata[15: 8];
78      func_reg[4*cmd_cnt  ]  <= m_addr_wdata[7 : 0];
79  `endif
80    end
81  end
82
83  `ifdef BIG_ENDIAN
84  assign s_rdata = srd_f ? {func_reg[4*cmd_cnt]  ,
85                            func_reg[4*cmd_cnt+1],
86                            func_reg[4*cmd_cnt+2],
87                            func_reg[4*cmd_cnt+3]} : 32'h0 ;
88  `elsif LITTLE_ENDIAN
89  assign s_rdata = srd_f ? {func_reg[4*cmd_cnt+3],
90                            func_reg[4*cmd_cnt+2],
91                            func_reg[4*cmd_cnt+1],
92                            func_reg[4*cmd_cnt  ]} : 32'h0 ;
93  `endif
```

```
94
95 assign s_vld   = swd_f | srd_f ;
96 endmodule
```

The FSMD integrates the slave datapath outlined in lines 59–95. Below, we illustrate memory access using both big-endian and little-endian formats:

- **Big-Endian Write:** It is important to note that the memory write following a big-endian operation is shown in lines 70–73, which means that the most significant byte on the write data bus ("m_addr_wdata[31:24]") is written to the lowest register address ("func_reg[4*cmd_cnt]"), while the least significant byte ("m_addr_wdata[7:0]") is written to the highest register address ("func_reg[4*cmd_cnt]+3") within each word-sized bus access. For instance, when the "cmd_cnt" is decimal 0, it indicates a write operation targeting the initial four bytes. As shown in lines 70–73, the most significant byte on the write data bus ("m_addr_wdata[31:24]") will be stored in func_reg[0], the least significant byte ("m_addr_wdata[7:0]") will be stored in func_reg[3], and the two middle bytes ("m_addr_wdata[23:16]" and ("m_addr_wdata[15:8]") will be stored in func_reg[1] and func_reg[2], respectively. This byte ordering within the register file aligns with the big-endian format.

- **Little-Endian Write**: The memory write operation following a little-endian operation is explained in lines 75–78. In this scenario, the most significant byte on the write data bus ("m_addr_wdata[31:24]") is written to the highest register address ("func_reg[4*cmd_cnt+3]"), while the least significant byte ("m_addr_wdata[7:0]") is written to the lowest register address ("func_reg[4*cmd_cnt]") within each word-sized bus access.

- **Big-Endian Read:** During each MSBUS read operation, four data bytes can be transmitted on the read data bus ("s_rdata"). Consistent with the big-endian format, the lowest register address ("func_reg[4*cmd_cnt]") will be represented by the most significant byte on the read data bus ("s_rdata[31:24]"), while the highest register address ("func_reg[4*cmd _cnt+3]") will be represented by the least significant byte on the read data bus ("s_rdata[7:0]"). This arrangement is described in lines 84–87, ensuring the correct byte ordering when reading data via the MSBUS.

- **Little-Endian Read:** The memory read operation in the little-endian format is explained in lines 89–92. In this case, the lowest register address ("func_reg[4*cmd_cnt]") will be represented by the least significant byte on the read data bus ("s_rdata[7:0]"), while the highest register address ("func_reg[4*cmd_cnt+3]") will be represented by the most significant byte on the read data bus ("s_rdata[31:24]").

cur_state	nxt_state		mwc_f	mrc_f	mwd_f
	~s_vld	s_vld			
INI	WC	WC	0	0	0
WC	WD	WD	1	0	0
WD	WD	RC	0	0	1
RC	RD	RD	0	1	0
RD	RD	WC	0	0	0

(a) State Table of MSBUS Master

cur_state {q2,q1,q0}	nxt_state {q2+,q1+,q0+}		mwc_f	mrc_f	mwd_f
	~s_vld	s_vld			
000	001	001	0	0	0
001	010	010	1	0	0
010	010	011	0	0	1
011	100	100	0	1	0
100	100	001	0	0	0

INI=3'b000, WC=3'b001, WD=3'b010,
RC=3'b011, RD=3'b100

(b) Transition Table of MSBUS Master

FIGURE 8.26
MSBUS master state and transition table.

The slave output signal, "s_vld", serves as the validator for write/read data transfers on the MSBUS. As depicted in line 95, this signal is asserted whenever the write or read data flag becomes active.

8.4.5 MSBUS master-slave circuit analysis

In this section, the FSMD circuit, consisting of both the master and slave designs, is analyzed using K-map optimization. The state table and transition table for the MSBUS master design are illustrated in Figure 8.26. As a Moore machine, the outputs of the FSM are all dependent on the current states, as indicated in the last three columns. Similarly, the state table and transition table for the MSBUS slave design are illustrated in Figure 8.27.

The K-map simplification process is omitted in this section. As a result, Figure 8.28 illustrates the synthesized circuit after the optimization. The FSMD design comprises two Moore FSMs and a combined datapath for master-slave operations. The master FSM generates flag signals ("mwc_f", "mrc_f", and "mwd_f") via the combinational output circuit within the state machine. Additionally, it incorporates a combinational circuit to determine

cur_state	nxt_state		swd_f	srd_f
	~m_ce	m_ce		
INI	WC	WC	0	0
WC	WC	WD	0	0
WD	RC	RC	1	0
RC	RC	RD	0	0
RD	WC	WC	0	1

(a) State Table of MSBUS Slave

cur_state {q2,q1,q0}	nxt_state {q2+,q1+,q0+}		swd_f	srd_f
	~m_ce	m_ce		
000	001	001	0	0
001	001	010	0	0
010	011	011	1	0
011	011	100	0	0
100	001	001	0	1

(b) Transition Table of MSBUS Slave

FIGURE 8.27
MSBUS slave state and transition table.

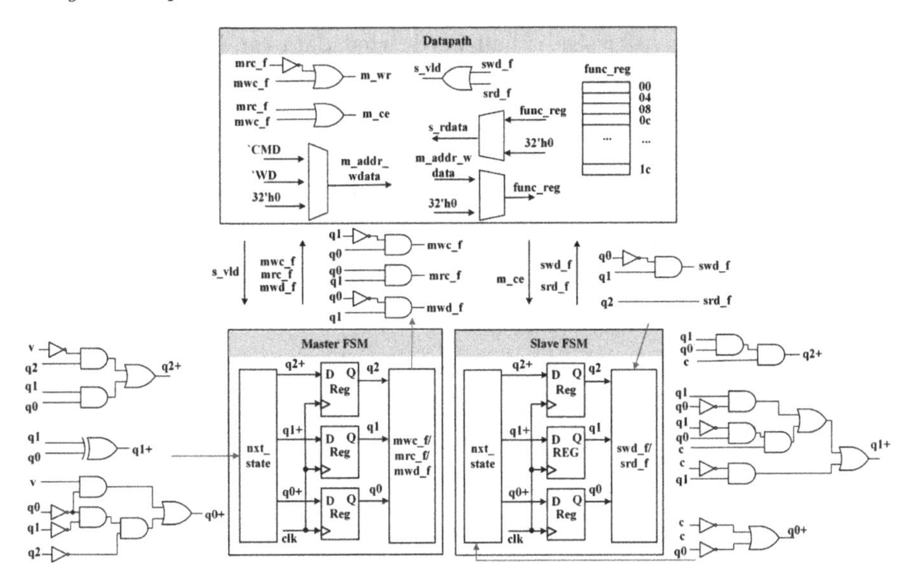

FIGURE 8.28
MSBUS master-slave circuit.

the next state transition. The sequential circuit comprises state registers that maintain the machine's current state. Likewise, the slave FSM generates flag signals ("swd_f" and "srd_f") through the combinational output circuit within the state machine and employs a combinational circuit for determining the next state transition. The sequential circuit is composed of state registers responsible for storing the current state.

Within the datapath, the master-slave communication adheres to the MSBUS protocol. The master datapath is responsible for transmitting essential command and data signals, which encompass "m_ce", "m_wr", and "m_addr_wdata" to the slave. Once these commands are received, the slave datapath issues corresponding signals, including "s_vld" and "s_rdata". It's worth noting that the slave datapath oversees the management of the register file, facilitating both data writing and reading via the MSBUS.

Exercises

Problem 8.1. A 16-byte memory block is depicted in Figure 8.29, showing both addresses and data bytes. Assuming the use of a big-endian mode data bus to read the memory block, please consider the following questions:

- What data can be read out on the data bus corresponding to the configured memory addresses $0x00$, $0x01$, $0x02$, and $0x03$ when the bus transfer size is one byte?

- When the bus transfer size is half-word, what data can be read out on the data bus corresponding to the configured memory addresses $0x04$, $0x05$, $0x06$, and $0x07$?

- If the bus transfer size is a word, what data can be read out on the data bus corresponding to the configured memory addresses $0x08$, $0x09$, $0x0a$, and $0x0b$?

Assuming the use of a little-endian mode data bus to access the memory block, please consider the three questions again.

Memory address:	00	01	02	03	04	05	06	07	08	09	0a	0b	0c	0d	0e	0f
Memory data:	a0	a1	a2	a3	b0	b1	b2	b3	c0	c1	c2	c3	d0	d1	d2	d3

FIGURE 8.29
Memory block access through data bus.

FP Design Projects

In this textbook, all single-precision FP operators, such as FP adders and multipliers, are readily accessible as pre-designed intellectual properties (IPs). These RTL designs are generated using the Chisel hardware description language. Two versions of the design are provided:

- **Basic design version:** These designs offer single-clock cycle latency between inputs and outputs.

- **Pipelined design version:** The pipelined designs can achieve a higher operational clock frequency compared to the basic designs, although they require more clock cycles for each data computation.

Hint: 1) It's crucial to recognize that the design IPs are activated by the rising clock edge and rising reset edge. Consequently, it's advisable to integrate the IPs within the system constructed with a rising clock edge and rising reset edge for optimal performance and compatibility. 2) Moreover, when dealing with FP operations in digital circuits, it is essential to conform to the IEEE 754 format for the representation of FP numbers. To ensure precise representation and conversion, please consult Table 8.2 for the numbers referenced.

TABLE 8.2
IEEE 754 format for FP numbers.

FP	1.0	2.0	3.0	4.0	5.0	6.0
HW	3f800000	40000000	40400000	40800000	40a00000	40c00000
FP	7.0	8.0	9.0	10.0	11.0	12.0
HW	40e00000	41000000	41100000	41200000	41300000	41400000
FP	-1.0	-2.0	-3.0	-4.0	-5.0	-6.0
HW	bf800000	c0000000	c0400000	c0800000	c0a00000	c0c00000
FP	-7.0	-8.0	-9.0	-10.0	-11.0	-12.0
HW	c0e00000	c1000000	c1100000	c1200000	c1300000	c1400000

PBL 18: FP Add-Mul-Sub Design with FSMD

1) Design an FSMD circuit according to the block diagram depicted in Figure 8.30(a). This FSMD circuit features a datapath comprising three single-precision FP design elements: an FP adder, an FP multiplier, and an FP subtractor arranged in a series configuration. With the assumption that each FP component operates within a single clock cycle, the final output, denoted as "g", will be produced after three clock cycles, following the provision of all four inputs "a-d".

   To coordinate the functioning of the datapath, a FSM serves as the timing controller. The FSM is activated by an enable signal, denoted

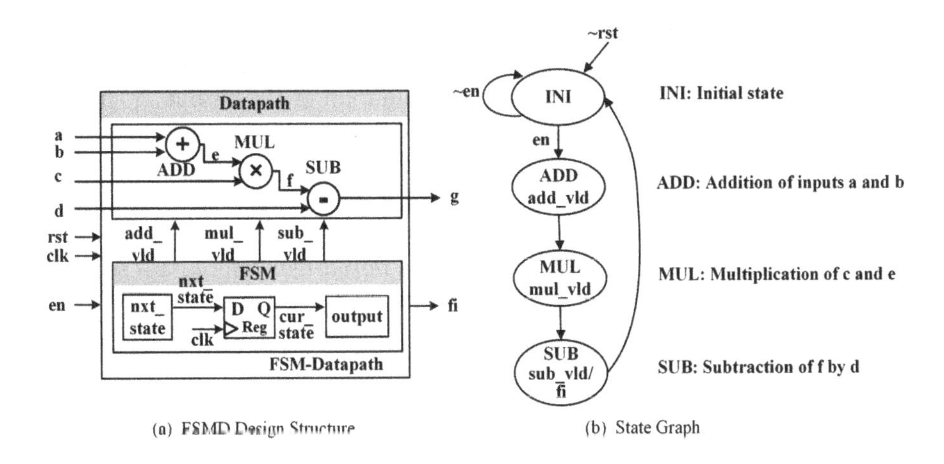

(a) FSMD Design Structure (b) State Graph

FIGURE 8.30
FP Add-Mul-Sub design with FSMD.

TABLE 8.3
IOs description of FP Add-Mul-Sub design with FSMD.

Name	Direction	Bit Width	Description
clk	Input	1	Clock, 100 MHz
rst	Input	1	Asynchronous reset, 0 valid
en	Input	1	Enable signal
a-d	Input	32	Input data
g	Output	32	Output data
fi	Output	1	Data processing completion indicator

as "en", and produces the completion of data processing utilizing a finish indicator, labeled as "fi". The "en-fi" mechanism follows the "enable-finish" handshaking protocol, facilitating the activation and termination of the FPU engine. A comprehensive list of all inputs and outputs for the circuit is provided in Table 8.3.

According to the design specifications, the state graph of the FSMD is illustrated in Figure 8.30(b). The design engine begins by resetting to the initial state (INI). Upon enabling the circuit, the FSM progresses into the first computational state, dedicated to addition (ADD). Subsequently, it proceeds to the second computational state, focused on multiplication (MUL). Finally, it transitions into the last computational state, which handles subtraction (SUB). During each of these computational states, the validation signals are activated as the outputs of the Moore machine, indicating the validity of the outputs produced by the three individual FP operators. To be specific, the signal "add_vld" confirms the validity of the output "e" from the adder, while the signal "mul_vld" signifies the validity of the output "f" from the multiplier. Furthermore, the signal "sub_vld" denotes the validation of the output "g" from the subtractor.

2) Create a testbench to perform simulations of the FSMD design, and assess its functionality using the provided direct test case illustrated in Figure 8.31. The FSMD is initialized and enabled during the initial clock cycle. Concurrently, the initial data inputs are set as follows: "a" to FP 1.0, "b" to FP 2.0, "c" to FP 3.0, and "d" to FP 0.0. Upon detecting the first asserted finish indicator "fi", the testbench modifies the inputs to the values FP 2.0, 3.0, 2.0, and 1.0 for "a", "b", "c", and "d", respectively. Similarly, subsequent updates to the third and fourth data inputs occur upon the appearance of the

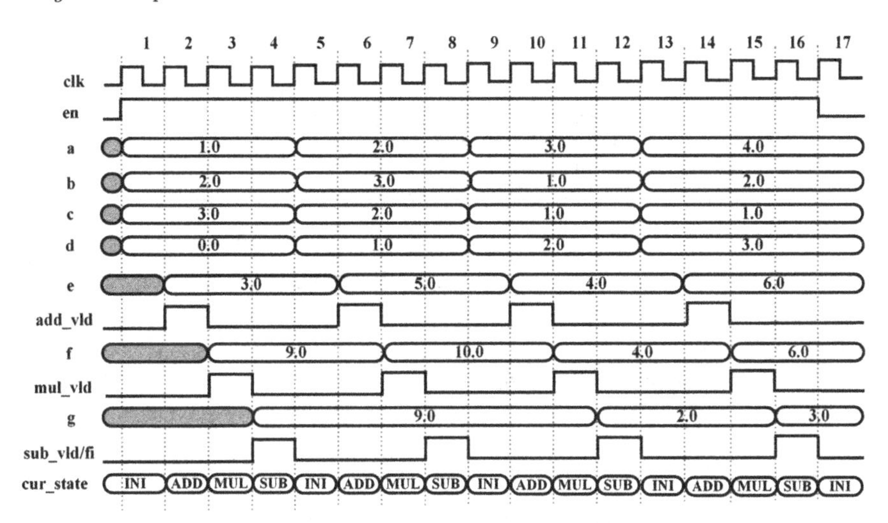

FIGURE 8.31
Timing diagram of FP Add-Mul-Sub with FSMD.

asserted "fi" output. It's important to note that the enable signal intentionally deasserts after receiving the fourth finish indicator, signaling the completion of the final round of operations.

The figure can be further clarified with additional details. The enable signal, represented as "en", signifies the readiness of input data for the FP design engine. This process initiates in the first clock cycle and persists until the completion of the final data package processing. Despite detecting three "fi" signals in clock cycles 4, 8, and 12, indicating the completion of data processing, the "en" signal remains active due to the transmission of four consecutive data packages on the input bus. Furthermore, it's essential to underscore that input data updates only occur upon detecting the finish indicator for each data package processing cycle. Specifically, these updates coincide with clock cycles 5, 9, and 13.

Starting from the first clock cycle, the state machine enters its initial state, INI, and then proceeds sequentially through ADD, MUL, and SUB, covering the entire datapath. This step-by-step process unfolds over clock cycles 1 to 4, resulting in the production of the final output for the first data packet in clock cycle 4. This same procedural pattern is applied to subsequent data packets. For each new data packet, the state machine cycles through its initial states, and the validation signals indicate the progression of operations across clock cycles. Consequently, the validation of the second, third, and fourth data outputs occurs on output "g" during clock cycles 8, 12, and 16, respectively.

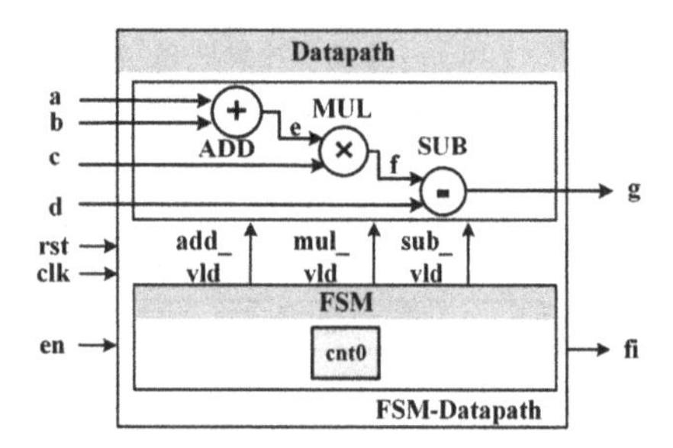

FIGURE 8.32
FP Add-Mul-Sub design with pipeline FP operators.

PBL 19: FP Add-Mul-Sub Design with Pipeline FP Operators

1) As illustrated in Figure 8.32, redesign the circuit in PBL 18 by integrating the pipeline-designed FP operators provided in this book.

   To elaborate, employ pipelined versions of the FP adder, multiplier, and subtractor to process data over the following clock cycles: 13 clock cycles for addition, 10 clock cycles for multiplication, and 13 clock cycles for subtraction, all the way from inputs to outputs. In the timing diagram depicted in Figure 8.33, therefore, the datapath for each set of data inputs spans 37 clock cycles, including one cycle for data input, 13 cycles for FP addition, 10 cycles for FP multiplication, and 13 cycles for FP subtraction. The complete data processing for four sets of data inputs encompasses 148 clock cycles. Rather than utilizing an FSM, you can efficiently control timing by implementing a counter design since the timing control is primarily determined by clock latency.

2) Create a testbench for conducting simulations of the circuit design, and evaluate its functionality using the provided direct test case involving four sets of data inputs in Figure 8.33.

PBL 20: Pipeline FP Add-Mul-Sub Design

1) Redesign the circuit in PBL 19 by implementing a pipeline design structure as depicted in Figure 8.34. Specifically, utilize a pipeline approach to feed four data sets into the design engine, following the timing diagram illustrated in Figure 8.35. The datapath for each

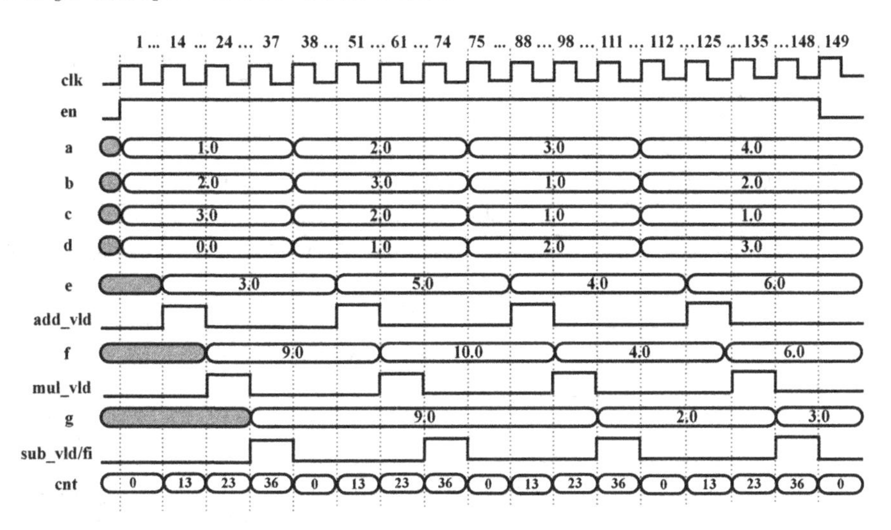

FIGURE 8.33
Timing diagram of FP Add-Mul-Sub design with pipeline FP operators.

set of data inputs spans 37 clock cycles, comprising 1 cycle for data input, 13 cycles for FP addition, 10 cycles for FP multiplication, and 13 cycles for FP subtraction. The complete data processing for these four sets of data inputs extends over 40 clock cycles, with an additional three clock cycles in a pipelined output manner. To design the timing controller effectively, it will be necessary to employ a minimum of four counters, with each counter responsible for tracking one data set.

2) Create a testbench for conducting simulations of the circuit design, and evaluate its functionality using the provided direct test case involving four sets of data inputs in Figure 8.35.

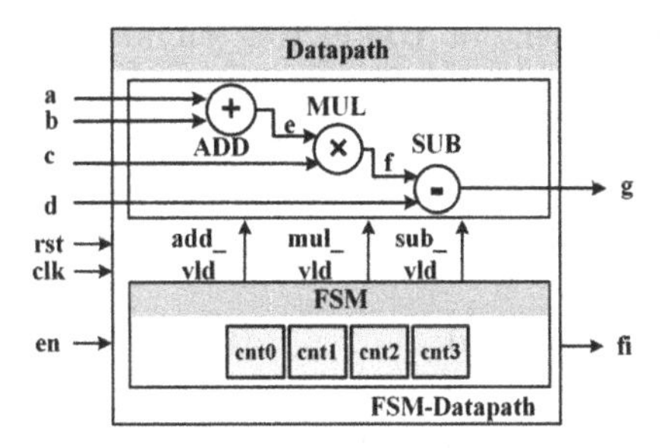

FIGURE 8.34
Pipeline FP Add-Mul-Sub design.

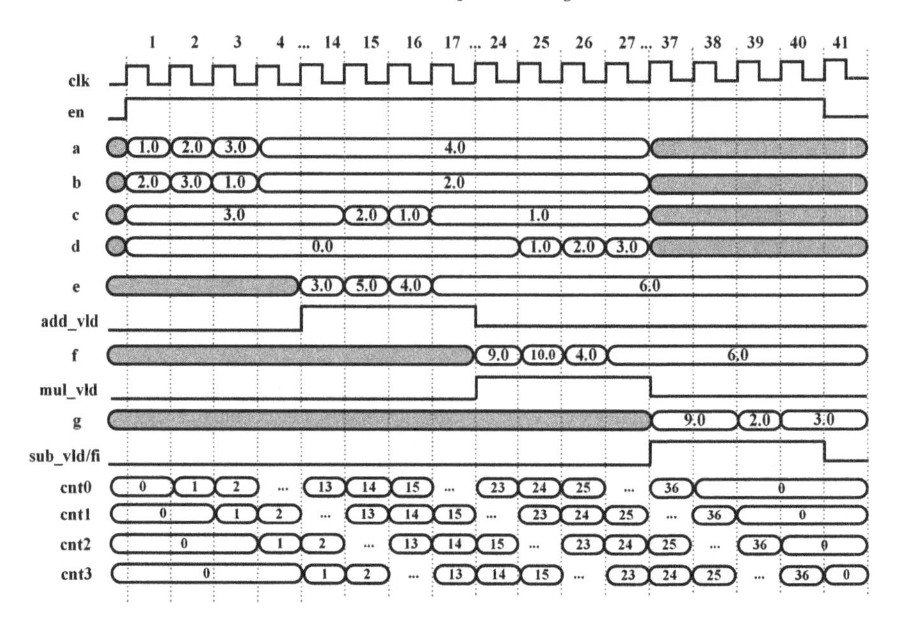

FIGURE 8.35
Timing diagram of pipeline FP Add-Mul-Sub design.

PBL 21: Parallel and Pipeline FP Add-Mul-Sub Design

1) Redesign the circuit in PBL 20 by implementing a parallel design structure. In more detail, employ the dual-core design approach to feed the four data sets into the design engine, as exemplified in the design structure shown in Figure 8.36(a). When considering the timing diagram, assess the number of clock cycles required for data processing in this parallel setup.

2) Create a testbench for conducting simulations of the dual-core design, and evaluate its functionality using the provided direct test case involving four sets of data inputs in Figure 8.35. It's worth noting that the dual-core design requires two clock cycles to input the four sets of data, in contrast to the four clock cycles needed in PBL 20.

3) Utilize the quad-core design approach to feed the four data sets into the design engine, as exemplified in the design structure shown in Figure 8.36(b). When considering the timing diagram, assess the number of clock cycles required for data processing in the quad-core setup.

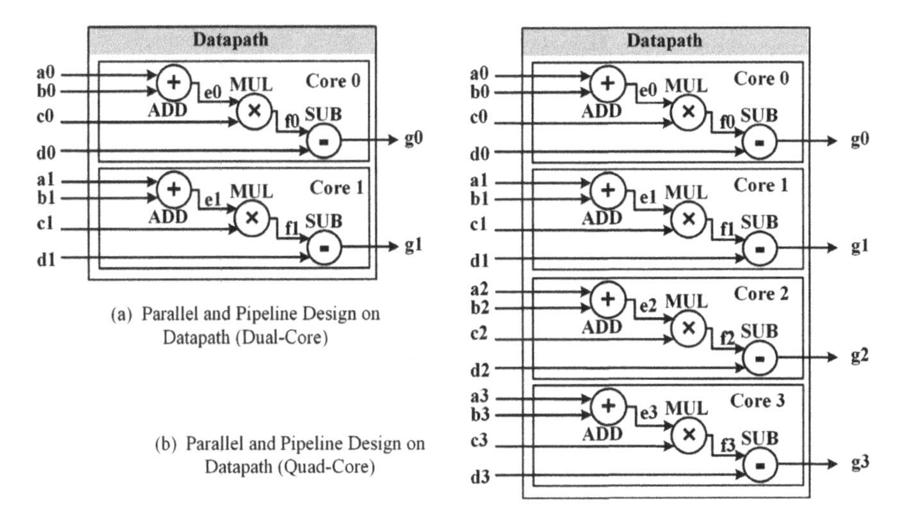

FIGURE 8.36
Parallel FP datapath.

4) Create a testbench for conducting simulations of the quad-core design and evaluate its functionality using the provided direct test case, which involves four sets of data inputs as shown in Figure 8.35. It's noteworthy that the quad-core design requires only one clock cycle to input the four sets of data, in contrast to the four clock cycles needed in PBL 20.

Part II

Advanced IC Design and Integration

9

Numerical Hardware Design and Integration

In the post-Moore's Law era, specialized hardware design emerges as a pivotal pathway for continual performance enhancements in digital electronics and high-performance computing. Therefore, Chapter 9 introduces basic hardware accelerator designs tailored for fundamental mathematical tasks and numerical operations. Optimizing hardware performance requires careful consideration of various factors and trade-offs, including latency, chip size, and power consumption. These constraints necessitate designers to customize and adapt the design structure as needed to meet specific requirements.

PBL 22–23 demonstrate memory access using both signal-port and dual-port register files. PBL 24–27 showcase a range of numerical hardware designs employing floating-point (FP) operators. These hardware implementations encompass matrix-matrix subtraction, matrix-matrix multiplication, and matrix power calculation.

9.1 Design Template for Numerical Hardware

A. Design Architecture

Figure 9.1 illustrates a general design architecture for numerical hardware integration. The design consists of two main components: a datapath and a timing controller. Within the datapath, two memory blocks, labeled as "mem0" and "mem1", are utilized to store the input and output data frames, respectively. The input data stream is stored in "mem0" via the "st_data_in" bus, while the final result is stored in "mem1" and can be accessed through the "st_data_out" bus. Situated between the two memory blocks is the data processing unit constructed using multiple FP operators implemented in parallel and pipeline structures, leading to a lengthy data path. The inputs to this data processing unit are the data read out from "mem0", and the output data from the unit is written back into "mem1".

In the design structure, the timing controller plays a central role in coordinating the timing of the entire datapath. It carries out various critical functions, including the generation of memory read and write commands ("mem_rd_cmd" and "mem_wr_cmd"), initiation of floating-point operations ("datapath_ctl"), monitoring the status of the datapath

DOI: 10.1201/9781003187080-9

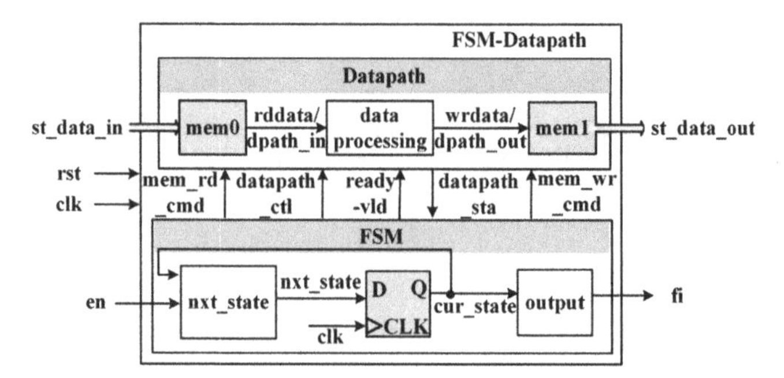

FIGURE 9.1
Design structure of numerical hardware.

("datapath_sta"), and enabling the essential "ready-valid" mechanism ("ready-vld") to signal data input and output within specific clock cycles. This design template showcases the implementation of a finite-state machine (FSM) as the timing controller. However, it's important to note that alternative timing-related designs, such as counters, can also be employed to oversee data processing across clock cycles. Additionally, to facilitate the entire FSM-Datapath circuit, an "enable-finish" handshaking signal ("en-fi") is generated to initiate and conclude the data processing.

B. Timing Diagram

Figure 9.2 provides a typical timing diagram featuring three timing control operations or data channels.

- The **Mem0 Read & Datapath In** channel serves a dual purpose: it reads data from the memory block labeled "mem0" through the "rddata" output and simultaneously transfers this data into the data processing unit via the "dpath_in" bus. To initiate the memory read operation, the "mem_rd_cmd" command is activated over a span of four command cycles, denoted as "C0-C3", allowing the retrieval of four data beats during clock cycles 1–4. Subsequently, one cycle later, the corresponding data, labeled "I0-I3", becomes accessible via the "rddata/dpath_in" bus and is ready to serve as input for the data processing unit.
- During the data processing phase, the results are generated and outputted across several clock cycles. Concurrently, these output data points are stored in memory labeled "mem1" via the **Mem1 Write & Datapath Out** channel. In this specific scenario, the lengthiest data path within the data processing unit requires one clock cycle. Consequently, the results of data processing are produced during clock cycles 3–6, which adds one clock cycle to the data input. During this period, the output data is also simultaneously written into "mem1". The commands for writing data are designated

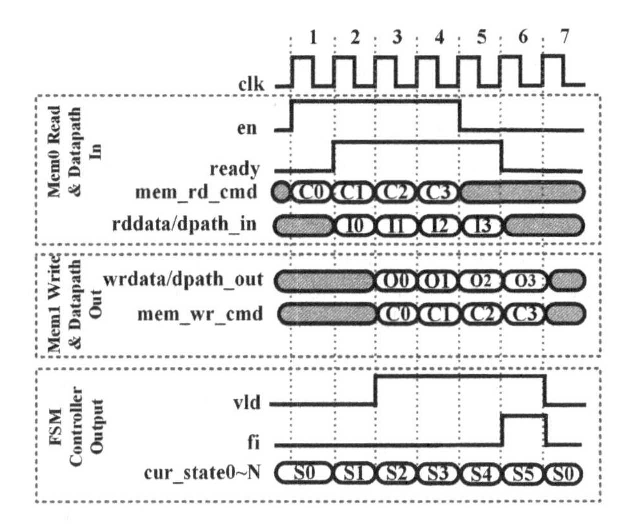

FIGURE 9.2
Timing diagram of numerical hardware.

as "C0-C3" and are transmitted through the "mem_wr_cmd" bus. Correspondingly, the data being written into memory is denoted as "O0-O3" and is conveyed via the "wrdata/dpath_out" bus.

- The FSM controller's responsibilities encompass updating read and write commands for the memories, managing the "ready-vld" mechanism for the datapath, and signaling the completion of data processing using the "fi" indicator. To break this down further, in states "S0-S3", the controller sends memory read commands through the "mem_rd_cmd" bus, specifically using commands "C0-C3". During states "S1-S4", it indicates the readiness of data input to the datapath. In states "S2-S5", the controller sends memory write commands via the "mem_wr_cmd" bus, while also activating the "vld" signal. Finally, in state "S5", the controller concludes the data processing by asserting the "fi" signal. In summary, the FSM controller proceeds through six machine states denoted as "S0-S5" to complete the entire data processing cycle, as illustrated in the **FSM Controller Output** channel.

9.2 Register Files

9.2.1 RTL design on register files

On-chip memories play a crucial role in system design and integration, serving as storage repositories for data across various applications. Among the available memory options, custom register files hold a significant position in RTL design. They function both as simulation models for random access memories (RAMs) and as synthesizable implementations of register arrays. However, it's essential to recognize that register files come at a high cost in terms of silicon real estate and are not intended for storing large data sets on ASICs. Consequently, once a fabrication technology is established, it becomes feasible to replace large-sized register files with synthesizable RAM blocks. This transition allows for their seamless integration into the final system-on-chip (SoC) design and simplifies synthesis alongside other system components.

When dealing with FPGAs, the use of large-sized register files can present challenges because of constraints on available resources like flip-flops and registers within the FPGA device. To address this issue, FPGA vendors offer pre-designed Block RAM or Ultra RAM IP cores. These IP cores can be readily integrated into SoC designs, simplifying the development process and ensuring efficient utilization of available FPGA resources.

In conclusion, within the domain of IC design and verification, register files play a crucial role in emulating memory behavior and ensuring its compliance with the specified timing and functional requirements. This section has introduced three distinct register file designs, which can be synthesized into register arrays or substituted with Block RAMs during the synthesis phase. This synthesis process can be carried out using tools like AMD Vivado® for FPGA design or Synopsys Design Compiler® for ASIC design.

9.2.2 Single-port register file

A. Design Specification

Figure 9.3 illustrates a typical design block of a single-port register array, commonly used for memory storage in digital circuits. The IO ports of the register array consist of the following:

- **Memory write/read enable signal "en":** This signal enables the write (binary one) or read (binary zero) operation of the register array.
- **Write enable signal "we":** This signal controls whether the corresponding data byte within the write data bus (multi-byte) is written into the register array.

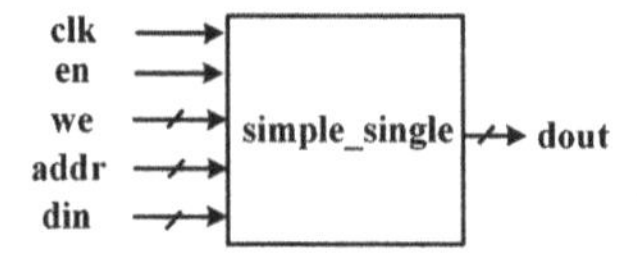

FIGURE 9.3
Block diagram of single-port register file.

- **Write/read address "addr":** This signal specifies the address at which the data is to be written or read.
- **Data input buses "din":** This signal carries the input data to be written into the register array.
- **Data output buses "dout":** This signal provides the output data read from the specified address.

It is important to highlight that the bit width of the write enable, write/read address, data input, and data output buses can be parameterized to accommodate various memory designs. This flexibility allows for customization to meet specific design requirements and optimize memory usage.

Figure 9.4 depicts write operations performed on a register array comprising four 32-bit registers. Each of these registers is accessible via the write/read address ("addr"), which takes on hexadecimal values of 2'h0, 2'h1, 2'h2, and 2'h3. It's important to note that these addresses correspond to the register file's internal addressing and do not increment by four, as each register size is configured to one word. These examples serve to illustrate how the write enable signal "we" plays a pivotal role in selectively determining which bytes of the write data bus are written into the register array, contingent upon their corresponding address.

- In Figure 9.4(a), we observe that the 4-bit write enable signal, denoted as "we[3:0]", has been configured to the hexadecimal value 4'hf or its binary equivalent 4'b1111. Each bit within this signal serves to enable a written byte on the write data bus. This specific configuration allows for the simultaneous writing of the entire data word into the designated register.
- In Figure 9.4(b), we observe that the write enable signal "we" has been set to the hexadecimal value 4'h7 or its binary representation, 4'b0111. This particular configuration enables the lower three bytes of the write data bus

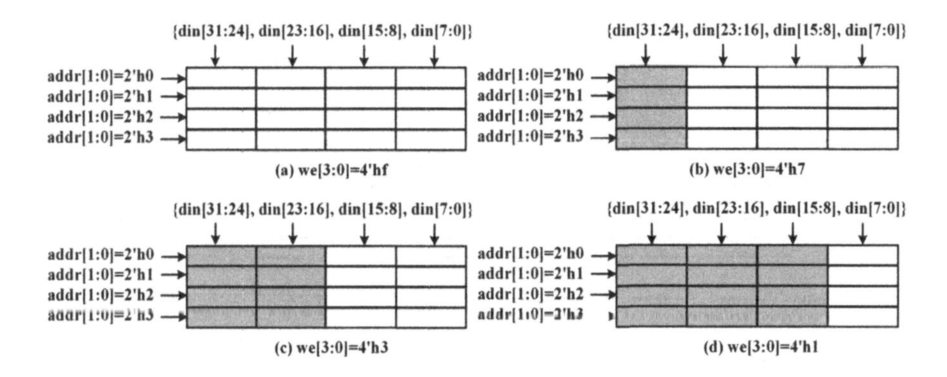

FIGURE 9.4
Register file access utilizing write enable signal (register bytes in white boxes indicate data writing enabling).

for writing. As a result, only the lower three bytes of the write data bus are writable into the register corresponding to the address.

- Similarly, in Figure 9.4(c), the write enable signal "we" is configured to the hexadecimal value 4'h3 or its binary equivalent, 4'b0011. This setting grants write access exclusively to the lower two bytes of the write data bus, allowing only those lower two bytes to be written into the register at the corresponding address.

- Lastly, in Figure 9.4(d), the write enable signal "we" is set to the hexadecimal value 4'h1 or binary 4'b0001. In this setup, only the least-significant byte of the write data bus is enabled for writing. Consequently, solely this particular byte can be written into the register at the corresponding address.

B. Timing Diagram

Figure 9.5 presents two timing diagrams illustrating write operations on a register file. In the first example, depicted in Figure 9.5(a), write operations are enabled during clock cycles 1–4, with corresponding write addresses of 2'h0-2'h3. Four data words, specifically 32'h0123, 32'h4567, 32'h89ab, and 32'hcdef, are written into the registers. Each write operation activates all four bytes of the write data bus ("we=4'hf"). In clock cycles 5–8, read addresses of

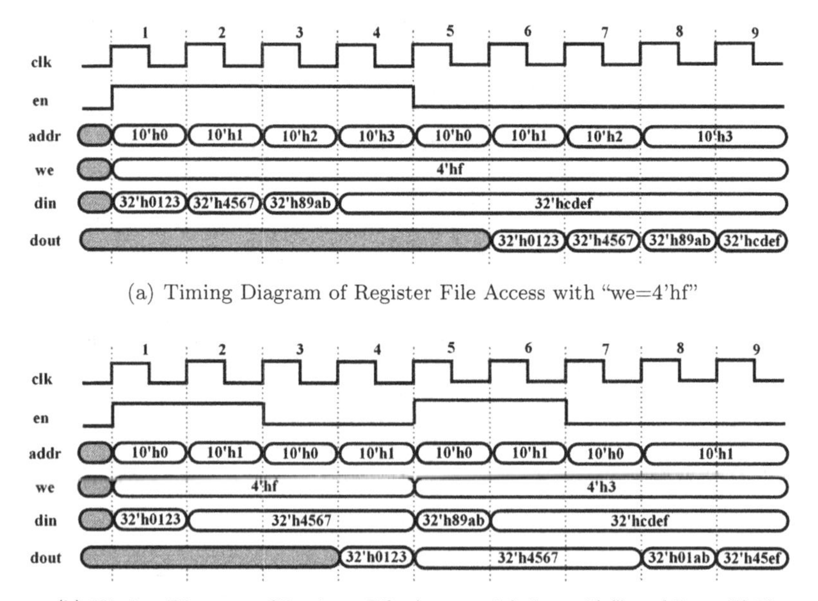

(a) Timing Diagram of Register File Access with "we=4'hf"

(b) Timing Diagram of Register File Access with "we=4'hf" and "we=4'h3"

FIGURE 9.5
Timing diagram of register file access utilizing write enable controlling.

2'h0-2'h3 are generated to verify the successful data write. It's important to note that there is a one-clock cycle delay between the read addresses and the availability of the read data, resulting in the data being read out during clock cycles 6–9.

The second example, depicted in Figure 9.5(b), demonstrates write operations performed in words and half-words to showcase the difference. In clock cycles 1–2, the write enable signal and write addresses are activated. The write data, 32'h0123 and 32'h4567, are written into the registers with the corresponding addresses 2'h0 and 2'h1, respectively, as the write enable signal enables all the write data bytes ("we=4'hf"). This can be observed in the subsequent data read addresses in clock cycles 3–4 and the corresponding data read out on the read data bus in clock cycles 4–5.

In the following sequence, during clock cycles 5–6, another round of data writing takes place. However, this time, only the lower two bytes are enabled with the write enable signal set to hexadecimal 4'h3. Despite having write data values of 32'h89ab and 32'hcdef, only the lower two bytes, 16'hab and 16'hef, are actually written into the register arrays. The higher two bytes, 16'h89 and 16'cd, are disregarded by the register operations.

Subsequently, as illustrated in the ensuing read operations spanning clock cycles 7–8, the corresponding data is retrieved from the memory by transmitting addresses 2'h0 and 2'h1 once again. During clock cycles 8-9, the data retrieved from the register arrays are 32'h01ab and 32'h45ef, with the lower two bytes updated, while the higher two bytes remain unchanged, preserving the data previously stored in the registers.

C. Verilog Design

The following Verilog code represents a register file design for a basic single-port memory. The line **`define WIDTH 32** specifies the data width of each memory location, which is 32 bits or 4 bytes. The line **`define DEPTH 1024** indicates the number of memory cells in the block RAM, resulting in a memory size of $DEPTH \times WIDTH = 1$ K Words. Additionally, the line **`define ADDR_WIDTH 10** specifies the bit width of the memory address. The memory depth and memory address follow the expression: $DEPTH = 1 << ADDR\_WIDTH$, which means that the memory depth $DEPTH$ is determined by left-shifting a binary one by $ADDR\_WIDTH$ bits. In this case, the memory depth is 1024, requiring a memory address width of 10 bits.

In the subsequent lines, the **`define WE_WIDTH 4** indicates that the write enable signal width is four bits. The write enable signals include **`define WE4 `WE_WIDTH'hf** (enabling all four bytes), **`define WE3 `WE_WIDTH'h7** (enabling the lower three bytes), **`define WE2 `WE_WIDTH'h3** (enabling the lower two bytes), and **`define WE1 `WE_WIDTH'h1** (enabling the least-significant byte).

```
1  `define WIDTH 32
2  `define DEPTH 1024
3  `define ADDR_WIDTH 10
4  `define WE_WIDTH 4
5  `define WE4 `WE_WIDTH'hf
6  `define WE3 `WE_WIDTH'h7
7  `define WE2 `WE_WIDTH'h3
8  `define WE1 `WE_WIDTH'h1
9  module simple_single(input                          clk  ,
10                     input                          en   ,
11                     input      [`WE_WIDTH-1:0]    we   ,
12                     input      [`ADDR_WIDTH-1:0]  addr ,
13                     input      [`WIDTH-1:0]       din  ,
14                     output reg [`WIDTH-1:0]       dout);
15 reg [`WIDTH-1:0] ram[0:`DEPTH-1];
16
17 always @(posedge clk) begin
18   if (en) begin
19     case(we)
20     `WE4: ram[addr]<=din;
21     `WE3: ram[addr]<={ram[addr][`WIDTH-1:`WIDTH-32*1],
22                      din[`WIDTH-32*1-1:0]};
23     `WE2: ram[addr]<={ram[addr][`WIDTH-1:`WIDTH-32*2],
24                      din[`WIDTH-32*2-1:0]};
25     `WE1: ram[addr]<={ram[addr][`WIDTH-1:`WIDTH-32*3],
26                      din[`WIDTH-32*3-1:0]};
27     default: ram[addr]<=`WIDTH'h0;
28     endcase
29   end else begin
30     dout <= ram[addr];
31   end
32 end
33 endmodule
```

In line 15, it specifies that each register location has a size of `WIDTH (32 bits or 4 bytes), and the total number of registers is `DEPTH (from "ram[0]" to "ram[`DEPTH-1]").

The description of write operations is outlined in lines 18–29. When the write enable signal "en" is asserted, as indicated in line 18, it signifies that the corresponding write data bytes present on the "din" bus are eligible for writing into the register file. The determination of which specific bytes to write is contingent upon the state of the write enable signal "we", which is covered in line 19.

If "we" is equivalent to `WE4, as noted in line 20, it implies that all four bytes can be written into the register. On the other hand, if "we" is set to `WE3, only the lower three bytes are writable, while the most significant byte retains the data stored in the register. This conditional scenario is detailed in lines 21–22. Similarly, when "we" is equal to `WE2, outlined in lines 23–24, or `WE1, as described in lines 25–26, only the lower two bytes or the least significant byte can be written into the register, respectively. This approach allows for precise control over which portions of the register are modified during the write operation.

The memory address labeled as "addr" is utilized for both the write and read channels, given that only a single write or read operation can be executed at any given time. In lines 29–31, the description for register reading is presented, including the associated address transmitted on the "addr" bus.

It's crucial to emphasize that input data can be written into the memory precisely one clock cycle following the issuance of the write command. Similarly, output data can be read from the output bus exactly one clock cycle after the read command is initiated.

9.2.3 Dual-port register file

A. Design Specification

Figure 9.6 illustrates a typical design block of a dual-port register array. This type of memory allows for simultaneous read and write operations through separate IOs: port "a" for write operations and port "b" for read operations. The IOs of the register array include:

- **Memory write enable signal "ena":** This signal enables the write operation of the register array.
- **Write enable signal "wea":** This signal controls whether the corresponding data byte within the write data bus (multi-byte) is written into the register array.
- **Write address "addra":** This signal specifies the address at which the data is to be written.

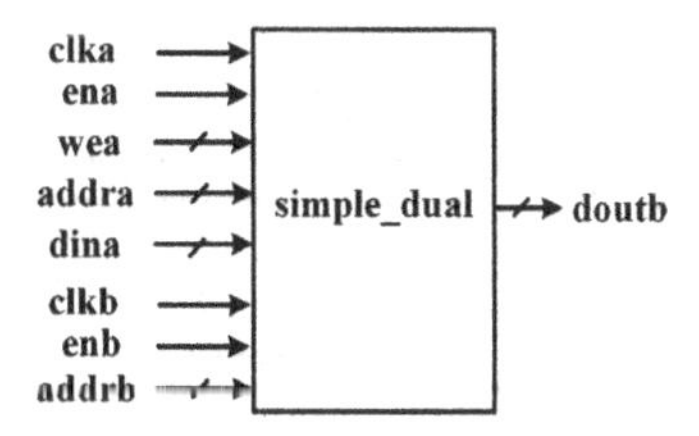

FIGURE 9.6
Block diagram of dual-port register file.

- **Data input buses "dina":** This signal carries the input data to be written into the register array.
- **Memory read enable signal "enb":** This signal enables the read operation of the register array.
- **Write/read address "addrb":** This signal specifies the address from which the data is to be read.
- **Data output buses "doutb":** This signal provides the output data read from the specified address.

These signals allow for independent read and write operations on the dual-port register array, enabling efficient data storage and retrieval.

B. Timing Diagram

Figure 9.7 presents a timing diagram that showcases the utilization of a dual-port register array. This diagram emphasizes the simultaneous read and write operations made possible by the dual-port architecture. The "clka" and "clkb" signals represent synchronous clock signals for simplicity.

During clock cycles 1 to 4, data is sequentially written into memory addresses 2'h0 to 2'h3. Each clock cycle corresponds to a write operation, enabling multiple data words to be simultaneously written into the register array. Subsequently, in clock cycles 3 to 6, the corresponding data previously written into the register file is read out. Thanks to the utilization of a dual-port memory architecture, the read operations can be performed in parallel with the ongoing write operations, with two clock cycles 3–4 overlapped for both write and read accesses.

Importantly, it should be noted that in the specific scenario depicted in the timing diagram, the read operation for address 2'h0 occurs two clock cycles after the write operation to the same address. This delay is attributed to the

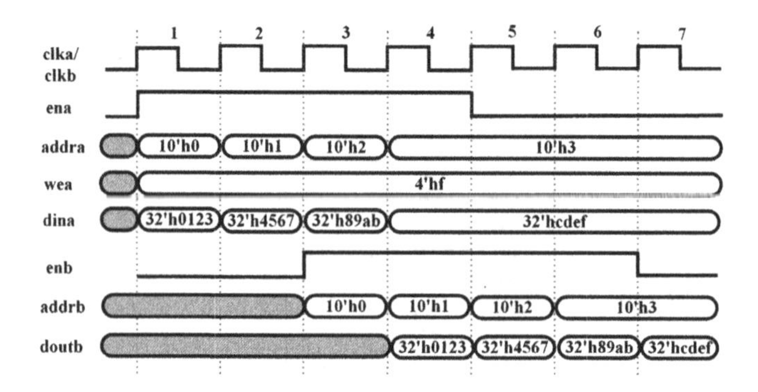

FIGURE 9.7
Timing diagram of dual-port register file access.

internal operation of the register array, which necessitates at least one clock cycle for the data to be written and stored before it can be successfully read out.

Additionally, it is crucial to highlight that simultaneous write and read operations to the same address within the same clock cycle are not permitted in a dual-port register array. This restriction is in place to prevent address collision issues that could result in unpredictable data corruption or erroneous behavior.

C. Verilog Design

The following Verilog code represents a register file design for a basic dual-port memory. In this design, the write-related IOs are defined in lines 9–13 using the port name "a". These signals include the clock signal "clka", port "a" enable signal "ena", write enable signal "wea", write address signal "addra", and data input signal "dia". Similarly, the read-related IO signals are declared in lines 14–17 using the port name "b". These signals consist of the clock signal "clkb", port "b" enable signal "enb", read address signal "addrb", and data output signal "dob". It's noteworthy that the clocks "clka" and "clkb" may be asynchronous, allowing the two ports to operate at different data speeds. This flexibility enables the dual-port memory to accommodate varying data processing requirements for each port independently.

```
1  `define WIDTH 32
2  `define DEPTH 1024
3  `define ADDR_WIDTH 10
4  `define WEA_WIDTH 4
5  `define WEA4 `WEA_WIDTH'hf
6  `define WEA3 `WEA_WIDTH'h7
7  `define WEA2 `WEA_WIDTH'h3
8  `define WEA1 `WEA_WIDTH'h1
9  module simple_dual(input                                clka ,
10                    input                                ena  ,
11                    input      [`WEA_WIDTH-1:0]          wea  ,
12                    input      [`ADDR_WIDTH-1:0]         addra,
13                    input      [`WIDTH-1:0]              dina ,
14                    input                                clkb ,
15                    input                                enb  ,
16                    input      [`ADDR_WIDTH-1:0]         addrb,
17                    output reg [`WIDTH-1:0]              doutb);
18 reg [`WIDTH-1:0] ram[0:`DEPTH-1];
19
20 always @(posedge clka) begin
21   if (ena) begin
22     case(wea)
23     `WEA4: ram[addra]<=dina;
```

```
24        `WEA3: ram[addra]<={ram[addra][`WIDTH-1:`WIDTH-32*1],
25                            dina[`WIDTH-32*1-1:0]};
26        `WEA2: ram[addra]<={ram[addra][`WIDTH-1:`WIDTH-32*2],
27                            dina[`WIDTH-32*2-1:0]};
28        `WEA1: ram[addra]<={ram[addra][`WIDTH-1:`WIDTH-32*3],
29                            dina[`WIDTH-32*3-1:0]};
30        default: ram[addra]<=`WIDTH'h0;
31        endcase
32    end
33  end
34
35  always @(posedge clkb) begin
36    if (enb) begin
37        doutb <= ram[addrb];
38    end
39  end
40  endmodule
```

The dual-port memory facilitates simultaneous write and read operations through two distinct IO ports, "a" and "b". The write operation is implemented within the *always* block from lines 20 to 33, while the read operation is carried out in the *always* block from lines 35 to 39. Address collisions can pose a significant concern when both *always* blocks attempt to access the same memory location concurrently, resulting in unpredictable outcomes during read operations. Careful synchronization and control mechanisms should be employed to avoid any potential conflicts in accessing the same memory addresses at the same time.

9.2.4 Ping-pong buffer

A. Design Specification

To enhance memory access efficiency and prevent address collisions, the ping-pong buffer architecture can be employed in data processing units. This architecture comprises two single-port memories that can perform read and write operations concurrently by alternating between them. The switching between these memories allows for transitions between input and output modes.

Figure 9.8 presents a block diagram illustrating a ping-pong buffer design comprising two single-port memories. The control signals, namely "u0_wr" and "u0_rd", are responsible for overseeing write and read access to the "u0_mem" memory. Correspondingly, the signals "u1_wr" and "u1_rd" govern write and read access to the "u1_mem" memory. Typically, the generation of these control signals falls under the purview of memory controller designs, which govern the timing and synchronization of read and write operations.

When "u0_rd" is asserted, indicating active reading or data processing in "u0_mem", it facilitates concurrent execution of "u1_wr" to update data in

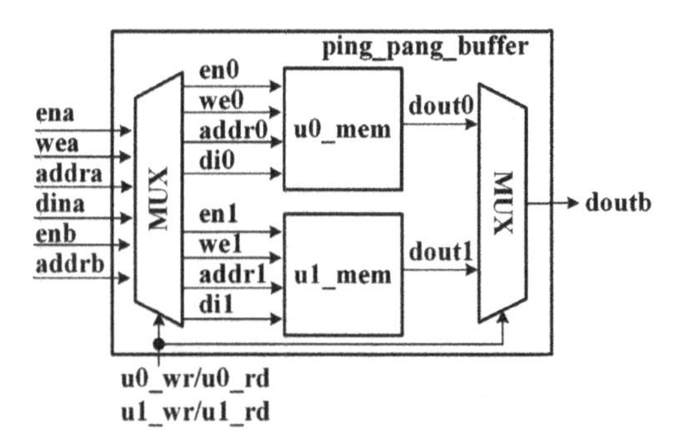

FIGURE 9.8
Block diagram of ping-pong buffer.

"u1_mem". Conversely, when "u1_rd" is asserted, signaling a read operation and data processing in "u1_mem", it enables simultaneous data feeding into "u0_mem" with an asserted "u0_wr". This ping-pong buffer design optimizes memory access management for data processing units, effectively mitigating conflicts during read and write operations.

B. Verilog Design

The provided Verilog code represents a ping-pong buffer design that uses two single-port memories to perform concurrent read and write operations. The first-stage multiplexer (lines 14–24) determines which input signals are directed to "u0_mem" and "u1_mem" based on the control signals "u0_wr", "u0_rd", "u1_wr", and "u1_rd". When a memory write or read operation occurs, the corresponding signals (memory write enable "ena", write data enable "wea", memory write address "addra", and memory write data "dina", memory read enable "enb", and memory read address "addra") are selected and routed to the appropriate memory module.

The second-stage multiplexer (lines 26–29) selects the output "doutb" from either the "dout0" bus or the "dout1" bus depending on whether "u0_mem" or "u1_mem" is chosen. The "simple_single" modules (instantiated in lines 31–42) represent the two single-port memory modules, "u0_mem" and "u1_mem", which operate with the selected signals accordingly.

```
1  `include "simple_single.v"
2  module ping_pong_buffer (input                      clk   ,
3                          input                      u0_wr ,
4                          input                      u1_wr ,
5                          input                      u0_rd ,
6                          input                      u1_rd ,
7                          input                      ena   ,
```

```
8                              input   [`WE_WIDTH-1:0]    wea   ,
9                              input   [`ADDR_WIDTH-1:0]  addra ,
10                             input   [`WIDTH-1:0]       dina  ,
11                             input                      enb   ,
12                             input   [`ADDR_WIDTH-1:0]  addrb ,
13                             output  [`WIDTH-1:0]       doutb);
14 wire en0 = (u0_wr&ena) | ~(u0_rd&enb) ;
15 wire en1 = (u1_wr&ena) | ~(u1_rd&enb) ;
16 wire [`WE_WIDTH-1:0] we0 = u0_wr ? wea : `WEA4;
17 wire [`WE_WIDTH-1:0] we1 = u1_wr ? wea : `WEA4;
18
19 wire [`ADDR_WIDTH-1:0] addr0 = u0_wr ? addra :
20                                u0_rd ? addrb : `ADDR_WIDTH'b0;
21 wire [`ADDR_WIDTH-1:0] addr1 = u1_wr ? addra :
22                                u1_rd ? addrb : `ADDR_WIDTH'b0;
23 wire [`WIDTH-1:0]      din0 = u0_wr ? dina  : `WIDTH'b0    ;
24 wire [`WIDTH-1:0]      din1 = u1_wr ? dina  : `WIDTH'b0    ;
25
26 wire [`WIDTH-1:0]      dout0 ;
27 wire [`WIDTH-1:0]      dout1 ;
28 assign doutb = u0_rd ? dout0 :
29                        u1_rd ? dout1 : `ADDR_WIDTH'h0;
30
31 simple_single u0_mem (.clk (clk   ),
32                       .en  (en0   ),
33                       .we  (we0   ),
34                       .addr(addr0),
35                       .din (din0 ),
36                       .dout(dout0));
37 simple_single u1_mem (.clk (clk   ),
38                       .en  (en1   ),
39                       .we  (we1   ),
40                       .addr(addr1),
41                       .din (din1 ),
42                       .dout(dout1));
43 endmodule
```

The ping-pong buffer design effectively manages memory access and mitigates address collisions by leveraging two single-port memories in an alternating fashion. This design enables simultaneous read and write operations and offers flexibility to data processing units, facilitating efficient data transfer and storage. The ping-pong buffer has found extensive applications in bus wrappers, data movement utilizing DMA (Direct Memory Access), and various data processing units.

9.3 Design Example: FP Matrix-Matrix Adder

In this section, our primary focus is on designing numerical hardware by leveraging FP operators and memory blocks. We will commence by exploring the foundational concept of FP matrix addition, which can be constructed using the design template outlined in Section 9.1. This design integrates memory blocks, FP adders, and a timing controller to achieve its objectives.

Mathematically, the equation for a 4×4 matrix addition can be expressed as follows:

$$\begin{aligned}
\begin{pmatrix} z_{00} & z_{01} & z_{02} & z_{03} \\ z_{10} & z_{11} & z_{12} & z_{13} \\ z_{20} & z_{21} & z_{22} & z_{23} \\ z_{30} & z_{31} & z_{32} & z_{33} \end{pmatrix} &= \begin{pmatrix} x_{00} & x_{01} & x_{02} & x_{03} \\ x_{10} & x_{11} & x_{12} & x_{13} \\ x_{20} & x_{21} & x_{22} & x_{23} \\ x_{30} & x_{31} & x_{32} & x_{33} \end{pmatrix} + \begin{pmatrix} y_{00} & y_{01} & y_{02} & y_{03} \\ y_{10} & y_{11} & y_{12} & y_{13} \\ y_{20} & y_{21} & y_{22} & y_{23} \\ y_{30} & y_{31} & y_{32} & y_{33} \end{pmatrix} \quad (9.1) \\
&= \begin{pmatrix} x_{00}+y_{00} & x_{01}+y_{01} & x_{02}+y_{02} & x_{03}+y_{03} \\ x_{10}+y_{10} & x_{11}+y_{11} & x_{12}+y_{12} & x_{13}+y_{13} \\ x_{20}+y_{20} & x_{21}+y_{21} & x_{22}+y_{22} & x_{23}+y_{23} \\ x_{30}+y_{30} & x_{31}+y_{31} & x_{32}+y_{32} & x_{33}+y_{33} \end{pmatrix}
\end{aligned}$$

In this equation, each element from matrix x is added to its corresponding element from matrix y, and the sum is stored in the same location in the new matrix z. From a hardware perspective, matrices x and y can be stored in two memory blocks, and the resulting matrix z can be written into another memory block.

To implement this functionality, three register arrays are implemented to store the matrices. As two arrays are used only for reading, and one is used only for writing, single-port memory blocks are utilized. This configuration ensures that read and write operations do not overlap, preventing any potential address collision issues and enabling concurrent access to the matrices with high efficiency.

9.3.1 Design structure of FP matrix-matrix adder

As previously mentioned, the design structure is depicted in Figure 9.9, featuring three separate register arrays designated for the storage of matrices x, y, and z. These matrices adhere to an organization in which columns are stored in memory row-wise, meaning the first matrix column resides in the first memory row, the second matrix column in the second memory row, and so forth. Given a data bus width of 128 bits, or four words, one memory row can be read out from "mem0" and/or "mem1" in each clock cycle. Likewise, the associated summation result of each memory row can be efficiently written into "mem2" within each clock cycles. This arrangement enables parallel

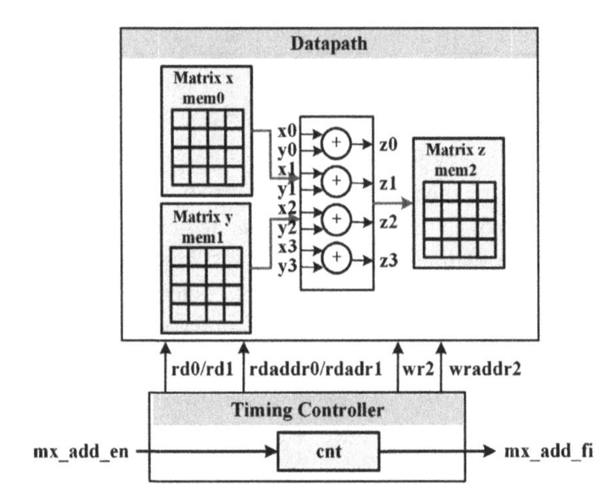

FIGURE 9.9
Design structure of FP matrix-matrix adder.

and concurrent read operations from "mem0" and "mem1" and simultaneous write operations to "mem2", maximizing the memory access efficiency during matrix addition.

The data processing unit partitions each row of data (128-bit) from both "mem0" and "mem1" into four words, with each word routed to an individual FP adder. The data is segmented in such a way that the most significant word is directed to the first FP adder via the "x0/y0" bus, the least significant word via the "x3/y3" bus, and the two intermediary words via the "x1/y1" and "x2/y2" buses. Conversely, the four FP outputs produced by the four FP adders are combined into a single 128-bit data stream, which is then intended to be written back into each respective row within "mem2". Executing a 4×4 matrix addition necessitates four iterations of this process. Each iteration encompasses memory read operations from "mem0" and "mem1", FP additions carried out in parallel, and memory write operation into "mem2".

For the data processing over multiple clock cycles, a timing controller is required to ensure that memory accesses and FP operations are executed in the correct time slots. This can be accomplished by utilizing either an FSM or a simple counter. In the upcoming section, the timing diagram for the matrix addition will be discussed, which involves a total of six clock cycles. These time slots include reading four data rows from memories, performing four rounds of 128-bit FP additions, and writing the results back into memory. To regulate the timing over these six clock cycles, a three-bit counter labeled "cnt" is employed. This counter assists in coordinating the sequential execution of various operations and ensures proper synchronization between the memory accesses and FP additions, enabling a successful matrix addition process.

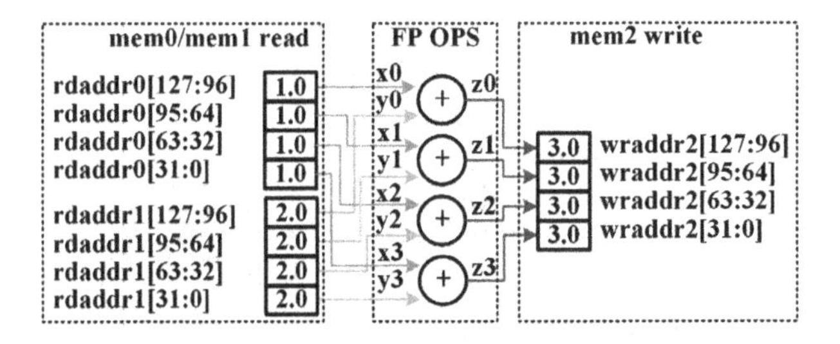

FIGURE 9.10
Data processing example of FP matrix-matrix adder.

9.3.2 Timing diagram of FP matrix-matrix adder

A. Assumption for Test Cases

Before we proceed with the timing diagram, let's simplify matters by assuming that all four word-sized data within each memory row are identical. For instance, in Figure 9.10, the four word-sized data in "rddata0[127:96]", "rddata0[95:64]", "rddata0[63:32]", and "rddata0[31:0]" are all identical (all set to FP 1.0 in this example), as are the four word-sized data in "rddata1[127:96]", "rddata1[95:64]", "rddata1[63:32]", and "rddata1[31:0]" (all set to FP 2.0 in the figure). Consequently, the data on the "x0-x3" buses for all four adders are identical, and similarly, the data on the "y0-y3" buses are also identical. This results in the summations "z0-z3" being the same as well. Therefore, all four word-sized data on "wrdata2[127:96]", "wrdata2[95:64]", "wrdata2[63:32]", and "wrdata2[31:0]" will also be identical. By making this assumption, we can simplify the timing diagram to display only one set of data buses: "rddata0[127:96]/x0", "rddata1[127:96]/y0", and "wrdata2[127:96]/z0". This simplification streamlines the diagram while preserving the core concept.

Additionally, in the realm of hardware design, it's imperative to represent single-precision numbers according to the IEEE 754 format. In this context, FP 1.0 is appropriately represented as 32'h3F800000, FP 2.0 as 32'h40000000, and FP 3.0 as 32'h40400000. To ensure clarity and simplicity, all FP numbers in the ensuing timing diagrams are illustrated using real data types, directly explaining FP computations.

Throughout all forthcoming examples of numerical hardware design, we maintain this consistent assumption, as it exerts no impact on the functionality of the hardware design. This simplification, however, serves to facilitate a more straightforward and uncomplicated analysis of the timing diagrams.

B. Timing Diagram of FP Matrix-Matrix Adder

Figure 9.11 illustrates the timing diagram for the FP 4×4 matrix addition process. In the **Mem0&1 Read** channel, the signals "rd0" and "rd1" are asserted during clock cycles 1–4 to initiate memory read operations. The memory

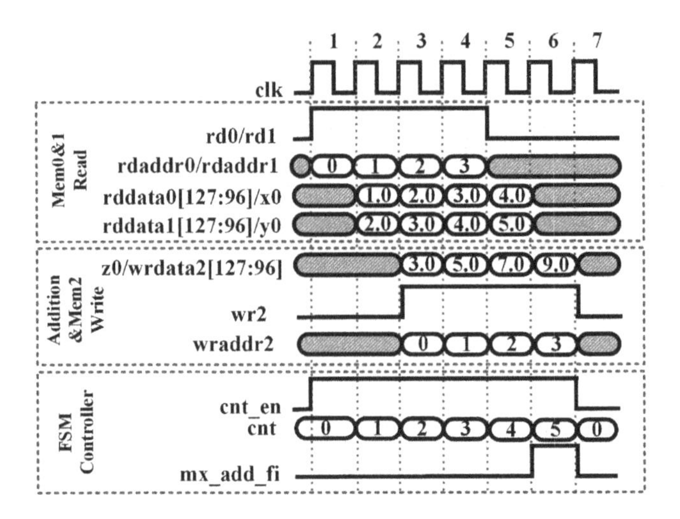

FIGURE 9.11
Timing diagram of FP matrix-matrix adder.

addresses for reading are specified as "rdaddr0" and "rdaddr1", with hexadecimal values ranging from 2'h0 to 2'h3, allowing access to four rows in "mem0" and "mem1", respectively. Following a one-clock-cycle delay, the corresponding data becomes available on the read data buses, namely "rddata0" for "mem0" and "rddata1" for "mem1". These identical data values are then directed into the FP operators via the "x0-x3" and "y0-y3" buses.

For the sake of simplifying timing analysis, we assume that each FP operator introduces a one-cycle delay. In the **Addition&Mem2 Write** channel, the summations are ready to be transmitted via the "z0-z3" buses in clock cycles 3–6. These four FP data results are combined to create a four-word-sized data on the "wrdata2" bus, which is subsequently stored in "mem2". For the memory write operation to proceed correctly, it is essential that both the write enable signal "wr2" and the write address "wraddr2" are updated simultaneously during the same clock cycles.

In the final **FSM Controller** channel, timing control is achieved through the use of a counter that counts from decimal 3'd0 to 3'd5. During the last cycle of this counting sequence, the timing controller sets the "mx_add_fi" indicator to signal the completion of the matrix addition process. This indicator plays a crucial role in signaling the status of the operation and ensuring synchronization with subsequent processes or applications that rely on the results of the matrix addition.

9.3.3 Verilog code for FP matrix-matrix adder

The Verilog description for the FP matrix-matrix adder design is provided below. It includes the clock "clk" and reset "rst", along with the input "mx_add_en" to enable the matrix addition and the output "mx_add_fi" to indicate its completion.

Lines 15–38 instantiate four FP adders. In lines 40–60, three memory blocks are utilized to store the matrices: "u0_mem" for matrix x, "u1_mem" for matrix y, and "u2_mem" for matrix z. Each 128-bit FP data read from "u0_mem" is divided into four words, shown in line 46, which are then used as inputs "x0", "x1", "x2", and "x3" for the four 32-bit FP adders. Similarly, each 128-bit FP data read from "u1_mem" is divided into four words, illustrated in line 53, serving as inputs "y0", "y1", "y2", and "y3" for the 32-bit FP adders. The outputs of the four 32-bit FP adders, "z0-z3", are combined to form the 128-bit input for "u2_mem2" as depicted in line 59.

```
1  module mx_add (input          clk        ,
2                 input          rst        ,
3                 input          mx_add_en  ,
4                 output         mx_add_fi  );
5  //*******************************
6  //****    Datapath Design *********
7  //*******************************
8  wire [31:0] x0, x1, x2, x3;
9  wire [31:0] y0, y1, y2, y3;
10 wire [31:0] z0, z1, z2, z3;
11
12 wire        wr2, rd0, rd1;
13 reg  [1:0]  wraddr2, rdaddr0, rdaddr1;
14
15 // FP adders instantiation
16 FP_adder u0_fp_add(.clock    (clk  ),
17                    .reset    (rst  ),
18                    .io_in_a  (x0   ),
19                    .io_in_b  (y0   ),
20                    .io_out_s (z0   ));
21
22 FP_adder u1_fp_add(.clock    (clk  ),
23                    .reset    (rst  ),
24                    .io_in_a  (x1   ),
25                    .io_in_b  (y1   ),
26                    .io_out_s (z1   ));
27
28 FP_adder u2_fp_add(.clock    (clk  ),
29                    .reset    (rst  ),
30                    .io_in_a  (x2   ),
```

```
31                     .io_in_b (y2   ),
32                     .io_out_s(z2   ));
33
34 FP_adder u3_fp_add(.clock   (clk  ),
35                     .reset   (rst  ),
36                     .io_in_a (x3   ),
37                     .io_in_b (y3   ),
38                     .io_out_s(z3   ));
39
40 // memory instantiation
41 simple_single u0_mem(.clk  (clk          ) ,
42                       .en   (rd0          ) ,
43                       .we   (`WEA4        ) ,
44                       .addr (rdaddr0      ) ,
45                       .di   (128'h0       ) ,
46                       .dout ({x0,x1,x2,x3}));
47
48 simple_single u1_mem(.clk  (clk          ) ,
49                       .en   (rd1          ) ,
50                       .we   (`WEA4        ) ,
51                       .addr (rdaddr1      ) ,
52                       .di   (128'h0       ) ,
53                       .dout ({y0,y1,y2,y3}));
54
55 simple_single u2_mem(.clk  (clk          ) ,
56                       .en   (wr2          ) ,
57                       .we   (`WEA4        ) ,
58                       .addr (wraddr2      ) ,
59                       .di   ({z0,z1,z2,z3}) ,
60                       .dout (             ));
61
62 //**************************************
63 //****    FSM Controller Design *********
64 //**************************************
65 reg  [2:0]  cnt;
66 wire cnt_en = mx_add_en | |cnt;
67 wire [2:0] nxt_cnt=(cnt==3'd5) ? 3'd0 :
68                                  cnt_en ? (cnt+3'd1) : cnt;
69 always @(posedge clk) begin
70   if(rst) begin
71     cnt<=3'd0     ;
72   end else begin
73     cnt<=nxt_cnt;
74   end
75 end
76
77 assign rd0 = mx_add_en ;
78 assign rd1 = mx_add_en ;
79 wire [1:0] nxt_rdaddr0 = rd0 ? rdaddr0+2'h1 : rdaddr0;
```

```
80  wire [1:0] nxt_rdaddr1 = rd1 ? rdaddr1+2'h1 : rdaddr1;
81  always @(posedge clk) begin
82    if(rst) begin
83      rdaddr0 <=2'h0    ;
84      rdaddr1 <=2'h0    ;
85    end else begin
86      rdaddr0 <=nxt_rdaddr0 ;
87      rdaddr1 <=nxt_rdaddr1 ;
88    end
89  end
90
91  assign wr2 = cnt>=3'd2 & cnt<=3'd5 ;
92  wire [1:0] nxt_wraddr2 = wr2 ? wraddr2+2'h1 : 2'h0;
93  always @(posedge clk) begin
94    if(rst) begin
95      wraddr2 <=2'h0         ;
96    end else begin
97      wraddr2 <=nxt_wraddr2 ;
98    end
99  end
100
101 assign mx_add_fi = &wraddr2;
102 endmodule
```

The timing controller design is illustrated in lines 62–101. Initially, a standard counter design that counts from decimal 3'd0 to 3'd5 is presented in lines 65–75. Upon reaching the maximum value of 3'd5, the counter resets to decimal 3'd0. The counter controls the entire datapath over six clock cycles, from 3'd0 to 3'd5. The counter enable signal "cnt_en" is asserted when the matrix-matrix adder is enabled or when the timing controller is "busy", as shown in line 66. The Reduction OR gate, expressed as "|*cnt*", indicates that the counter is non-zero or in a "busy" status.

Memory read and write operations are explained in lines 77–99. The input "mx_add_en" initiates the commencement of memory read commands, transmitted through the "rd0" and "rd1" buses. Simultaneously, it updates the corresponding read addresses using the "rdaddr0" and "rdaddr1" buses during the cycles when the commands are asserted. In clock cycles where the counter values fall between decimal 3'd2 and 3'd5, write operations are carried out. In each of these clock cycles, the write addresses are updated using the "wraddr2" bus.

The completion of the matrix addition is indicated by the "mx_add_fi" output in line 101. This output is asserted when the last write address is reached, where all the bits of "wraddr2" are binary ones. To determine the maximum value of the write address "wraddr2", a Reduction AND gate can be employed.

9.4 Design Example: FP AXPY Calculation

As another design example, we explore the implementation of the mathematical expression $y = a \times x + y$ in hardware using multiple FP adders and multipliers. The equation involves 4×4 FP matrices x and y, along with an FP number a. The computation is performed in hardware, and we can represent the *axpy* expression as follows:

$$\begin{pmatrix} y_{00} & y_{01} & y_{02} & y_{03} \\ y_{10} & y_{11} & y_{12} & y_{13} \\ y_{20} & y_{21} & y_{22} & y_{23} \\ y_{30} & y_{31} & y_{32} & y_{33} \end{pmatrix} = a \begin{pmatrix} x_{00} & x_{01} & x_{02} & x_{03} \\ x_{10} & x_{11} & x_{12} & x_{13} \\ x_{20} & x_{21} & x_{22} & x_{23} \\ x_{30} & x_{31} & x_{32} & x_{33} \end{pmatrix} + \begin{pmatrix} y_{00} & y_{01} & y_{02} & y_{03} \\ y_{10} & y_{11} & y_{12} & y_{13} \\ y_{20} & y_{21} & y_{22} & y_{23} \\ y_{30} & y_{31} & y_{32} & y_{33} \end{pmatrix} \tag{9.2}$$

$$= \begin{pmatrix} ax_{00} + y_{00} & ax_{01} + y_{01} & ax_{02} + y_{02} & ax_{03} + y_{03} \\ ax_{10} + y_{10} & ax_{11} + y_{11} & ax_{12} + y_{12} & ax_{13} + y_{13} \\ ax_{20} + y_{20} & ax_{21} + y_{21} & ax_{22} + y_{22} & ax_{23} + y_{23} \\ ax_{30} + y_{30} & ax_{31} + y_{31} & ax_{32} + y_{32} & ax_{33} + y_{33} \end{pmatrix}$$

The equation signifies that each element of matrix y is updated by multiplying the corresponding element of matrix x with the constant a, and then adding the result to the previous value of the corresponding element in matrix y. This operation is executed for all elements in the 4×4 matrices, resulting in an updated matrix y after the computation.

To implement this functionality, two dual-port memory blocks are utilized with register arrays, allowing simultaneous reading and writing operations to the matrix y. Additionally, we employ a parallel design structure with four FP adders and four FP multipliers to efficiently implement the $a \times x + y$ computation.

9.4.1 Design structure of FP AXPY calculation

As previously mentioned, Figure 9.12 depicts the design structure, featuring two dual-port register arrays and four 32-bit FP multipliers followed by four 32-bit FP adders. The organization of matrix columns in memory follows a row-wise pattern, where each subsequent column is stored in the successive memory rows. With a memory data bus width of 128 bits, a single memory access permits the reading or writing of an entire memory row. The "mem0" and "mem1" read accesses occur in distinct time slots: data retrieved from "mem0" acts as input for the parallel multipliers ("x0-x3"), while data from "mem1" serves as input for the parallel adders ("y0-y3"). To coordinate read and write commands across different clock cycles, a timing controller is created as shown in the figure.

Within the data processing unit, the row data sourced from "mem0" is divided into four segments, each transmitted along the "x0-x3" buses. These

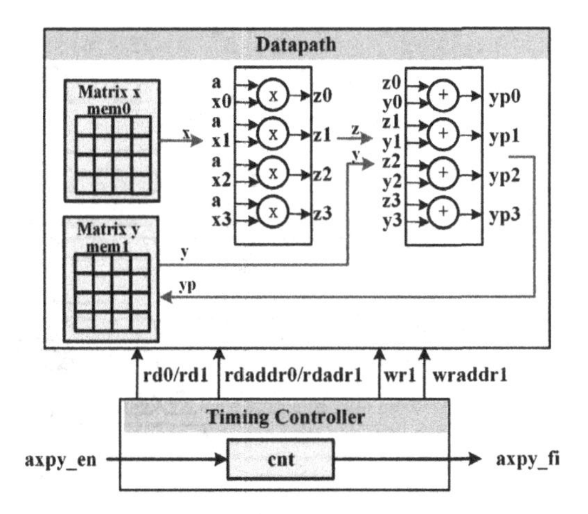

FIGURE 9.12
Design structure of FP AXPY.

individual segments are subsequently fed into separate 32-bit FP multipliers. These multipliers execute FP multiplications utilizing the FP number "a", yielding resultant products conveyed through the "z0-z3" buses.

Similarly, the row data originating from "mem1" is partitioned into four segments, transmitted along the "y0-y3" buses. Each of these segments is directed into an individual 32-bit FP adder. These adders perform the operation of adding the values from the "z0-z3" buses with those from the "y0-y3" buses, culminating in summations denoted as "yp0-yp3".

Finally, the four FP summations, namely "yp0-yp3", are concatenated into a 128-bit dataset propagated through the "yp" bus. This concatenated data is subsequently written back into "mem1", serving as the repository for the ultimate outcome of the *axpy* computation. The computation itself encompasses four iterations, each involving a sequence of parallel 32-bit FP operations for both multiplication and addition. The concluding *axpy* outcomes, comprising four sets of 128-bit FP data, are stored within "mem1" and occupy a span of four clock cycles.

9.4.2 Timing diagram of FP AXPY calculation

The progression of events is illustrated through the timing diagram in Figure 9.13. In the **Mem0&1 Read** channel, the process of reading "mem0" is initiated during the first clock cycle, followed by the initiation of "mem1" read during the subsequent clock cycle. In each clock cycle, a single row is fetched from "mem0", routed to the "x0-x3" buses, serving as inputs for the four 32-bit FP multipliers. Similarly, during each clock cycle, an equivalent row from "mem1" is retrieved, channeled into the "y0-y3" buses, functioning as inputs for the four 32-bit FP adders. The operation of reading the complete matrix

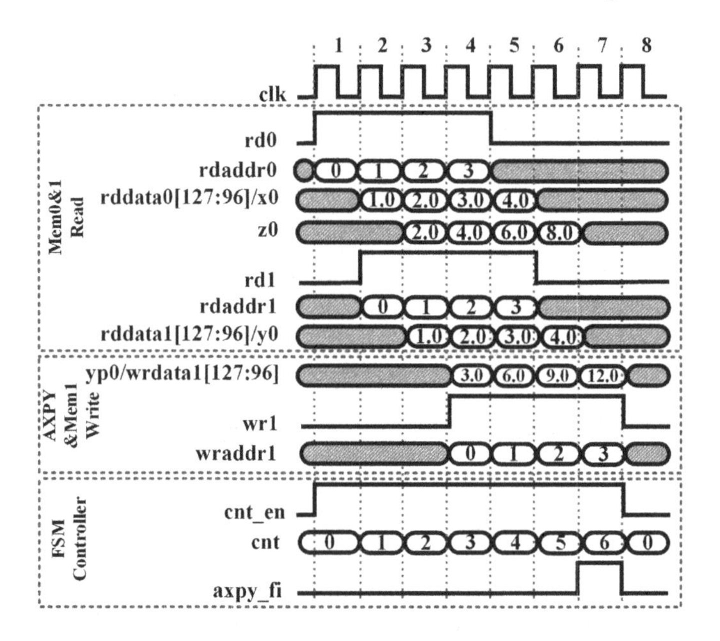

FIGURE 9.13
Timing diagram of FP AXPY.

from "mem0" spans four clock cycles, occurring over cycles 1–4. The same process for "mem1" takes place over cycles 2–5, encompassing a total of four cycles.

In line with the assumption articulated in Section 9.3.2, where all four word-sized data within a memory row are identical, only the most significant word on the "rddata0[127:96]/x0" bus is presented in the diagram. Specifically, FP 1.0 is showcased in the second clock cycle, FP 2.0 in the third, FP 3.0 in the fourth, and FP 4.0 in the fifth. Upon multiplication with the constant "a=FP 2.0", the resultant values "z0" emerge as FP 2.0, 4.0, 6.0, and 8.0, respectively, spanning clock cycles 3 through 6. Simultaneously, the 128-bit FP data within "mem1" is accessed and displayed on the "rddata1[127:96]/y0" buses. These values encompass FP 1.0 in the third clock cycle, FP 2.0 in the fourth, FP 3.0 in the fifth, and FP 4.0 in the sixth, following a sequential order.

By conducting the addition of FP data found in "z0" and "y0" during clock cycles 3, 4, 5, and 6, the outcomes of the *axpy* operation can be propagated onto the "yp0" bus with a single clock cycle delay, precisely in clock cycles 4, 5, 6, and 7. In the **AXPY&Mem1 Write** channel, this implies that the resultant values are inscribed into "mem1" spanning clock cycles 4 through 7. Simultaneously, the corresponding write enable signal denoted as "wr1" and the write address "wraddr1" are both activated within the same clock cycles.

In the **FSM Controller channel**, a counter is initiated, commencing from the decimal value 3'd0 and progressing up to 3'd6. This counter monitors the progression of the *axpy* computation and governs the corresponding memory access. As the counter attains the decimal value 3'd6 during the last clock cycle, it triggers the activation of the indicator "axpy_fi", serving as a signal to indicate the successful completion of the *axpy* processing.

9.4.3 Verilog code for FP AXPY calculation

The provided Verilog design is presented below. In line 8, the FP number "a" is assigned the hexadecimal value 32'h4000000 (corresponding to FP 2.0). The instantiation of four parallel FP multipliers is depicted in lines 17 to 40, while lines 42 to 65 instantiate four parallel FP adders. In the FP multipliers, one input receives the FP number "a", while the other inputs "x0-x3" are retrieved from "mem0". The outputs "z0-z3" from the FP multipliers are interconnected with the inputs of the four FP adders. Concurrently, the additional inputs "y0-y3" of the adders are sourced from "mem1". The ultimate outcomes of the adders, i.e., the summations, are conveyed through the "yp0-yp3" buses.

The instantiation of two dual-port memories, "u0_mem" for matrix x and "u1_mem" for matrix y, is elucidated in lines 67–86. Each memory port possesses a write operation port "a" and a read operation port "b". Notably, "u0_mem" is solely employed for read operations, prompting the deactivation of its "a" port by setting the enable signal "ena" to binary 1'b0.

```
1  module axpy (input              clk       ,
2               input              rst       ,
3               input              axpy_en   ,
4               output             axpy_fi   );
5  //********************************
6  //****    Datapath Design  *********
7  //********************************
8  wire [31:0] a=32'h40000000; //FP 2.0
9  wire [31:0] x0, x1, x2, x3;
10 wire [31:0] y0, y1, y2, y3;
11 wire [31:0] z0, z1, z2, z3;
12 wire [31:0] yp0, yp1, yp2, yp3;
13
14 wire        wr1, rd0, rd1;
15 reg  [1:0]  wraddr1, rdaddr0, rdaddr1;
16
17 // FP multipliers instantiation
18 FP_multiplier u0_fp_mul(.clock    (clk  ),
19                         .reset    (rst  ),
20                         .io_in_a  (a    ),
21                         .io_in_b  (x0   ),
22                         .io_out_s (z0   ));
23
```

```
24 FP_multiplier u1_fp_mul(.clock    (clk  ),
25                         .reset    (rst  ),
26                         .io_in_a (a    ),
27                         .io_in_b (x1   ),
28                         .io_out_s(z1   ));
29
30 FP_multiplier u2_fp_mul(.clock    (clk  ),
31                         .reset    (rst  ),
32                         .io_in_a (a    ),
33                         .io_in_b (x2   ),
34                         .io_out_s(z2   ));
35
36 FP_multiplier u3_fp_mul(.clock    (clk  ),
37                         .reset    (rst  ),
38                         .io_in_a (a    ),
39                         .io_in_b (x3   ),
40                         .io_out_s(z3   ));
41
42 // FP adders instantiation
43 FP_adder u0_fp_add(.clock    (clk  ),
44                    .reset    (rst  ),
45                    .io_in_a (z0   ),
46                    .io_in_b (y0   ),
47                    .io_out_s(yp0  ));
48
49 FP_adder u1_fp_add(.clock    (clk  ),
50                    .reset    (rst  ),
51                    .io_in_a (z1   ),
52                    .io_in_b (y1   ),
53                    .io_out_s(yp1  ));
54
55 FP_adder u2_fp_add(.clock    (clk  ),
56                    .reset    (rst  ),
57                    .io_in_a (z2   ),
58                    .io_in_b (y2   ),
59                    .io_out_s(yp2  ));
60
61 FP_adder u3_fp_add(.clock    (clk  ),
62                    .reset    (rst  ),
63                    .io_in_a (z3   ),
64                    .io_in_b (y3   ),
65                    .io_out_s(yp3  ));
66
67 // memory instantiation
68 simple_dual u0_mem(.clka  (clk          ) ,
69                    .ena   (1'b0         ) ,
70                    .wea   (`WEA4        ) ,
71                    .addra(2'h0         ) ,
72                    .dia   (128'h0       ) ,
```

```
73                      .clkb (clk          ) ,
74                      .enb  (rd0          ) ,
75                      .addrb(rdaddr0      ) ,
76                      .dob  ({x0,x1,x2,x3}));
77
78 simple_dual u1_mem(.clka (clk               ) ,
79                      .ena  (wr1               ) ,
80                      .wea  (`WEA4             ) ,
81                      .addra(wraddr1           ) ,
82                      .dia  ({yp0,yp1,yp2,yp3}) ,
83                      .clkb (clk               ) ,
84                      .enb  (rd1               ) ,
85                      .addrb(rdaddr1           ) ,
86                      .dob  ({y0,y1,y2,y3}     ));
87
88 //************************************
89 //****    FSM Controller Design ********
90 //************************************
91 reg [2:0] cnt;
92 wire cnt_en = axpy_en | |cnt ;
93 wire [2:0] nxt_cnt = &cnt[2:1] ? 3'd0 :
94                                  cnt_en ? (cnt+3'd1) : cnt;
95 always @(posedge clk) begin
96   if(rst) begin
97     cnt<=3'd0     ;
98   end else begin
99     cnt<=nxt_cnt ;
100   end
101 end
102
103 assign rd0 = axpy_en ;
104 assign rd1 = cnt>=3'd1 & cnt<=3'd4;
105 wire [1:0] nxt_rdaddr0 = rd0 ? rdaddr0+2'h1 : 2'h0;
106 wire [1:0] nxt_rdaddr1 = rd1 ? rdaddr1+2'h1 : 2'h0;
107 always @(posedge clk) begin
108   if(rst) begin
109     rdaddr0<=2'h0     ;
110     rdaddr1<=2'h0     ;
111   end else begin
112     rdaddr0<=nxt_rdaddr0 ;
113     rdaddr1<=nxt_rdaddr1 ;
114   end
115 end
116
117 assign wr1 = cnt>=3'd3 & cnt<=3'd6;
118 wire [1:0] nxt_wraddr1 = wr1 ? wraddr1+2'h1 : 2'h0;
119 always @(posedge clk) begin
120   if(rst) begin
121     wraddr1<=2'h0          ;
```

```
122     end else begin
123       wraddr1<=nxt_wraddr1 ;
124     end
125   end
126
127   assign axpy_fi  = &wraddr1;
128   endmodule
```

Lines 88–127 describe the functioning of the timing controller, employing a straightforward counter architecture. Upon activation of the circuit through the "axpy_en" signal, the counter enable signal "cnt_en" is triggered, persisting until the counter attains a zero value. The counter status, checked for non-zero condition using a Reduction OR operation ($|cnt$) in line 92, dictates the behavior. Upon reaching the maximum decimal value of 3'd6 (binary 3'b110), with the two most significant bits set to binary ones, the counter is reset. The condition for this is verified in line 93, utilizing a Reduction AND operation ($\&cnt[2:1]$) to confirm the binary one status of the two most significant bits.

In lines 103–115, the logic for memory read commands and addresses is outlined. The read command "rd0" for "u0_mem0" is activated upon enabling the *axpy* module (line 103). The address for "u0_mem" is sequentially updated as long as the *axpy* operation is in progress. Correspondingly, the read command "rd1" for "u1_mem" is activated within the counter range of decimal values 3'd1 to 3'd4 (line 104), and the address for "u1_mem" follows a similar incremental pattern during these cycles (line 106).

Subsequently, lines 117–125 detail the protocol for write commands and addresses concerning "u1_mem". The write command is activated when the counter registers values ranging from decimal 3'd3 to 3'd6 (line 117), and the associated write address evolves during these clock cycles (line 118). Ultimately, in line 127, the indicator "axpy_fi" is asserted as the write address attains its maximum decimal value of 2'b11. This state is determined through a Reduction AND expression ($\&wraddr1$).

9.5 Design Example: Basic Design on FP DDOT

Concluding the last case study, the mathematical expression governing the FP *ddot* calculation takes the form of $z = x^T \times y$, where x and y denote single-precision FP vectors each consisting of eight elements, and z symbolizes the FP outcome. This can be expressed mathematically as:

$$z = \begin{pmatrix} x0 & x1 & x2 & x3 & x4 & x5 & x6 & x7 \end{pmatrix} * \begin{pmatrix} y0 \\ y1 \\ y2 \\ y3 \\ y4 \\ y5 \\ y6 \\ y7 \end{pmatrix} \tag{9.3}$$

$$= x0 \cdot y0 + x1 \cdot y1 + x2 \cdot y2 + x3 \cdot y3 + x4 \cdot y4 + x5 \cdot y5 + x6 \cdot y6 + x7 \cdot y7$$

This mathematical expression encapsulates the concept of vector multiplication and addition. In a direct implementation, the process would entail employing eight parallel FP multipliers to compute the individual products. Subsequently, seven consecutive FP adders would be utilized to aggregate these products, culminating in the final outcome, denoted as z.

9.5.1 Design structure of basic design on FP DDOT

Figure 9.14 shows a design tailored for the FP *ddot* calculation, coordinating the multiplication-addition task denoted as $z = x^T \times y$. The underlying datapath configuration encompasses several stages. Firstly, there are eight parallel FP multipliers forming the initial stage, dedicated to generating the products labeled as "p0-p7". Subsequently, the ensuing stages feature the following: four FP adders in the second stage, two FP adders in the third stage, and culminating with a single FP adder in the final stage. The collective purpose of these stages is to consolidate the eight products, eventually yielding the conclusive result "z". This resultant value is released across a span of four clock cycles.

To oversee and regulate the operations of this datapath structure, the inclusion of a timing controller becomes indispensable. Let's consider each FP operator to consume one clock cycle, spanning from input to output. With this assumption, the latency of the multiplication-addition circuit can be computed using the formula $1 + \log_2 N$, wherein N denotes the streaming width of the design mechanism. This formula signifies that a single clock cycle is needed for the parallel multiplications, and an additional $\log_2 N$ clock cycles are essential for the sequential aggregation of the N multiplication products. In the context of this design instance, where the streaming width is set at 8, the latency for data processing is determined as $1 + \log_2 8 = 4$ clock cycles.

In the *ddot* design, the timing controller delineates the distinct states of data processing. The FSM states are labeled "S0" to represent the initial phase of parallel multiplications, while "S1-S3" represent the subsequent three cascading stages of additions. Additionally, the timing controller introduces a "ready-vld" handshake mechanism. Here, the input "ready" signifies the preparedness of data on the "x0-x7" and "y0-y7" inputs to enter the *ddot* datapath.

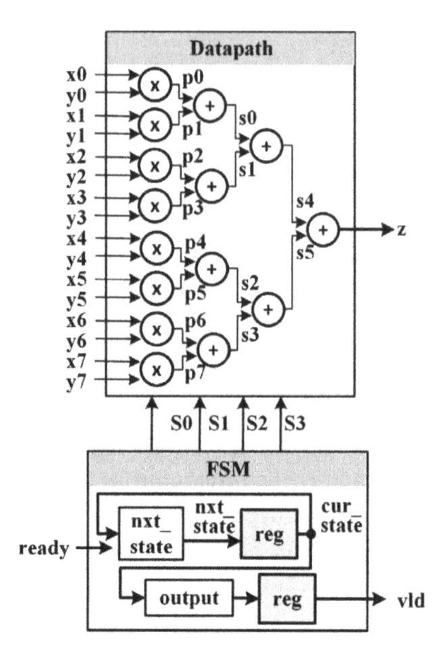

FIGURE 9.14
Design structure of basic design on FP DDOT.

Conversely, the output "vld" conveys the validity of the data present on the "z" output.

9.5.2 Timing diagram of basic design on FP DDOT calculation

The timing diagram for the *ddot* design is depicted in Figure 9.15. As with the previous examples, for the sake of clarity and simplification, only the data on the "x0" and "y0" buses are showcased in the diagram. This approach is underpinned by the assumption that all the data contained within the "x0-x7" buses are identical, and correspondingly, all the data within the "y0-y7" buses share the same uniformity.

During the initial clock cycle, the "ready" input is asserted, introducing "x0 = FP 1.0" and "y0 = FP 2.0" into the *ddot* processing unit. At this point, the FSM state is labeled "S0", thereby activating the eight parallel multipliers to generate the products during the subsequent clock cycle. In clock cycles 2, 3, and 4, the states transition to "S1-S3", signifying the presence of valid data on the "p0-p7", "s0-s3", and "s4-s5" buses correspondingly, referencing the signals showcased in Figure 9.14.

As clock cycle 5 commences, the "vld" indicator is asserted, validating the final output, which is "z = FP 16.0". This action also triggers a handshake with the "ready" input, effectively initiating a reset of the state machine for

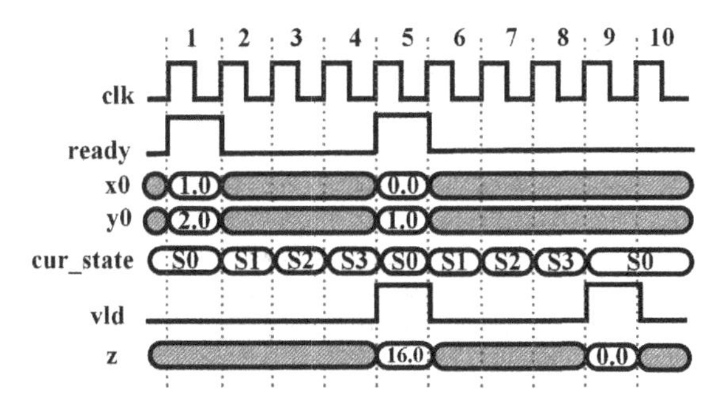

FIGURE 9.15
Timing diagram of basic design on FP DDOT.

subsequent *ddot* calculations. Subsequently, this handshake prepares the ground for the input of another data frame into the processing unit. As illustrated in the figure, during clock cycle 5, the second inputs "x0 = FP 0.0" and "y0 = FP 1.0" stand ready on the respective buses. Eventually, in clock cycle 9, the ultimate output, "z = FP 0.0", is propagated out, culminating in the completion of the entire process.

9.5.3 Verilog code for basic design on FP DDOT

Below is the Verilog code for the foundational FP *ddot* design. The design employs parallel instantiation of eight FP multipliers, demonstrated in lines 14–61. The ensuing multiplication products are linked to the "p0-p7" buses. Subsequently, these products are directed into the second-stage FP adders, outlined in lines 63–86, where the resulting summations are channeled through the "s0-s3" buses. The third stage encompasses the instantiation of FP adders, executed in lines 88–99. These adders utilize "s0-s3" as inputs and yield outputs on "s4" and "s5" buses. The ultimate stage is comprised of a single FP adder, instantiated in lines 101–106. This adder functions to aggregate the values from "s4" and "s5", thereby producing the conclusive output "z".

```
1  module basic_ddot (input              clk   ,
2                     input              rst   ,
3                     input              ready ,
4                     input       [31:0] x0,x1,x2,x3,x4,x5,x6,x7,
5                     input       [31:0] y0,y1,y2,y3,y4,y5,y6,y7,
6                     output reg         vld   ,
7                     output      [31:0] z     );
8  //********************************
9  //****    Datapath Design  ********
10 //********************************
```

```
11 wire [31:0] p0, p1, p2, p3, p4, p5, p6, p7;
12 wire [31:0] s0, s1, s2, s3, s4, s5;
13
14 // The first stage of multipliers
15 FP_multiplier u0_fp_mul(.clock    (clk  ),
16                         .reset    (rst  ),
17                         .io_in_a  (x0   ),
18                         .io_in_b  (y0   ),
19                         .io_out_s (p0   ));
20
21 FP_multiplier u1_fp_mul(.clock    (clk  ),
22                         .reset    (rst  ),
23                         .io_in_a  (x1   ),
24                         .io_in_b  (y1   ),
25                         .io_out_s (p1   ));
26
27 FP_multiplier u2_fp_mul(.clock    (clk  ),
28                         .reset    (rst  ),
29                         .io_in_a  (x2   ),
30                         .io_in_b  (y2   ),
31                         .io_out_s (p2   ));
32
33 FP_multiplier u3_fp_mul(.clock    (clk  ),
34                         .reset    (rst  ),
35                         .io_in_a  (x3   ),
36                         .io_in_b  (y3   ),
37                         .io_out_s (p3   ));
38
39 FP_multiplier u4_fp_mul(.clock    (clk  ),
40                         .reset    (rst  ),
41                         .io_in_a  (x4   ),
42                         .io_in_b  (y4   ),
43                         .io_out_s (p4   ));
44
45 FP_multiplier u5_fp_mul(.clock    (clk  ),
46                         .reset    (rst  ),
47                         .io_in_a  (x5   ),
48                         .io_in_b  (y5   ),
49                         .io_out_s (p5   ));
50
51 FP_multiplier u6_fp_mul(.clock    (clk  ),
52                         .reset    (rst  ),
53                         .io_in_a  (x6   ),
54                         .io_in_b  (y6   ),
55                         .io_out_s (p6   ));
56
57 FP_multiplier u7_fp_mul(.clock    (clk  ),
58                         .reset    (rst  ),
59                         .io_in_a  (x7   ),
```

```
60                          .io_in_b (y7    ),
61                          .io_out_s(p7    ));
62
63  // The second stage of adders
64  FP_adder u0_fp_add(.clock   (clk  ),
65                     .reset   (rst  ),
66                     .io_in_a (p0   ),
67                     .io_in_b (p1   ),
68                     .io_out_s(s0   ));
69
70  FP_adder u1_fp_add(.clock   (clk  ),
71                     .reset   (rst  ),
72                     .io_in_a (p2   ),
73                     .io_in_b (p3   ),
74                     .io_out_s(s1   ));
75
76  FP_adder u2_fp_add(.clock   (clk  ),
77                     .reset   (rst  ),
78                     .io_in_a (p4   ),
79                     .io_in_b (p5   ),
80                     .io_out_s(s2   ));
81
82  FP_adder u3_fp_add(.clock   (clk  ),
83                     .reset   (rst  ),
84                     .io_in_a (p6   ),
85                     .io_in_b (p7   ),
86                     .io_out_s(s3   ));
87
88  // The third stage of adders
89  FP_adder u4_fp_add(.clock   (clk  ),
90                     .reset   (rst  ),
91                     .io_in_a (s0   ),
92                     .io_in_b (s1   ),
93                     .io_out_s(s4   ));
94
95  FP_adder u5_fp_add(.clock   (clk  ),
96                     .reset   (rst  ),
97                     .io_in_a (s2   ),
98                     .io_in_b (s3   ),
99                     .io_out_s(s5   ));
100
101 // The fourth stage of adders
102 FP_adder u6_fp_add(.clock   (clk  ),
103                    .reset   (rst  ),
104                    .io_in_a (s4   ),
105                    .io_in_b (s5   ),
106                    .io_out_s(z    ));
107
108 //****************************************
```

```
109 //****    FSM Controller Design *********
110 //************************************
111 parameter S0 = 2'h0 ;
112 parameter S1 = 2'h1 ;
113 parameter S2 = 2'h2 ;
114 parameter S3 = 2'h3 ;
115
116 reg [1:0] nxt_state;
117 reg [1:0] cur_state;
118 always @(cur_state,ready) begin
119   case(cur_state)
120     S0: if(ready) nxt_state <= S1;
121     S1:           nxt_state <= S2;
122     S2:           nxt_state <= S3;
123     S3:           nxt_state <= S0;
124     default:      nxt_state <= S0;
125   endcase
126 end
127
128 always @(posedge clk) begin
129   if(rst) begin
130     cur_state <= S0       ;
131   end else begin
132     cur_state <= nxt_state;
133   end
134 end
135
136 wire nxt_vld = cur_state==S3;
137 always @(posedge clk) begin
138   if(rst) begin
139     vld <= 1'b0   ;
140   end else begin
141     vld <= nxt_vld;
142   end
143 end
144 endmodule
```

The subsequent lines, 108–134, delineate the FSM responsible for governing the timing controller's four distinct states. The state "S0" denotes the initial phase, wherein eight parallel multiplications are conducted. Following that, the FSM transitions through "S1-S3" states, which respectively encapsulate the second, third, and fourth stages of additions.

Following the conclusion of the final state "S3", a single clock cycle elapses, culminating in the availability of the ultimate outcome on the "z" output. This accomplishment is accompanied by the activation of the "vld" indicator, which is demonstrated in lines 136–143.

PBL 22: Design Simulation on Single-Port Register File

1) Create a testbench to simulate the single-port register file introduced in Section 9.2.2.
2) Create the test case illustrated in Figure 9.5(a), which involves writing four words into the register file followed by reading them out. The write enable signal "we=4'hf" ensures that all four bytes in each data burst are enabled.
3) Create the test case depicted in Figure 9.5(b), which entails writing two words into the register file with "we=4'hf", followed by reading them out. Additionally, conduct writing two words into the register file with "we=4'h3", followed by reading them out.

PBL 23: Design Simulation on Dual-Port Register File

1) Create a testbench to simulate the dual-port register file introduced in Section 9.2.3.
2) Create the test case illustrated in Figure 9.7, which involves writing four words into the register file followed by reading them out. The write enable signal "wea=4'hf" ensures that all four bytes in each data burst are enabled. It's essential to emphasize that the write and read operations need to overlap to illustrate the concurrent operations between write and read accesses for the dual-port register file.

PBL 24: FP Matrix-Matrix Subtractor

1) Design a FP matrix-matrix subtractor utilizing the signal-clock-cycle FP operators, denoted as $z = x - y$, where x, y, and z are 4×4 FP matrices. The FP subtractor can be used as the design IP. Three single-port memories can be used to store the input matrices

x and y and the final output matrix z.

$$\begin{pmatrix} z_{00} & z_{01} & z_{02} & z_{03} \\ z_{10} & z_{11} & z_{12} & z_{13} \\ z_{20} & z_{21} & z_{22} & z_{23} \\ z_{30} & z_{31} & z_{32} & z_{33} \end{pmatrix} = \begin{pmatrix} x_{00} & x_{01} & x_{02} & x_{03} \\ x_{10} & x_{11} & x_{12} & x_{13} \\ x_{20} & x_{21} & x_{22} & x_{23} \\ x_{30} & x_{31} & x_{32} & x_{33} \end{pmatrix} - \begin{pmatrix} y_{00} & y_{01} & y_{02} & y_{03} \\ y_{10} & y_{11} & y_{12} & y_{13} \\ y_{20} & y_{21} & y_{22} & y_{23} \\ y_{30} & y_{31} & y_{32} & y_{33} \end{pmatrix} \tag{9.4}$$

$$= \begin{pmatrix} x_{00}-y_{00} & x_{01}-y_{01} & x_{02}-y_{02} & x_{03}-y_{03} \\ x_{10}-y_{10} & x_{11}-y_{11} & x_{12}-y_{12} & x_{13}-y_{13} \\ x_{20}-y_{20} & x_{21}-y_{21} & x_{22}-y_{22} & x_{23}-y_{23} \\ x_{30}-y_{30} & x_{31}-y_{31} & x_{32}-y_{32} & x_{33}-y_{33} \end{pmatrix}$$

2) Create a testbench to simulate the designed circuit and verify its functionalities using the provided direct test case below. Represent the FP data using the IEEE-754 format depicted in Table 8.2. Store the columns of each matrix in memory in a row-wise manner.

$$\begin{pmatrix} x_{00} & x_{01} & x_{02} & x_{03} \\ x_{10} & x_{11} & x_{12} & x_{13} \\ x_{20} & x_{21} & x_{22} & x_{23} \\ x_{30} & x_{31} & x_{32} & x_{33} \end{pmatrix} = \begin{pmatrix} 1.0 & 2.0 & 3.0 & 4.0 \\ 1.0 & 2.0 & 3.0 & 4.0 \\ 1.0 & 2.0 & 3.0 & 4.0 \\ 1.0 & 2.0 & 3.0 & 4.0 \end{pmatrix} \tag{9.5}$$

$$\begin{pmatrix} y_{00} & y_{01} & y_{02} & y_{03} \\ y_{10} & y_{11} & y_{12} & y_{13} \\ y_{20} & y_{21} & y_{22} & y_{23} \\ y_{30} & y_{31} & y_{32} & y_{33} \end{pmatrix} = \begin{pmatrix} 0.0 & 1.0 & 2.0 & 3.0 \\ 0.0 & 1.0 & 2.0 & 3.0 \\ 0.0 & 1.0 & 2.0 & 3.0 \\ 0.0 & 1.0 & 2.0 & 3.0 \end{pmatrix} \tag{9.6}$$

PBL 25: FP Multiplication-Addition Circuit

1) Design a FP multiplication-addition circuit (MAC) utilizing the signal-clock-cycle FP operators, denoted as $z = x^T \times y$, where x and y are FP vectors and z is a FP number. The FP multiplier and adder can be used as design IPs.

$$\begin{aligned} z &= \begin{pmatrix} x0 & x1 \end{pmatrix} \times \begin{pmatrix} y0 \\ y1 \end{pmatrix} \\ &= x0 \cdot y0 + x1 \cdot y1 \end{aligned} \tag{9.7}$$

2) Create a testbench to simulate the designed circuit and verify its functionalities using the provided direct test case below. Represent the FP data using the IEEE-754 format depicted in Table 8.2.

$$\begin{pmatrix} x0 \\ x1 \end{pmatrix} = \begin{pmatrix} 1.0 \\ 2.0 \end{pmatrix} \tag{9.8}$$

$$\begin{pmatrix} y0 \\ y1 \end{pmatrix} = \begin{pmatrix} 3.0 \\ 4.0 \end{pmatrix} \tag{9.9}$$

PBL 26: FP Matrix-Matrix Multiplier

1) Design a FP matrix-matrix multiplier utilizing the signal-clock-cycle FP operators, denoted as $z = x \times y$, where x, y, and z are 2×2 FP matrices. The multiplication-addition circuit in PBL 25 can be used as the design IP.

$$\begin{pmatrix} z_{00} & z_{01} \\ z_{10} & z_{11} \end{pmatrix} = \begin{pmatrix} x_{00} & x_{01} \\ x_{10} & x_{11} \end{pmatrix} \times \begin{pmatrix} y_{00} & y_{01} \\ y_{10} & y_{11} \end{pmatrix} \tag{9.10}$$
$$= \begin{pmatrix} MAC(x_{00} \sim x_{01}, y_{00} \sim y_{10}) & MAC(x_{00} \sim x_{01}, y_{01} \sim y_{11}) \\ MAC(x_{10} \sim x_{11}, y_{00} \sim y_{10}) & MAC(x_{10} \sim x_{11}, y_{01} \sim y_{11}) \end{pmatrix}$$

2) Create a testbench to simulate the designed circuit and verify its functionalities using the provided direct test case below. Represent the FP data using the IEEE-754 format depicted in Table 8.2.

$$\begin{pmatrix} x_{00} & x_{01} \\ x_{10} & x_{11} \end{pmatrix} = \begin{pmatrix} 1.0 & 3.0 \\ 2.0 & 4.0 \end{pmatrix} \tag{9.11}$$

$$\begin{pmatrix} y_{00} & y_{01} \\ y_{10} & y_{11} \end{pmatrix} = \begin{pmatrix} 0.0 & 2.0 \\ 1.0 & 3.0 \end{pmatrix} \tag{9.12}$$

PBL 27: Power of FP Matrix

1) Design a circuit to compute the power of an FP matrix utilizing the signal-clock-cycle FP operators, denoted as $y = x^2$, where x and y are 2×2 FP matrices. The FP multiplication-addition circuit in PBL 25 can be used as the design IP.

$$\begin{pmatrix} y_{00} & y_{01} \\ y_{10} & y_{11} \end{pmatrix} = \begin{pmatrix} x_{00} & x_{01} \\ x_{10} & x_{11} \end{pmatrix} * \begin{pmatrix} x_{00} & x_{01} \\ x_{10} & x_{11} \end{pmatrix} \tag{9.13}$$
$$= \begin{pmatrix} MAC(x_{00} \sim x_{01}, x_{00} \sim x_{10}) & MAC(x_{00} \sim x_{01}, x_{01} \sim x_{11}) \\ MAC(x_{10} \sim x_{11}, x_{00} \sim x_{10}) & MAC(x_{10} \sim x_{11}, x_{01} \sim x_{11}) \end{pmatrix}$$

2) Create a testbench to simulate the designed circuit and verify its functionalities using the provided direct test case below. Represent the FP data using the IEEE-754 format depicted in Table 8.2.

$$\begin{pmatrix} x_{00} & x_{01} \\ x_{10} & x_{11} \end{pmatrix} = \begin{pmatrix} 1.0 & 3.0 \\ 2.0 & 4.0 \end{pmatrix} \tag{9.14}$$

10

Streaming and Iterative Design on Numerical Hardware

The *ddot* design, as elaborated in Chapter 9, stands as a foundational example of numerical hardware design and integration. Furthermore, Chapter 10 introduces two advanced design architectures for the same *ddot* operation: the streaming design and iterative design. The objective is to present diverse hardware architectures aimed at improving data processing efficiency and/or providing cost-effective solutions to meet a range of design requirements.

PBL 28 extends the *ddot* design into a sequential implementation with a streaming width of one. PBL 29–30 demonstrate hardware designs tailored for Fourier transform applications, encompassing sizes two and four.

10.1 Streaming and Iterative Design and Integration

10.1.1 Streaming and iterative designs

An effective strategy for achieving high-performance computing involves maximizing the parallelization of floating-point (FP) operations within the hardware design. This approach, known as the streaming design paradigm, emphasizes achieving the highest possible floating-point operations per second (FLOPS) in data processing. Within this context, the streaming width aligns with the size of the data frame, enabling each frame to serve as both input and output in a single clock cycle. While this approach offers exceptional throughput, it demands a substantial number of hardware resources, leading to significant hardware costs in terms of IOs, gate count, and memory. These costs can present challenges for FPGAs due to their limited hardware resources or for ASICs due to power constraints. Consequently, the expenses associated with the streaming design may outweigh its benefits when dealing with data packages containing extensive content.

Given these considerations, the iterative design emerges as a more viable alternative for processing large-sized data frames. In this design scheme, the goal is to reduce hardware costs by making a trade-off in data bandwidth. This trade-off is achieved by limiting the streaming width (STW) to a value significantly smaller than the data frame size (N). Consequently, the entire

DOI: 10.1201/9781003187080-10

data frame must be incrementally fed into the design circuit over multiple clock cycles. Mathematically, it takes $\lceil N/STW \rceil$ clock cycles to progressively input the entire data frame, where $\lceil\ \rceil$ denotes the mathematical ceiling function. This function rounds a number up to the nearest integer greater than or equal to the provided number. For instance, consider a data frame consisting of 96 words and a hardware streaming width of two words. In this scenario, it requires $\lceil 96/2 \rceil = 48$ clock cycles to fully input the entire data frame.

10.1.2 Multi-controller design with pipelined and parallel framework

A. High-Performance Design Analysis

The *ddot* example depicted in Figure 9.14 in Chapter 9 showcased a basic design that employed a solitary timing controller to govern the datapath. However, the present section embarks on an exploration of high-performance designs that necessitate the involvement of multiple timing controllers for parallel data frame processing.

As an illustration, let's consider a data frame containing eight bytes, with a design streaming width of four bytes. The timing analysis depicted in Figure 10.1 outlines the data operations. For the first data frame, encompassing hexadecimal values "x00-x07", a span of two clock cycles is required to feed the data into the processing unit due to the streaming width of four. Similarly, the second data frame (hexadecimal "x10-x17") necessitates clock cycles 3 and 4. Subsequently, the third data frame (hexadecimal "x20-x27") aligns with clock cycles 5 and 6, and finally, the last data frame (hexadecimal "x30-x37") unfolds across clock cycles 7 and 8. To maintain the unbroken flow of data streams, it becomes evident that a network of several timing controllers is essential to oversee the parallel processing of these four distinct data frames.

Assume that the longest data processing path spans four clock cycles. The output of the initial data frame becomes evident in clock cycle 5, indicating a 4-clock-cycle delay from the clock cycle of data input. This data operation is governed by a single state machine (FSM0), responsible for timing control across all four clock cycles. Following the completion of processing for the initial data frame, FSM0 returns to an idle state, ready for subsequent utilization in processing other data frames, either in the fifth clock cycle or beyond.

The second data stream enters during clock cycle 3, culminating in clock cycle 7 due to the encompassing 4-clock-cycle path. Since FSM0 remains engaged until clock cycle 4, an additional timing controller (FSM1) becomes necessary to oversee the processing of the second data frame from cycles 3 through 6. Subsequently, the third data stream is activated from clock cycle 5 to clock cycle 8. Given FSM0's availability within and after clock cycle 5, it can be promptly repurposed for supervising the processing of the third data frame. Similarly, FSM1's availability facilitates the administration of the

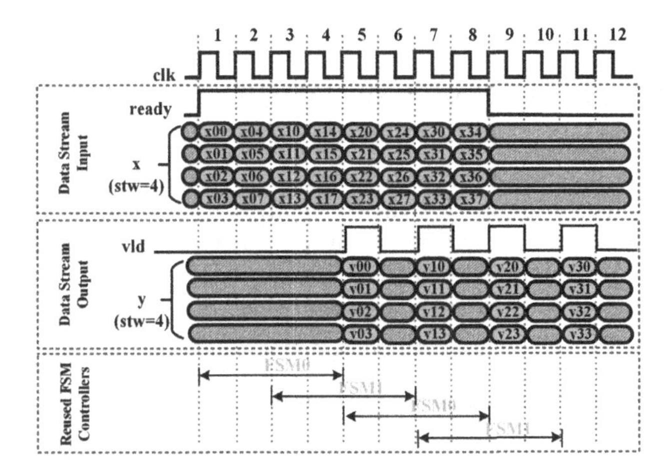

FIGURE 10.1
Timing analysis for iterative design.

fourth data frame, which commences from clock cycle 7 and concludes in clock cycle 10.

B. Multi-Controller Structural Assessment

Efficient reuse of timing controllers can be achieved through rotation, provided they are unoccupied. The number of required timing controllers depends on an assessment of the longest data path and the streaming width inherent in the data processing unit. As previously mentioned, the number of clock cycles needed to feed each data frame into the design engine can be calculated as $NO\_CY = \lceil N/STW \rceil$. Consequently, the formula for determining the necessary number of timing controllers can be expressed as follows:

$$NO\_TCs = \lceil Longest\_Path/NO\_CY \rceil. \tag{10.1}$$

Here, NO_TCs denotes the count of timing controllers needed for seamless data processing continuity. In our illustrative example, the longest path spans four clock cycles. Therefore, the calculation becomes $\lceil 4/2 \rceil = 2$ controllers, which are essential for consecutively processing data streams within a fully pipelined and parallel framework. The practical implementation of these timing controllers can take the form of FSMs or streamlined design counters.

It's worth noting that these controllers are equipped with distinct priority levels for reuse. FSM0 takes precedence with the highest priority, serving as the primary option when it's in an idle state. In instances where FSM0 is already engaged, FSM1, holding the second-highest priority, steps in to assume control. This systematic rotation of FSMs enables the seamless execution of continuous data stream processing, effortlessly achieved within a fully pipelined and parallel framework.

10.2 Streaming Design on DDOT

The foundational building block for computing *ddot* involves multiplication and addition operations. In the fundamental design framework outlined in Section 9.5, data frames enter the FP multipliers in parallel, but they must await the completion of previous multiplication-addition operations. To coordinate the synchronization of each data frame's processing, a handshaking protocol is used, employing the signals "ready" and "vld" to indicate status. This basic design structure requires four clock cycles per computation of each data frame, resulting in a total of $4n$ clock cycles needed for processing n data frames. To support the uninterrupted processing of consecutive data frames, this section introduces a pipeline configuration within a streaming design scenario.

10.2.1 Design structure of streaming design on DDOT

The structure for processing the *ddot* computation through a pipeline and parallel manner is depicted in Figure 10.2. The underlying datapath retains its similarity to the basic *ddot* design. The streaming width is configured at eight, enabling the input of eight FP data sets "x0-x7" and eight FP data sets "y0-y7" into the design engine during each clock cycle. To effectively manage this high data throughput, multiple timing controllers are employed. Each timing controller is dedicated to processing one data frame, overseeing the journey from inputs to the ultimate output "z".

The number of timing controllers required for the streaming design can be determined using Equation 10.1. In this illustration, the streaming width

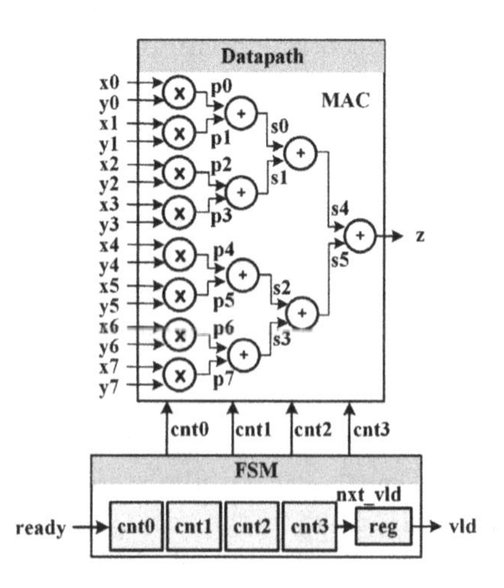

FIGURE 10.2
Streaming design on DDOT.

corresponds to the data frame size, denoted as $STW = N = 8$. This implies that each data frame can be fed to the datapath within a single clock cycle, representing as $NO\_CY = \lceil N/STW \rceil = \lceil 8/8 \rceil = 1$. The longest path within the design encompasses four clock cycles, as depicted in Figure 10.2, which includes one cycle for parallel FP multiplications and three cycles for successive additions. Consequently, the number of timing controllers can be computed as $NO\_TCs = \lceil Longest\_Path/NO\_CY \rceil = \lceil 4/1 \rceil = 4$. In this design, counters are utilized to serve as the timing controllers.

10.2.2 Timing diagram of streaming design on DDOT

For illustrative purposes, consider Figure 10.3, which presents a timing diagram demonstrating the progression of eight data frames through the *ddot* design engine. In each clock cycle, the engine can accept eight sets of FP data via the "x0-x7" inputs and an additional eight sets of FP data via the "y0-y7" inputs. The readiness of FP data on the input buses is denoted by the "ready" signal during clock cycles 1 to 8.

For the sake of simplicity, let's assume uniform data on the "x0-x7" buses within each data frame and uniform data on the "y0-y7" buses as well. For example, the initial data frame comprises eight FP 0.0 on the "x0-x7" buses and eight FP 1.0 on the "y0-y7" buses, entering in the first clock cycle. Subsequently, the second data frame contains eight FP 1.0 on the "x0-x7" buses and eight FP 2.0 on the "y0-y7" buses, entering in the second clock cycle. This sequence continues until the final data frame, composed of eight FP 7.0 on the "x0-x7" buses and eight FP 8.0 on the "y0-y7" buses, enters in the eighth clock cycle. For the sake of clarity, the timing diagram portrays only the FP data on the "x0" bus and the "y0" bus.

As elaborated in Section 10.2.1, the processing of each data frame necessitates four clock cycles to complete the *ddot* computation. Accordingly, the

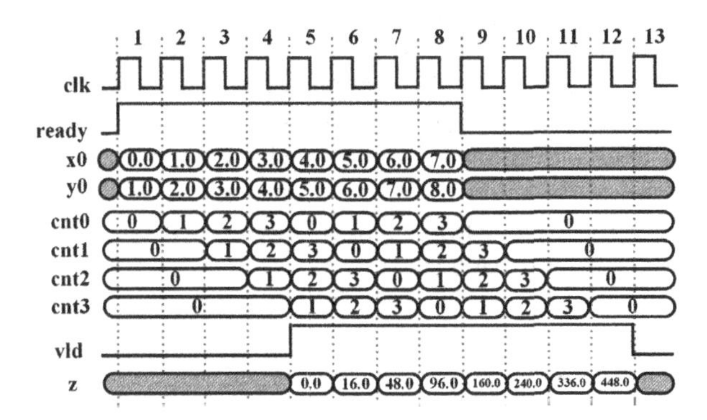

FIGURE 10.3
Timing diagram of streaming design on DDOT.

initial *ddot* result is generated in the fifth clock cycle, as depicted in the timing diagram. The counter labeled "cnt0" serves to monitor the processing of the first data frame across clock cycles 1–4. Starting at decimal value 2'd0, it denotes the introduction of the first data frame input and the parallel multiplications in the first clock cycle. Subsequently, it transitions to decimal value 2'd1 for the second stage of four parallel additions, 2'd2 for the third stage of two additions, and finally 2'd3 for the last clock cycle involving the final addition. The output "z" materializes in the fifth clock cycle, coinciding with the reset of "cnt0" to decimal value 2'd0. Within the fifth clock cycle, the data valid indicator "vld" is activated to signal the presence of the first valid data output. Similarly, the other three counters, namely "cnt1", "cnt2", and "cnt3", oversee the *ddot* processing of the subsequent three data frames. The *ddot* outputs for these three data frames emerge in clock cycles 6 to 8, with the corresponding valid indicators "vld" being asserted.

As for the fifth data frame, it is introduced to the *ddot* engine during clock cycle 5 ("x0 = FP 4.0" and "y0 = FP 5.0"). Given that the first counter "cnt0" has been initialized to decimal value 3'd0 (in idle status), it can be repurposed to control the processing of this data frame across clock cycles 5 to 8. The ultimate output, "z = FP 160.0", is presented in clock cycle 9, representing the *ddot* result for eight FP 4.0 ("x0-x7") and eight FP 5.0 ("y0-y7"). During the same clock cycle, the valid indicator "vld" is asserted. Employing a similar approach, the second, third, and fourth counters, "cnt1", "cnt2", and "cnt3" respectively, can be recycled to oversee the processing of the sixth, seventh, and final data frames across the corresponding clock cycles 6 to 12. The resulting outputs are pushed out, and the valid data indicators are activated at the conclusion of each data output cycle.

10.2.3 Timing controller of streaming design on DDOT

The process of constructing the datapath involves both connecting and instantiating all the FP design operators, which has been introduced in Figure 9.14 in Chapter 9. This section concentrates on the streaming design that employs four timing controllers to handle distinct data frames in a parallel pipeline manner. The primary challenge lies in coordinating the rotation of these four controllers in a predetermined order, based on dynamic statuses. The predefined priorities of these four counters, from highest to lowest, are labeled as "cnt0", "cnt1", "cnt2", and "cnt3". The status of each counter is indicated by its corresponding counter value.

The overall architecture of each timing controller is presented in Figure 10.4, involving a counter enable circuit and a sequential counter. The signals labeled with "x" in the names, such as "st$x$_rdy", "cnt$x$_en", "nxt_cnt$x$", and "cnt$x$", correspond to the control signals for all the counters: "cnt0" to "cnt3". For instance, when designing the counter labeled as "cnt0", the associated signals should be "st0_rdy", "cnt0_en", "nxt_cnt0", and "cnt0".

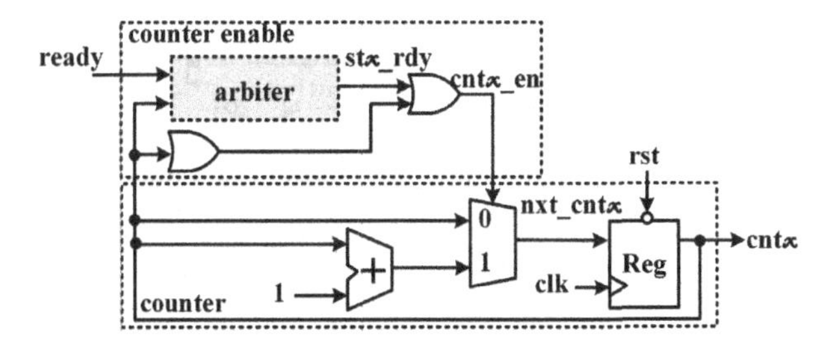

FIGURE 10.4
Timing controller of streaming design on DDOT.

In the counter enable circuit, the "cnt$x$_en" output is activated by the "st$x$_rdy" pulse signal and stays active as long as the counter is actively monitoring data. This condition is determined through the use of a Reduction OR gate. The "st$x$_rdy" trigger signal is controlled by the arbiter, a critical component responsible for ensuring the desired rotation sequence of counters within the counter enable circuit.

A. Arbiter Design

The arbiter's role encompasses selecting the highest-priority controller that currently resides in an idle status and initiating its operation by asserting the corresponding "st$x$_rdy" signal. To illustrate, if "cnt0" is in an idle state, it's activated using a designated pulse signal labeled as "st0_rdy". In cases where "cnt0" is occupied, but "cnt1" is idle, the arbiter triggers the start of "cnt1" via a pulse signal termed "st1_rdy". Similarly, when both "cnt0" and "cnt1" are occupied, yet "cnt2" is idle, the arbiter initiates "cnt2" using a pulse signal denoted as "st2_rdy". In a scenario where all counters from "cnt0" to "cnt2" are engaged, but "cnt3" is idle, the arbiter commences "cnt3" with a pulse signal named "st3_rdy".

To distinguish between idle and busy states, a NOT Reduction OR gate can be employed to represent the former (when a counter equals zero), while a standard Reduction OR gate serves to denote the latter (when a counter isn't zero). In particular, Figure 10.5(a) provides an illustration of the arbiter's design for "cnt0". In case "cnt0" is in an idle state (as signaled by a NOT Reduction OR gate) and a valid data frame is present on the bus (as indicated by the asserted "ready" input), the initiation of "cnt0" becomes feasible.

Moving to the second highest priority, "cnt1" can commence its operation if both "cnt0" is currently active (evident through a Reduction OR gate) and "cnt1" is idle (conveyed by a NOT Reduction OR gate). This specific hardware configuration is visualized in Figure 10.5(b). The designs for the arbiters controlling "cnt2" and "cnt3" are respectively depicted in Figure 10.5(c) and Figure 10.5(d). For "cnt2" to initiate, both "cnt0" and "cnt1" must be active, and "cnt2" itself should be in an idle state. As for "cnt3", being of the lowest

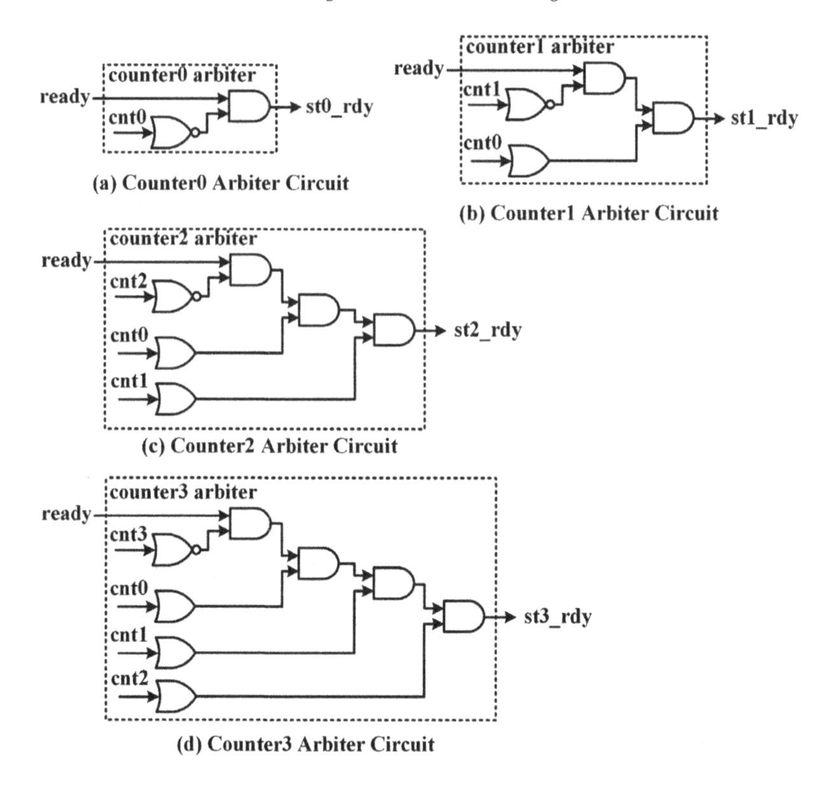

FIGURE 10.5
Arbiter design of timing controller (streaming design on DDOT).

priority, its activation is contingent upon all the other three counters being engaged.

B. Timing Controllers

Figure 10.6 illustrates the timing diagram encompassing the behavior of the four controllers across eight successive data frames. Commencing with the initial data frame, control is assumed by the "cnt0" spanning clock cycles 1 to 4. This is pursued by the second data frame regulated by "cnt1", operating from clock cycles 2 to 5, then the third data frame guided by "cnt2" spanning clock cycles 3 to 6. Subsequently, the fourth data frame is overseen by "cnt3" over clock cycles 4 to 7.

Upon the arrival of the fifth data frame within the *ddot* design, the availability of an idle "cnt0" enables its reutilization for managing the fifth data frame, spanning clock cycles 5 to 8. This rotational sequence proceeds, with "cnt1" controlling the sixth data frame over clock cycles 6 to 9, followed by "cnt2" overseeing the seventh data frame across clock cycles 7 to 10, and ultimately, "cnt3" governing the eighth data frame throughout clock cycles 8 to

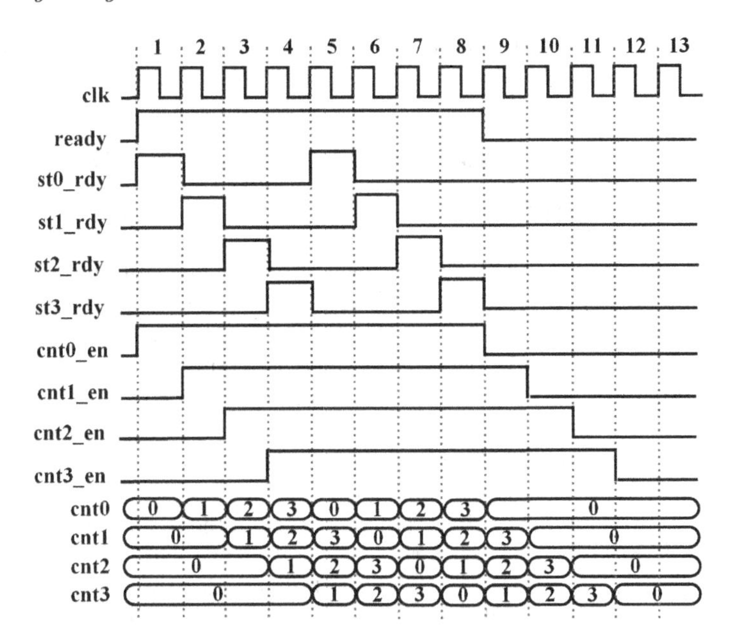

FIGURE 10.6
Timing diagram of timing controller (streaming design on DDOT).

11. This coordinated rotation of the four counters not only facilitates parallel processing of four data frames but also permits the sequential handling of consecutive data frames through the streaming design mechanism.

To elaborate, the activation of the counter enable signal is synchronized with the corresponding start indicator and remains active until the counter attains a value of 2'd3. To illustrate this, consider the first clock cycle, where the initiation signal "st0_rdy" is asserted, subsequently triggering the enablement of the counter through the signal "cnt0_en". This activation persists until the counter "cnt0" reaches the value of 2'd3 in the fourth clock cycle. As the fifth clock cycle arrives, the reactivation of "cnt0_en" coincides with the renewed assertion of "st0_rdy", and this persists until "cnt0" reaches the value of 2'd3 in the eighth clock cycle. As the ninth clock cycle dawns, the counter enable signal "cnt0_en" is then deactivated.

Likewise, the counter enable signals "cnt1_en", "cnt2_en", and "cnt3_en" are successively asserted during the second, third, and fourth clock cycles, respectively, in correspondence with their respective start indicators "st1_rdy", "st2_rdy", and "st3_rdy". These signals endure until their corresponding counters "cnt1", "cnt2", and "cnt3" each attain the maximum value of 2'd3 in clock cycles 9, 10, and 11. Subsequently, these enable signals are deasserted. Despite reaching their maximum values in clock cycles 6, 7, and 8, the counters "cnt1", "cnt2", and "cnt3" continue their operation. This continuation is enabled by the reassertion of their respective start indicators "st1_rdy", "st2_rdy", and "st3_rdy", prompted by the consecutive input of data frames.

10.2.4 Verilog code for streaming design on DDOT

Given that the datapath design is the same as the previous foundational structure outlined in Chapter 9, the Verilog code only for the timing controller is presented below. Specifically, lines 5–8 elucidate the description of the arbiters responsible for initiating the four counters. The highest-priority counter, labeled as "cnt0", is addressed first. Its start indicator, denoted as "st0_rdy", can be activated under the condition that the counter is idle (indicated by $\sim|cnt0$), and simultaneously, the "ready" input is asserted.

Moving on to the second counter, represented by "cnt1", its start indicator "st1_rdy" is considered only if the preceding "cnt0" counter is engaged (as indicated by $|cnt0$), and "cnt1" itself is in an idle state (as conveyed by $\sim|cnt1$). Continuing to the third indicator, designated as "st2_rdy", its activation hinges on the busyness of both "cnt0" and "cnt1", while "cnt2" is idle, and the "ready" input is asserted. Finally, the fourth indicator, termed "st3_rdy", can be set into motion if all three counters "cnt0", "cnt1", and "cnt2" are busy, while "cnt3" remains idle, and the "ready" input receives an assertion.

```
1  //************************************
2  //**** Timing Controller Design ********
3  //************************************
4  reg  [1:0] cnt0, cnt1, cnt2, cnt3;
5  wire st0_rdy = ready & ~|cnt0 ;
6  wire st1_rdy = ready & ~|cnt1 & |cnt0 ;
7  wire st2_rdy = ready & ~|cnt2 & |cnt0 & |cnt1 ;
8  wire st3_rdy = ready & ~|cnt3 & |cnt0 & |cnt1 & |cnt2 ;
9
10 wire cnt0_en = st0_rdy | |cnt0 ;
11 wire cnt1_en = st1_rdy | |cnt1 ;
12 wire cnt2_en = st2_rdy | |cnt2 ;
13 wire cnt3_en = st3_rdy | |cnt3 ;
14
15 wire [1:0] nxt_cnt0 = cnt0_en ? (cnt0+2'd1) : cnt0;
16 wire [1:0] nxt_cnt1 = cnt1_en ? (cnt1+2'd1) : cnt1;
17 wire [1:0] nxt_cnt2 = cnt2_en ? (cnt2+2'd1) : cnt2;
18 wire [1:0] nxt_cnt3 = cnt3_en ? (cnt3+2'd1) : cnt3;
19
20 always @(posedge clk) begin
21   if(rst) begin
```

```
22        cnt0<=2'd0     ;
23        cnt1<=2'd0     ;
24        cnt2<=2'd0     ;
25        cnt3<=2'd0     ;
26      end else begin
27        cnt0<=nxt_cnt0;
28        cnt1<=nxt_cnt1;
29        cnt2<=nxt_cnt2;
30        cnt3<=nxt_cnt3;
31      end
32    end
33
34    wire nxt_vld = &cnt0 | &cnt1 | &cnt2 | &cnt3;
35    always @(posedge clk) begin
36      if(rst) begin
37        vld <=1'b0     ;
38      end else begin
39        vld <=nxt_vld ;
40      end
41    end
```

Within lines 10–13, the counters initiate their operations upon the activation of the ready signals, denoted as "st0_rdy" through "st3_rdy", and persist as long as the counters remain engaged. The utilization of Reduction OR gates, specifically $|cnt0$, $|cnt1$, $|cnt2$, and $|cnt3$, facilitates the determination of the counter's busy status.

Moving to the combinational components of the counters, which are typically configured as outlined in lines 15–18, their behavior is expounded. When the enable signals are asserted, the counters experience an increment by one; conversely, in the absence of the enable signals, they maintain their present register output value. It's noteworthy that in this scenario, the counters are inherently initialized to 2'd0, since the maximum value of these 2-bit counters is 2'd3, resulting in the subsequent count rolling back to 2'd0. This aspect negates the need for manual initialization of the counters. Proceeding to lines 20–32, registers come into play for the purpose of storing the counter results across different clock cycles.

Lastly, lines 34–41 encapsulate the formulation of the output valid signal "vld". This signal's design is delineated, wherein its assertion hinges on any of the counters reaching their maximum value of 2'd3, thus triggering the activation of the "nxt_vld" signal. Subsequently, following a single clock cycle, the "vld" output is registered to signify the successful transmission of valid data.

10.3 Iterative design with streaming width four

The streaming design of the *ddot* engine featured eight sets of single-precision FP inputs, labeled as "x0-x7" for the "x" input and "y0-y7" for the "y" input. This particular configuration required a total of 512 single-bit IO connections (resulting from 16 × 32 connections), along with the incorporation of eight FP multipliers and seven FP adders. To strike a balance between resource utilization and meeting the specified requirements, an iterative approach was employed in this section. This approach enabled a reduction in the streaming width within the design engine, thereby decreasing the number of necessary IO connections and overall resource costs.

10.3.1 Iterative design structure with streaming width four

In an effort to reduce both the quantity of necessary IO connections and the associated resource expenses, an iterative design structure, depicted in Figure 10.7, is introduced. This new configuration employs a streaming width of four, enabling the input of a complete data frame, comprising eight FP data, within two clock cycles. Specifically, each cycle accommodates the feeding of four FP data into the design engine. By adopting this approach, the number of required IOs is effectively halved, involving four FP data inputs labeled "x0-x3" and another four labeled "y0-y3". Moreover, the design requisites are streamlined to encompass solely four FP multipliers and three consecutive cascading FP adders, collectively constructing the multiplication-addition circuit, denoted as *MAC* within the figure.

The *MAC* block is employed twice, resulting in two distinct outcomes, with each operation involving the multiplication and addition of four FP data inputs. To facilitate this process, the outcome of the initial round is stored

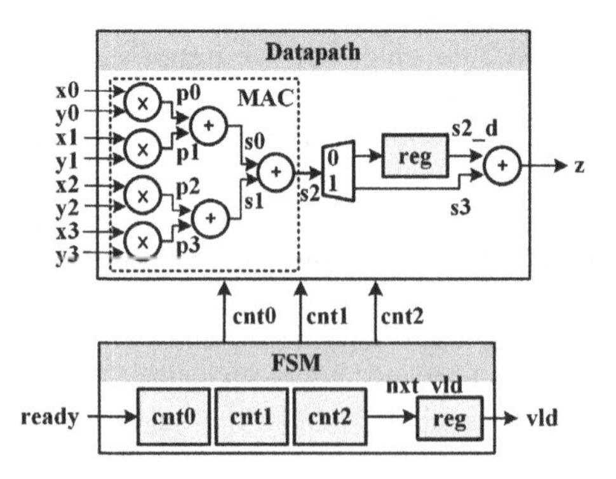

FIGURE 10.7
Iterative design structure with streaming width four.

using a register to introduce a single-clock delay. Subsequently, this result is accumulated with the second round MAC output, ultimately culminating in the production of the ultimate output labeled "z".

The necessary number of timing controllers can be determined through Equation 10.1. Given a streaming width of four, referred to as $STW = 4$, and a data frame size of eight, denoted as $N = 8$, the quantity of clock cycles needed to input each data frame is calculated as $NO\_CY = \lceil N/STW \rceil = \lceil 8/4 \rceil = 2$. In alignment with Figure 10.7, the most extended latency within the datapath encompasses five clock cycles. This entails one clock cycle for the parallel execution of four FP multiplications, two clock cycles dedicated to the sequential cascade of additions, another clock cycle for the registration of the interim outcome "s2_d", and finally, a concluding clock cycle for the summation "z". Consequently, the requisite tally of timing controllers can be calculated as $NO\_TC = \lceil Longest\_Path/NO\_CY \rceil = \lceil 5/2 \rceil = 3$. As a result shown in the figure, three counters are effectively employed as timing controllers, ensuring synchronization with the input of consecutive data frames and the subsequent processing stages.

10.3.2 Timing diagram of iterative design with streaming width four

Illustrated in Figure 10.8 is a comprehensive timing diagram that outlines the data processing of four successive data frames, each encompassing eight FP data. With a design embracing a streaming width of four, the input of each data frame can be accomplished over a span of two clock cycles. To streamline the depiction while retaining coherence, we adopt the same assumption to the one established in Section 10.2.2. In this context, it is presumed that all the data transmitted through the "x0-x3" buses remain uniform within a given data frame, and similarly, the contents conveyed via the "y0-y3" buses remain constant as well.

Notably, the initial data frame, housing eight FP 0.0 on the "x0-x3" buses and eight FP 1.0 on the "y0-y3" buses, enters the system across clock cycles 1 to 2. Subsequently, the second data frame, constituted of eight FP 1.0 on the "x0-x3" buses and eight FP 2.0 on the "y0-y3" buses, enters in clock cycles 3 to 4. This pattern persists until the fourth data frame, comprising eight FP 3.0 on the "x0-x3" buses and eight FP 4.0 on the "y0-y3" buses, is processed during clock cycles 7 to 8. In the interest of clarity and simplicity, the timing diagram solely illustrates the FP data on the "x0" bus and the "y0" bus.

As evident from the figure, the data processing coordination involves the utilization of three counters. The initial counter, referred to as "cnt0", boasts the highest priority and commences operation in the first clock cycle, precisely coinciding with the introduction of the first data frame into the engine. This counter persists through the fifth clock cycle, marking the conclusion of the longest data path, and recommences in the subsequent seventh clock cycle.

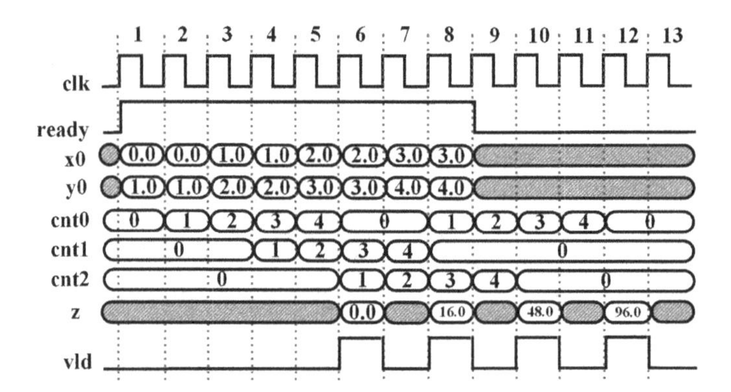

FIGURE 10.8
Timing diagram of iterative design with streaming width four.

Diving into specifics, during the first clock cycle, half of the initial data frame is channeled into the design engine, while the second half follows in the subsequent clock cycle. Referencing Figure 10.7, the products "p0-p3" can be obtained in the second clock cycle, the summations "s0-s1" can be reached by the third clock cycle, and the intermediate MAC result "s2" for the first half data frame can be received by the fourth clock cycle. Subsequently, this intermediate outcome is registered as "s2_d" within the fifth clock cycle. By the fifth clock cycle, the MAC outcome "s2" for the second half of the same data frame also comes to the fore, subsequently being channeled to the "s3" bus. Significantly, the ultimate outcome can be achieved by accumulating "s2_d" and "s3", effectuating the final summation "z = FP 0.0" within the sixth clock cycle.

Simultaneously, the counters "cnt1" and "cnt2" contribute to the concurrent management of the second and third data frames. Counter "cnt1" initiates its operation in the third clock cycle and concludes by the seventh clock cycle, while "cnt2" commences activity in the fifth clock cycle and operates until the ninth clock cycle. Capitalizing on a one-cycle delay, the second *ddot* output "z = FP 16.0" takes form during clock cycle 8, followed by the emergence of the third output "z = FP 48.0" in clock cycle 10. The fourth and final data frame is subsequently processed, employing the rotational usage of counter "cnt0". The final output "z = FP 96.0" comes to the output within clock cycle 12.

10.3.3 Timing controller of iterative design with streaming width four

The timing controller design for the iterative design structure closely resembles that of the streaming design, but with a minor variation. In the context of an iterative design, an additional consideration comes into play: the need to define the initial clock cycle for each data frame input, given that the iterative

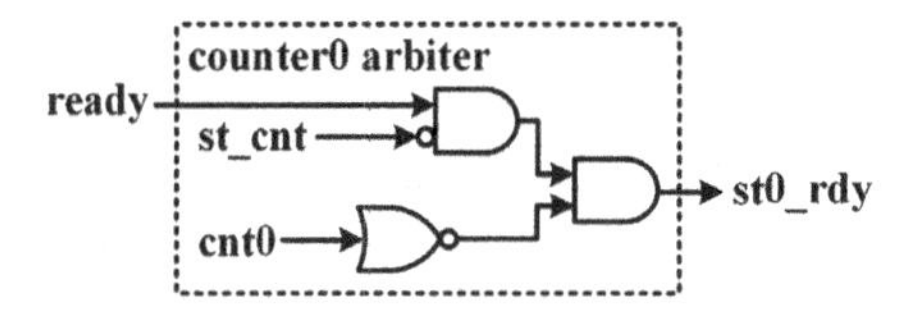

FIGURE 10.9
Arbiter design of timing controller (iterative design with streaming width four).

approach takes more than one clock cycle to fully input each data frame. This distinction is particularly relevant when considering different streaming widths. For instance, with a streaming width of four, it takes two clock cycles to input a data frame containing eight FP data points. If the streaming width is two, then it takes four clock cycles. To achieve this, the start indicator "st_rdy" must be asserted during the first clock cycle of each data frame input period, functioning as a pulse signal to signal the start of the input process. This requires an additional counter, referred to as "st_cnt", to track the clock cycles within each data frame input period.

The arbiter design for the first counter, "cnt0", is depicted in Figure 10.9. In an iterative design employing a streaming width of four, each data frame input takes two clock cycles. The "st_cnt" counter is instrumental in distinguishing these two clock cycles: the first clock cycle is marked when "st_cnt" is zero, leading to the use of an inverter output as the input of the AND gate. In contrast, the second clock cycle is identified when "st_cnt" equals one, causing the output of the AND gate to be driven to zero and subsequently deactivating the "st0_rdy" signal. To determine if "cnt0" is in an idle state, a NOT Reduction OR gate is employed, as depicted in the figure.

In summary, the initiation of the start indicator "st0_rdy" is exclusively allowed during the first clock cycle of each data frame input period, and only when "cnt0" is in an idle state. This approach ensures the synchronization of the iterative design structure with the input timing requirements.

The timing diagram illustrating the controller design is displayed in Figure 10.10. Much like the streaming design, the start indicators are sequentially activated in line with the priority of the counters, progressing from "st0_rdy" to "st1_rdy", and finally to "st2_rdy". It's crucial to bear in mind that these indicators are asserted every two clock cycles, as each data frame necessitates two clock cycles to be introduced into the design engine. The "st_cnt" counter is utilized to keep track of the 2-clock cycle duration of each data frame input.

The counter enable signals "cnt0_en", "cnt1_en", and "cnt2_en" are triggered the moment the corresponding start indicator is active. These counters commence counting from decimal 3'd0 to 3'd4, guided by the asserted enable signals, and reset when reaching the maximum count of decimal 3'd4. These counters coordinate data processing in a rotational fashion, thereby enabling parallel data streams. For instance, the initial counter "cnt0" oversees the

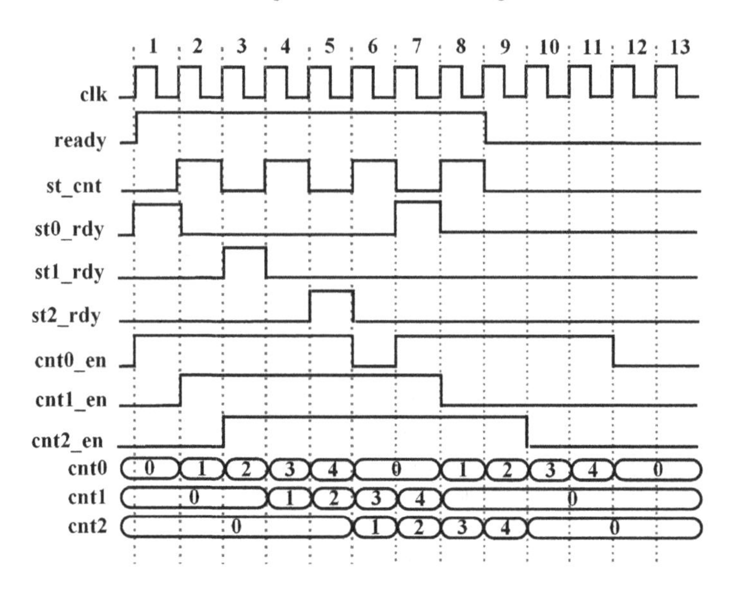

FIGURE 10.10
Timing diagram of timing controller (iterative design with streaming width four).

processing of the first data frame across clock cycles 1–5. The subsequent counter, "cnt1", takes the reins for the second data frame, spanning clock cycles 3–7. Meanwhile, the third counter, "cnt2", guides the processing of the third data frame within clock cycles 5–9. Upon the arrival of the fourth data frame in clock cycle 7, the first counter "cnt0" is freed up and can be repurposed to govern the fourth data frame's processing. Meanwhile, the other two counters can be cycled to manage the forthcoming data frames.

10.3.4 Verilog code for iterative design with streaming width four

Here's the Verilog description for the iterative design with a streaming width of four. The key difference in the datapath between this iterative design (lines 8–79) and the streaming design is how multipliers and adders are used, along with the inclusion of additional registers. These registers store and delay intermediate MAC results during multiple iterations.

In this illustration, intermediate results are stored using 32-bit registers (lines 61–70). A flag signal is employed to identify the fourth clock cycle within each data frame processing, allowing the "s2" result to be registered as "s2_d" and await the result of the next iteration, denoted as "s3". The second-iteration result becomes available in the fifth clock cycle within each data frame processing. During this fifth clock cycle, "s2_d" and "s3" are combined through the use of an FP adder, as shown in lines 75–79, to generate the final output "z"

```
module iterative_stw4_ddot (input                  clk   ,
                            input                  rst   ,
                            input                  ready ,
                            input         [31:0]   x0,x1,x2,x3,
                            input         [31:0]   y0,y1,y2,y3,
                            output reg             vld   ,
                            output        [31:0]   z     );
//********************************
//****    Datapath Design ********
//********************************
reg  [2:0]  cnt0, cnt1, cnt2;
wire [31:0] p0, p1, p2, p3;
wire [31:0] s0, s1, s2, s3;
wire        nxt_vld;

// The first stage of multipliers
FP_multiplier u0_fp_mul(.clock    (clk  ),
                        .reset    (rst  ),
                        .io_in_a  (x0   ),
                        .io_in_b  (y0   ),
                        .io_out_s (p0   ));

FP_multiplier u1_fp_mul(.clock    (clk  ),
                        .reset    (rst  ),
                        .io_in_a  (x1   ),
                        .io_in_b  (y1   ),
                        .io_out_s (p1   ));

FP_multiplier u2_fp_mul(.clock    (clk  ),
                        .reset    (rst  ),
                        .io_in_a  (x2   ),
                        .io_in_b  (y2   ),
                        .io_out_s (p2   ));

FP_multiplier u3_fp_mul(.clock    (clk  ),
                        .reset    (rst  ),
                        .io_in_a  (x3   ),
                        .io_in_b  (y3   ),
                        .io_out_s (p3   ));

// The second stage of adders
FP_adder u0_fp_add(.clock   (clk  ),
                   .reset   (rst  ),
                   .io_in_a (p0   ),
```

```
45                    .io_in_b (p1   ),
46                    .io_out_s(s0   ));
47
48 FP_adder u1_fp_add(.clock   (clk  ),
49                    .reset   (rst  ),
50                    .io_in_a (p2   ),
51                    .io_in_b (p3   ),
52                    .io_out_s(s1   ));
53
54 // The third stage of adders
55 FP_adder u2_fp_add(.clock   (clk  ),
56                    .reset   (rst  ),
57                    .io_in_a (s0   ),
58                    .io_in_b (s1   ),
59                    .io_out_s(s2   ));
60
61 wire s2_d_flag = &cnt0[1:0] | &cnt1[1:0] | &cnt2[1:0];
62 wire [31:0] nxt_s2_d = s2_d_flag ? s2 : 32'h0;
63 reg  [31:0] s2_d;
64 always @(posedge clk) begin
65   if(rst) begin
66     s2_d <= 32'h0    ;
67   end else begin
68     s2_d <= nxt_s2_d;
69   end
70 end
71
72 assign s3 = cnt0[2]|cnt1[2]|cnt2[2] ? s2 : 32'h0;
73
74 // The fourth stage of adders
75 FP_adder u3_fp_add(.clock   (clk  ),
76                    .reset   (rst  ),
77                    .io_in_a (s2_d ),
78                    .io_in_b (s3   ),
79                    .io_out_s(z    ));
80
81 //*************************************
82 //****    FSM Controller Design ********
83 //*************************************
84 reg  st_cnt;
85 wire nxt_st_cnt = ready ? st_cnt+1'b1 : st_cnt;
86 always @(posedge clk) begin
87   if(rst) begin
88     st_cnt<=1'd0      ;
89   end else begin
90     st_cnt<=nxt_st_cnt;
91   end
92 end
93
```

```
94  wire st0_rdy = ready & ~st_cnt & ~|cnt0 ;
95  wire st1_rdy = ready & ~st_cnt & ~|cnt1 & |cnt0;
96  wire st2_rdy = ready & ~st_cnt & ~|cnt2 & |cnt0 & |cnt1;
97
98  wire cnt0_en = st0_rdy | |cnt0 ;
99  wire cnt1_en = st1_rdy | |cnt1 ;
100 wire cnt2_en = st2_rdy | |cnt2 ;
101
102 wire [2:0] nxt_cnt0 = cnt0[2] ? 3'd0 :
103                       cnt0_en ? (cnt0+3'd1) : cnt0;
104 wire [2:0] nxt_cnt1 = cnt1[2] ? 3'd0 :
105                       cnt1_en ? (cnt1+3'd1) : cnt1;
106 wire [2:0] nxt_cnt2 = cnt2[2] ? 3'd0 :
107                       cnt2_en ? (cnt2+3'd1) : cnt2;
108 always @(posedge clk) begin
109   if(rst) begin
110     cnt0<=3'd0    ;
111     cnt1<=3'd0    ;
112     cnt2<=3'd0    ;
113   end else begin
114     cnt0<=nxt_cnt0;
115     cnt1<=nxt_cnt1;
116     cnt2<=nxt_cnt2;
117   end
118 end
119
120 assign nxt_vld = cnt0[2] | cnt1[2] | cnt2[2];
121 always @(posedge clk) begin
122   if(rst) begin
123     vld <=1'b0    ;
124   end else begin
125     vld <=nxt_vld ;
126   end
127 end
128 endmodule
```

The data stream counter "st_cnt" is established in lines 84–92 to keep track of the clock cycles associated with each data frame input. It begins counting as soon as the data input becomes available on the bus and resets to zero once it reaches its maximum value. Subsequently, the logic governing the initiation of start indicators, namely "st0_rdy" through "st2_rdy", is described in lines 94–96. These indicators follow a priority sequence: "st0_rdy" begins first, followed by "st1_rdy", and finally "st2_rdy". Here's how each of these initiations is determined: The counter "cnt0" becomes active in the initial clock cycle when it's idle, as indicated by $\sim |cnt0$ in line 94. Additionally, the data stream counter "st_cnt" must be equal to zero, denoted as $\sim st\_cnt$. The counter "cnt1" starts when "cnt0" is already active, as expressed by $|cnt0$ in line 95. Counter "cnt2" commences when both "cnt0" and "cnt1" are operational, as indicated by $|cnt0$ & $|cnt1$ in line 96.

The counter enable signals, "cnt0_en", "cnt1_en", and "cnt2_en", are discussed in lines 98–100. These signals are triggered by the start indicators, namely "st0_rdy", "st1_rdy", and "st2_rdy", and they are deactivated when the counters are reset to zero. The reset process is controlled through lines 102–107. Upon reaching the maximum value of decimal 3'd4, the counters are reset to decimal 3'd0. This maximum value determination is facilitated by examining the most significant bit – if it's set to true, the counter has reached its maximum value and resets to zero.

The validity indicator "vld" is also triggered by the maximum counter value, as evident in lines 120. One clock cycle later, as detailed in the register design from lines 121–127, the "vld" output is activated, signifying the validity of the *ddot* result on the "z" output.

10.4 Iterative Design with Streaming Width Two

In this section, we delve into the iterative design with a streaming width of two, highlighting its distinctions from the iterative design employing a streaming width of four. In contrast to the latter approach, the former demands a reduced number of FP operators and IOs. However, it comes with an extended latency due to the necessity of multiple iterations to recycle the multiplication-addition circuit.

10.4.1 Iterative design structure with streaming width two

Figure 10.11 illustrates the iterative design structure employing a streaming width of two. This configuration mandates four clock cycles for the input of each data frame comprising eight FP data. The design incorporates two FP multipliers and one FP adder to construct the multiplication-addition circuit, referred to as MAC in the figure. Additionally, it employs three FP adders to accumulate the outcomes of the 4-round MAC computation. Notably, the initial MAC output necessitates a one-clock cycle delay, leading to the insertion of a register after the MAC circuit. This MAC circuit is reused four times to generate the outputs of the first, second, third, and fourth rounds, namely "s0_d", "s1", "s2", and "s3".

To delve further into the specifics, the first-round MAC outcome, denoted as "s0", is selected through a multiplexer and subsequently stored in the register labeled "s0_d". This value is then subjected to addition with the second-round result, accessible via "s1", yielding the summation displayed on "s4". Following this, the third-round result, available on "s2", is combined with "s4",

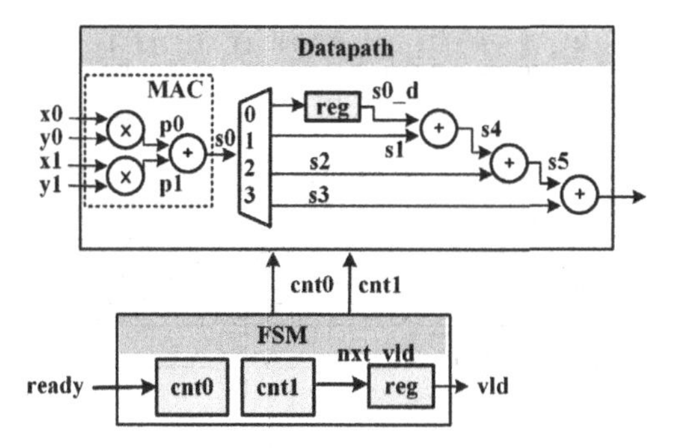

FIGURE 10.11
Iterative design structure with streaming width two.

resulting in the summation on "s5". Finally, the outcome of the last iterative round, attainable from "s3", is added to "s5", culminating in the ultimate output designated as "z". It is essential to highlight that each successive cascading FP adder inherently integrates a register-out, negating the need for supplementary registers for the sums of "s4" and "s5". The interconnected cascading adders play a pivotal role in upholding the pipeline design structure, enabling the seamless processing of consecutive data frames in a streaming fashion.

The calculation of required timing controllers can be accomplished using Equation 10.1. Given a streaming width of two, indicated as $STW = 2$, and a data frame size of eight, referred as $N = 8$, the number of clock cycles needed for each data frame input is determined as $NO\_CY = \lceil N/STW \rceil = 4$. Examining Figure 10.11, the most extensive route within the *ddot* design involves six clock cycles, encompassing a series of five FP operators and a register for "s0_d" output registration. Consequently, the count of required timing controllers can be expressed as $NO\_TC = \lceil Longest\_Path/NO\_CY \rceil = \lceil 6/4 \rceil = 2$.

10.4.2 Timing diagram of iterative design with streaming width two

Figure 10.12 presents a comprehensive timing diagram illustrating the data processing of four successive data frames, each encompassing eight FP data elements. To maintain consistency with the previous Sections 10.3.2 and 10.2.2, we adopt the same test case and make an assumption that all data on the "x0-x1" buses remains consistent within each data frame, as does the data on the "y0-y1" buses. To delve into specifics, let's consider each data frame's entry and progression.

Commencing with the initial data frame, comprised of eight FP 0.0 on the "x0-x1" buses and eight FP 1.0 on the "y0-y1" buses, this data sequence enters the design engine over the span of clock cycles 1–4. Subsequently, the second

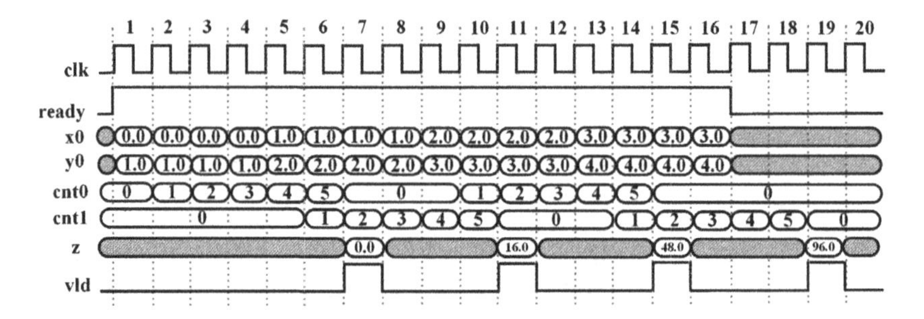

FIGURE 10.12
Timing diagram of iterative design with streaming width two.

data frame makes its ingress, housing eight FP 1.0 on the "x0-x1" buses and eight FP 2.0 on the "y0-y1" buses. This data sequence undergoes processing during clock cycles 5–8. Similarly, the third data frame, comprising eight FP 2.0 on the "x0-x1" buses and eight FP 3.0 on the "y0-y1" buses, enters the engine for computation over clock cycles 9–12. Lastly, the fourth data frame, encompassing eight FP 3.0 on the "x0-x1" buses and eight FP 4.0 on the "y0-y1" buses, initiates processing during clock cycles 13–16. In order to simplify the diagram, only the FP data on the "x0" and "y0" buses are showcased.

As depicted in the figure, the data processing unfolds through the utilization of two counters, namely, "cnt0" and "cnt1", operating in parallel. To delve into specifics, "cnt0" takes charge of overseeing the computation for the first data frame, spanning clock cycles 1–6. Referencing Figure 10.11, the initial *MAC* outcome surfaces during clock cycle 3, subsequently being registered as "s0_d" to accommodate the arrival of the second *MAC* outcome, "s1". Progressing through clock cycles 4–6, the second, third, and fourth *MAC* results materialize on the "s1", "s2", and "s3" buses, respectively. As clock cycles 5–7 ensue, the accumulations of "s4", "s5", and the ultimate output "z = FP 0.0" are dispatched, characterized by a one-clock cycle delay inherent to the cascading additions.

The second data frame falls under the control of "cnt1". Within clock cycle 11, the conclusive *ddot* result "z = FP 16.0" is unveiled. As the third data frame unfolds, "cnt0" can be reemployed, leading to the emergence of the *ddot* result "z = FP 48.0" at clock cycle 15. In culmination, "cnt1" steps in once again to oversee the fourth data frame, culminating in a *ddot* result "z = FP 96.0" during clock cycle 19.

10.4.3 Timing controller of iterative design with streaming width two

Similar to the timing controller design for a streaming width of four, a pulse starter, named "st_rdy", is needed to signal the commencement of each data

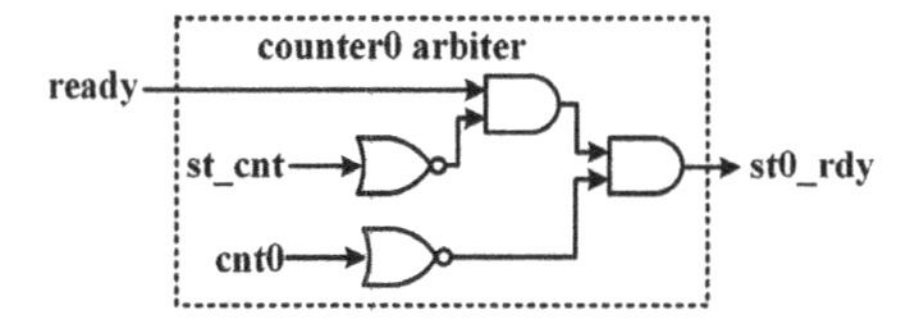

FIGURE 10.13
Arbiter design of timing controller (iterative design with streaming width two).

frame. To fulfill this role, a data stream counter, denoted as "st_cnt", is utilized to accurately track the progress of the four clock cycles required for each data frame input. This counter is designed with a bit size of two, encompassing the complete range from its minimum value of decimal 2'd0 to its maximum value of decimal 2'd3.

Figure 10.13 illustrates the arbiter design for the first timing controller. Within this circuit, a NOT Reduction OR gate is utilized to distinguish the instances when the "st_cnt" counter is decimal 2'd0. This condition signifies the initial clock cycle of data frame input.

The timing diagram detailing the controller operations is presented in Figure 10.14. An essential observation to make is the sequential assertion of the start indicators, namely "st0_rdy" and "st1_rdy", occurring at intervals of four clock cycles. This sequential triggering reflects their prioritized nature. With the activation of these start indicators, the corresponding counter enable signals, "cnt0_en" and "cnt1_en", experience activation. Given that the longest data path encompasses six clock cycles, the counters are set in motion, counting from the decimal 3'd0 to the decimal 3'd5 within the active enable signals.

In a practical scenario, the processing timeline unfolds as follows. The first data frame undergoes processing during clock cycles 1–6. Subsequently, the second data frame is processed spanning clock cycles 5–10. As the ninth clock

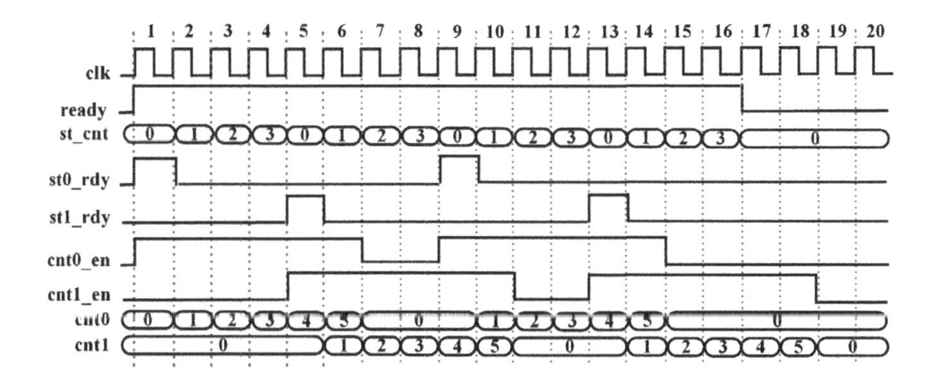

FIGURE 10.14
Timing diagram of timing controller (iterative design with streaming width two).

cycle begins, the third data frame enters the data engine. At this point, "cnt0" becomes available for reassignment and is repurposed to oversee the processing of this third data frame, following a rotational approach. Likewise, as the 13th clock cycle commences, the last data frame enters the processing pipeline. In this clock cycle, "cnt1" is idle while "cnt0" is already in use. Therefore, "cnt1" is repurposed to keep track of the processing for this final data frame.

10.4.4 Verilog code for iterative design with streaming width two

Below is the Verilog code illustrating the iterative design with a streaming width of two. In this implementation, only two FP multipliers and four FP adders are necessary, as demonstrated in lines 16–34 and 51–70. Furthermore, the code encompasses the declaration of a register for storing the intermediate sum "s0", as seen in lines 36–45. When the counter value is 3'd2, a deliberate one-cycle delay is applied to the initial *MAC* outcome "s0". This delay facilitates the subsequent accumulation of this result with the second iterative outcome "s1".

In relation to multiplexers, the code delineates the routing strategy for the "s0" output to various channels, as portrayed in lines 47–49. The sequential presentation of *MAC* results is governed by counter values 3'd3, 3'd4, and 3'd5. Accordingly, the output channels "s1", "s2", and "s3" sequentially receive these results. Subsequently, these results are fed into distinct FP adders in lines 51–70.

```
1  module iterative_st2_ddot (input                    clk   ,
2                             input                    rst   ,
3                             input                    ready ,
4                             input           [31:0]   x0,x1,
5                             input           [31:0]   y0,y1,
6                             output reg               vld   ,
7                             output          [31:0]   z     );
8  //********************************
9  //****    Datapath Design *********
10 //********************************
11 reg   [2:0]   cnt0, cnt1;
12 wire  [31:0]  p0, p1;
13 wire  [31:0]  s0, s1, s2, s3, s4, s5;
14 wire          nxt_vld;
15
16 // The first stage of multipliers
17 FP_multiplier u0_fp_mul(.clock    (clk  ),
18                         .reset    (rst  ),
19                         .io_in_a  (x0   ),
20                         .io_in_b  (y0   ),
```

```
21                              .io_out_s(p0   ));
22
23  FP_multiplier u1_fp_mul(.clock    (clk  ),
24                          .reset    (rst  ),
25                          .io_in_a  (x1   ),
26                          .io_in_b  (y1   ),
27                          .io_out_s(p1   ));
28
29  // The second stage of adder
30  FP_adder u0_fp_add(.clock    (clk  ),
31                     .reset    (rst  ),
32                     .io_in_a  (p0   ),
33                     .io_in_b  (p1   ),
34                     .io_out_s(s0   ));
35
36  wire s0_d_flag = cnt0==3'd2|cnt1==3'd2;
37  wire [31:0] nxt_s0_d = s0_d_flag ? s0 : 32'h0;
38  reg  [31:0] s0_d;
39  always @(posedge clk) begin
40    if(rst) begin
41      s0_d<=32'h0     ;
42    end else begin
43      s0_d<=nxt_s0_d ;
44    end
45  end
46
47  assign s1 = (cnt0==3'd3|cnt1==3'd3) ? s0 : 31'h0;
48  assign s2 = (cnt0==3'd4|cnt1==3'd4) ? s0 : 31'h0;
49  assign s3 = (cnt0==3'd5|cnt1==3'd5) ? s0 : 31'h0;
50
51  // The third stage of adder
52  FP_adder u1_fp_add(.clock    (clk  ),
53                     .reset    (rst  ),
54                     .io_in_a  (s0_d ),
55                     .io_in_b  (s1   ),
56                     .io_out_s(s4   ));
57
58  // The fourth stage of adder
59  FP_adder u2_fp_add(.clock    (clk  ),
60                     .reset    (rst  ),
61                     .io_in_a  (s4   ),
62                     .io_in_b  (s2   ),
63                     .io_out_s(s5   ));
64
65  // The fifth stage of adder
66  FP_adder u3_fp_add(.clock    (clk  ),
67                     .reset    (rst  ),
68                     .io_in_a  (s5   ),
69                     .io_in_b  (s3   ),
```

```
70                          .io_out_s(z    ));
71
72  //*************************************
73  //****    FSM Controller Design ********
74  //*************************************
75  reg  [1:0] st_cnt;
76  wire [1:0] nxt_st_cnt  = ready ? st_cnt+2'd1 : st_cnt;
77  always @(posedge clk) begin
78    if(rst) begin
79      st_cnt<=2'd0      ;
80    end else begin
81      st_cnt<=nxt_st_cnt;
82    end
83  end
84
85  wire st0_rdy = ready & ~|st_cnt & ~|cnt0;
86  wire st1_rdy = ready & ~|st_cnt & ~|cnt1 & |cnt0 ;
87
88  wire cnt0_en = st0_rdy | |cnt0 ;
89  wire cnt1_en = st1_rdy | |cnt1 ;
90
91  wire [2:0] nxt_cnt0 = cnt0==3'd5 ? 3'd0 :
92                                 cnt0_en ? cnt0+3'd1 : cnt0;
93  wire [2:0] nxt_cnt1 = cnt1==3'd5 ? 3'd0 :
94                                 cnt1_en ? cnt1+3'd1 : cnt1;
95  always @(posedge clk) begin
96    if(rst) begin
97      cnt0<=3'd0    ;
98      cnt1<=3'd0    ;
99    end else begin
100     cnt0<=nxt_cnt0;
101     cnt1<=nxt_cnt1;
102   end
103 end
104
105 assign nxt_vld = cnt0==3'd5|cnt1==3'd5;
106 always @(posedge clk) begin
107   if(rst) begin
108     vld <=1'b0    ;
109   end else begin
110     vld <=nxt_vld ;
111   end
112 end
113 endmodule
```

The timing controller design includes a 2-bit streaming counter named "st_cnt". As described in lines 75–83, this counter is responsible for accurately counting the four clock cycles associated with each incoming data frame. When "st_cnt" equals zero, it signifies the first clock cycle of each data frame input. During the initial clock cycle, the starter signal "st0_rdy" becomes active when "cnt0" is in an idle state (indicated as $\sim |cnt0$ in line 85). Conversely, the pulse starter signal "st1_rdy" becomes active when "cnt0" is active (expressed as $|cnt0$ in line 86) and "cnt1" is in an idle state (noted as $\sim |cnt1$ in line 86).

The design for generating counter enable signals, "cnt0_en" and "cnt1_en", is outlined in lines 88–89. These signals enable the counters when the corresponding start indicators are asserted. The counters are active until they are reset to zero. The counter designs themselves are described in lines 91–103, where they increment by decimal 3'd1 with each clock cycle when enabled. Upon reaching the maximum value of decimal 3'd5, the counters reset to decimal 3'd0.

The design for the validation signal, referred to "vld", is detailed in lines 105–112. The conclusion of data processing is signaled by the counters reaching their maximum value. Subsequently, after a single clock cycle, the valid *ddot* result is produced, and the "vld" signal is asserted, indicating the presence of valid data on the "z" output.

PBL 28: Iterative design on DDOT with streaming width one

1) Design a *ddot* circuit utilizing the signal-clock-cycle FP operators, symbolized as $z = x^T \times y$, where x and y represent FP vectors of size eight, and z signifies the resulting *ddot* value. The mathematical representation is as follows:

$$z = \begin{pmatrix} x0 & x1 & x2 & x3 & x4 & x5 & x6 & x7 \end{pmatrix} * \begin{pmatrix} y0 \\ y1 \\ y2 \\ y3 \\ y4 \\ y5 \\ y6 \\ y7 \end{pmatrix} \tag{10.2}$$

$$= x0 \cdot y0 + x1 \cdot y1 + x2 \cdot y2 + x3 \cdot y3 + x4 \cdot y4 + x5 \cdot y5 + x6 \cdot y6 + x7 \cdot y7$$

The design can employ the FP multiplier and FP adder as IPs. As the design structure depicted in Figure 10.15, the *ddot* design's streaming width is set to one, permitting the utilization of a single FP multiplier for the multiplication-addition procedure. The accumulation of products generated in each multiplication iteration requires the utilization of multiple following FP adders. Consider

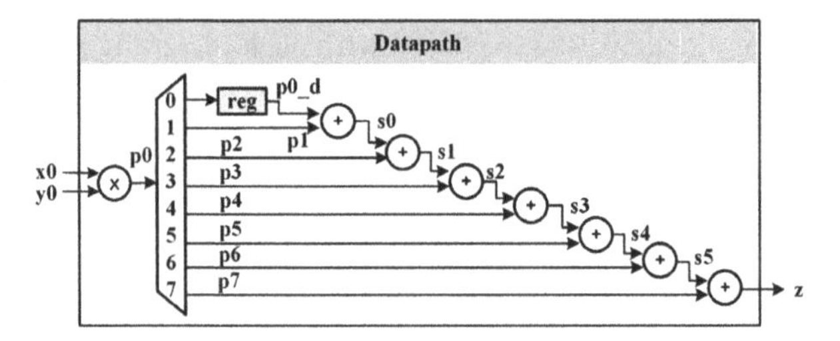

FIGURE 10.15
Iterative design datapath with streaming width one.

the significant demand for timing controllers to facilitate data processing efficiently in both a pipeline and parallel manner.

2) Create a testbench to facilitate the simulation of the *ddot* circuit. Validate the functionality of the circuit using the following specific test case: the FP vector x is composed of eight instances of FP 1.0, represented in IEEE-754 format as 32'h3f800000, and the FP vector y is comprised of eight instances of FP 2.0, represented in IEEE-754 format as 32'h40000000.

$$\begin{pmatrix} x0 \\ x1 \\ x2 \\ x3 \\ x4 \\ x5 \\ x6 \\ x7 \end{pmatrix} = \begin{pmatrix} 1.0 \\ 1.0 \\ 1.0 \\ 1.0 \\ 1.0 \\ 1.0 \\ 1.0 \\ 1.0 \end{pmatrix} \tag{10.3}$$

and

$$\begin{pmatrix} y0 \\ y1 \\ y2 \\ y3 \\ y4 \\ y5 \\ y6 \\ y7 \end{pmatrix} = \begin{pmatrix} 2.0 \\ 2.0 \\ 2.0 \\ 2.0 \\ 2.0 \\ 2.0 \\ 2.0 \\ 2.0 \end{pmatrix} \tag{10.4}$$

PBL 29: Streaming Design on DFT2

1) Design a hardware circuit for the discrete Fourier transform of size 2 ($DFT2$), utilizing the signal-clock-cycle FP operators. The general formula for a DFT is a linear transform represented as:

$$y[k] = \sum_{n=0}^{N-1} x[n]\omega_N^{kn} \tag{10.5}$$

Here, x and y represent one-dimensional input/output arrays of size N. The symbol $\omega_N = e^{-j\frac{2\pi}{N}}$, where $j^2 = -1$. For a specific DFT size of 2, the complex output y can be represented as follows:

$$\begin{aligned} y[0] &= x[0]\omega_2^0 + x[1]\omega_2^0 = x[0] + x[1] \\ y[1] &= x[0]\omega_2^0 + x[1]\omega_2^1 = x[0] - x[1] \end{aligned} \tag{10.6}$$

From a hardware perspective, the complex inputs $x[0]$ and $x[1]$ can be represented by the real parts "x0_re" and "x1_re", as well as the imaginary parts "x0_im" and "x1_im". Similarly, the complex outputs $y[0]$ and $y[1]$ can be represented by the real parts "y0_re" and "y1_re", and the imaginary parts "y0_im" and "y1_im". These inputs and outputs are illustrated in Figure 10.16(a). In addition to the data interfaces, the "ready-valid" bus protocol is employed to indicate when data on the input buses is ready for use and to validate output data on the output buses. The clock, input, and enable signals are also essential IOs of the sequential circuit. It is important to note that all the registers in the design are triggered by the rising clock edge and rising reset edge.

A specific simulation example is illustrated in the timing diagram in Figure 10.16(b). Four sets of data are input into the $DFT2$ core:

- First complex input set: $x[0]$ is $1.0+j\cdot 0.0$, and $x[1]$ is $2.0+j\cdot 0.0$;
- Second complex input set: $x[0]$ is $2.0+j\cdot 1.0$, and $x[1]$ is $2.0+j\cdot 4.0$;
- Third complex input set: $x[0]$ is $1.0+j\cdot 2.0$, and $x[1]$ is $3.0+j\cdot 4.0$;
- Fourth complex input set: $x[0]$ is $4.0+j\cdot 3.0$, and $x[1]$ is $2.0+j\cdot 1.0$.

After two clock cycles delay, the corresponding outputs are pushed out sequentially in a pipeline manner.

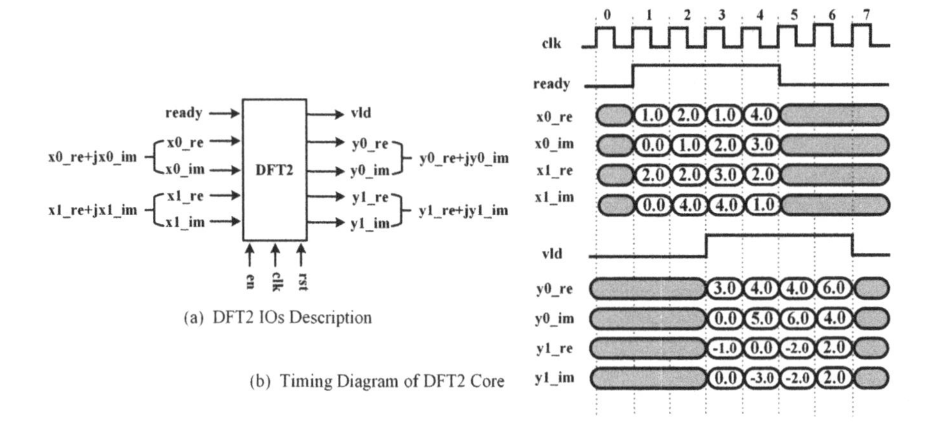

FIGURE 10.16
Design IOs and timing diagram of $DFT2$.

Design Hint: Following Equation 10.6, the $DFT2$ core can be implemented using the following FP operations:

$$\begin{aligned} y0\_re &= x0\_re + x1\_re \\ y0\_im &= x0\_im + x1\_im \\ y1\_re &= x0\_re - x1\_re \\ y1\_im &= x0\_im - x1\_im \end{aligned} \tag{10.7}$$

This design necessitates two FP additions and two FP subtractions, which can be constructed using the FP adders and FP subtractors as depicted in Figure 10.17.

It's worth noting that the ready-valid bus protocol can be devised with shift registers, aligning with the clock cycles required for the FP operational datapath. In addition, the multiplexers serve to diminish switching activities for input data to the FP operators, thereby reducing dynamic power consumption. Consequently, FP operations are triggered solely when the input data is ready for use on the bus. Finally, the last stage of registers is employed to register-out all the combinational outputs from the multiplexers.

2) Create a testbench to facilitate the simulation of the $DFT2$ core and validate its functionality using specific test cases shown in Figure 10.16(b). To clarify, the real part of complex input $x[0]$ can be driven on "x0_re" as 1.0, 2.0, 1.0, and 4.0 for the four input sets, and the imaginary part as "x0_im" with values 0.0, 1.0, 2.0, and 3.0. Similarly, "x1_re" and "x1_im" represent the real and imaginary parts of each complex number input for $x[1]$. It's important to note that they must be represented in IEEE-754 format.

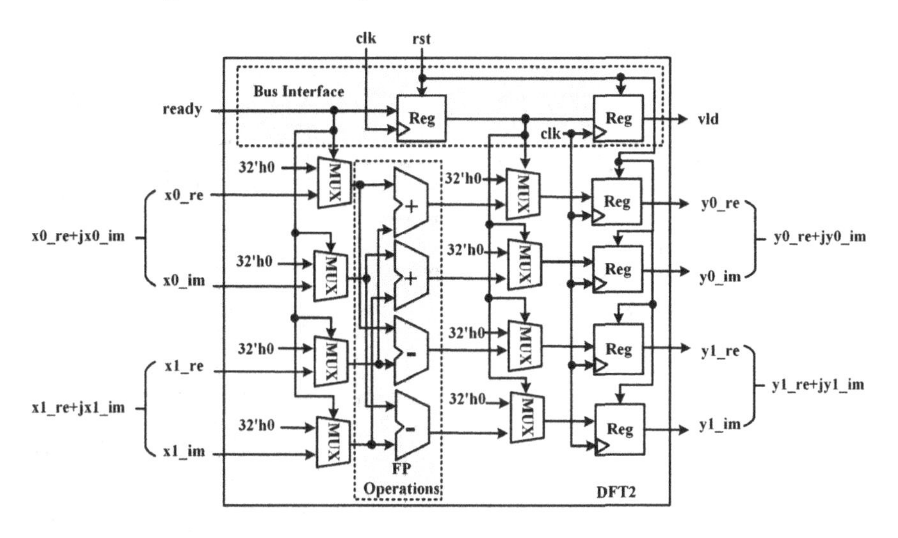

FIGURE 10.17
Block diagram of $DFT2$.

PBL 30: Streaming Design on DFT4

Design a hardware circuit for the discrete Fourier transform of size 4 ($DFT4$), utilizing the $DFT2$ core in PBL 29. Similar to PBL 29, the complex inputs and outputs, as well as the "ready-valid" buses are depicted in Figure 10.18(a). A specific simulation example is illustrated in the timing diagram in Figure 10.18(b). Two sets of data are input into the $DFT4$ core:

- First complex input set: $x[0]$ is $1.0+j\cdot 0.0$, $x[1]$ is $2.0+j\cdot 0.0$, $x[2]$ is $3.0+j\cdot 0.0$, $x[3]$ is $4.0 + j \cdot 0.0$;
- Second complex input set: $x[0]$ is $-1.0 + j \cdot 1.0$, $x[1]$ is $-2.0 + j \cdot 2.0$, $x[2]$ is $-3.0 + j \cdot 3.0$, $x[3]$ is $-4.0 + j \cdot 4.0$.

After four clock cycles delay, the corresponding outputs are pushed out sequentially in a pipeline manner.

Design Hint: Following the Cooley-Tukey algorithm [6], you can create the $DFT4$ core by integrating four $DFT2$ cores, as illustrated in Figure 10.19. It's essential to incorporate the permutation operation into the $DFT4$ design. Below is the detailed guidance related to the permutation for $DFT4$ computation.

- For the interleaved interfacing, e.g., between **u2** and **u3**, the complex output $u2.y[0]$ from **u2** connects to the complex input $u3.x[1]$ from **u3**. Hence, the

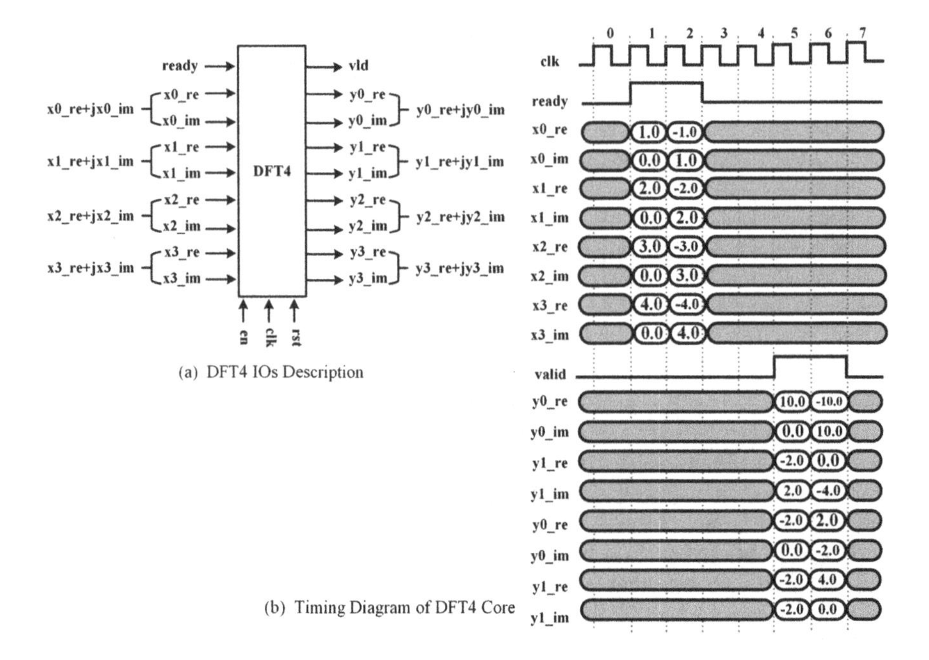

(a) DFT4 IOs Description

(b) Timing Diagram of DFT4 Core

FIGURE 10.18
Design IOs and timing diagram of $DFT4$.

hardware connections should be $u3.x1\_re = u2.y0\_re$ and $u3.x1\_im = u2.y0\_im$.

- For the interfacing between **u2** and **u4**, the complex output $u2.y[1]$ from **u2** multiplies the imaginary unit $-j$, resulting in $-j \times (y1\_re + j \cdot y1\_im) = -j \cdot y1\_re + y1\_im$, and then feed it into the following complex input $u4.x[1]$ from **u4**. Hence, the hardware connections should be $u4.x1\_re = u2.y1\_im$ and $u4.x1\_im = -u2.y1\_re$.

2) Create a testbench to facilitate the simulation of the $DFT4$ core and validate its functionality using the specific test cases shown in Figure 10.18(b).

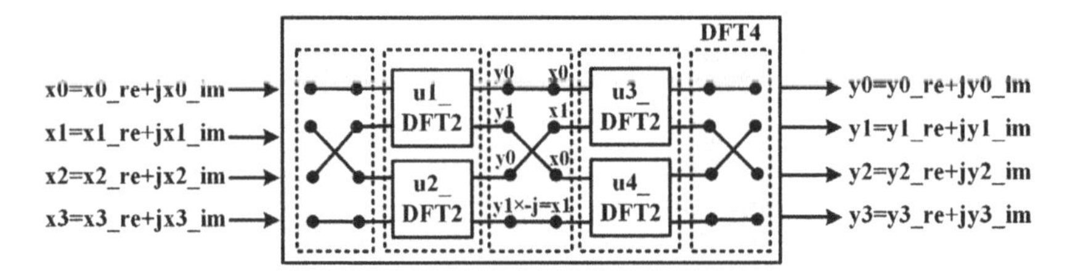

FIGURE 10.19
Block diagram of $DFT4$ using Cooley-Tukey algorithm.

11

Timing Constraints and High-Speed RTL Design

Chapter 11 offers an in-depth exploration and consideration of timing constraints in RTL design and their pivotal role in achieving high-speed ICs. In contrast to software programming, hardware design directly dictates and influences physical hardware, imposing stringent timing requirements and design rules. Failing to meet these constraints can result in issues throughout the IC design and manufacturing process, highlighting the critical need for comprehensive timing checks during the IC design flow.

Furthermore, Chapter 11 delves into the factors that contribute to enhancing IC performance. One of the key factors is the achievable operational clock frequency, typically measured in MHz or GHz in contemporary ICs. While fabrication technology does influence operational frequency, RTL designers can also optimize computational speed and efficiency by considering various design specifications. This chapter will provide valuable insights into such considerations for achieving high-performance designs at the register-transfer level.

PBL 31 is included in this chapter to approximate the **maximum operational frequency (MOF)** and to showcase the design factors that influence high-speed designs. PBL 32 introduces design rules for signals crossing different clock domains (CDC), while PBL 33 demonstrates a typical FIFO (First-In-First-Out) design. PBL 34-41 are provided in this chapter to exemplify high-speed and pipelined designs aimed at achieving higher MOFs.

11.1 Critical Path and True/False Path

This section discusses the concept of distinct data paths within a combinational circuit. These paths can be categorized as either true paths, which are executable, or false paths, which remain inactive during regular circuit operation.

DOI: 10.1201/9781003187080-11

11.1.1 True/false path for AND/NAND and OR/NOR gates

Before delving into the distinctions between true and false paths, let's first examine the control input of a basic AND/NAND gate. Taking the case of an AND gate, depicted in Figure 11.1(a), the controlling input holds a binary value of 1'b0. For instance, if input "b" for the AND gate is set to 1'b0, the output "c" must unequivocally be 1'b0, regardless of the value assigned to the other input, "a". Consequently, the data path from "a" to "c" remains inactive when "b=0". Such a data path that can't be executed is termed a false path. To transform the data path from "a" to "c" into a true path that can be executed, the input "b" must adopt a binary value of 1'b1, as depicted in Figure 11.1(b). Similar principles hold for a NAND gate, where the controlling input value of 1'b0 ensures the output to be 1'b1.

Conversely, in the realm of an OR gate, the controlling input value assumes the form of binary 1'b1. In practical terms, this signifies that when any one of the inputs is set to 1'b1, the output "c" is compelled to adopt the value of 1'b1, independent of the state of the other input. As depicted in Figure 11.1(c), if input "b=1", then the output "c" unmistakably assumes a value of "c=1". In this context, the data path from "a" to "c" becomes a false path, as its execution is precluded. For the data path from "a" to "c" to be designated as a true path, the input "b" must assume the binary value of 1'b0, as illustrated in Figure 11.1(d). Similar principles hold for a NOR gate, where the controlling input value of 1'b1 ensures the output to be 1'b0.

11.1.2 True/false path for combinational circuits.

A. Examples of True/False Path

Within a combinational circuit, the presence of multiple data paths is a common occurrence; however, not all of these paths qualify as true paths. This concept is aptly depicted in Figure 11.2(a), where the data path "a-c-d"

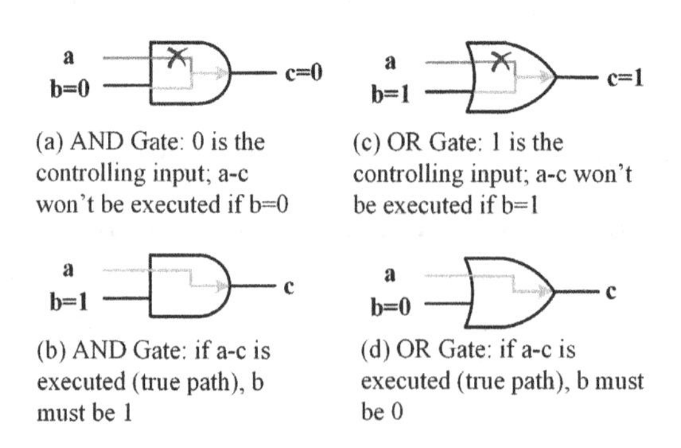

FIGURE 11.1
Controlling inputs of AND/NAND and OR/NOR gates.

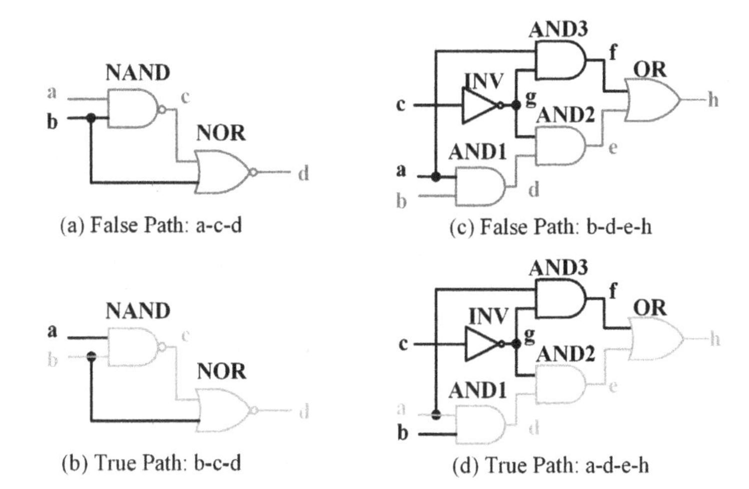

FIGURE 11.2
Two Examples of true/false path.

assumes the role of a false path. To elaborate, the input "b" of the NAND gate is necessitated to adopt the binary value 1'b1 for the "a-c" path to be substantiated as a true path. Interestingly, this same value of "b" results in the output "d" of the NOR gate assuming the binary value 1'b0, thereby rendering the "c-d" path as a false path. Consequently, the dual paths are inherently incompatible with simultaneous truth, thereby characterizing the overarching "a-c-d" path as a false path in its entirety.

On the contrary, Figure 11.2(b) showcases the "b-c-d" path, demonstrating the characteristics of a true path. To establish the truth of the "b-c" path, it's necessary for the input "a" of the NAND gate to adopt the binary value 1'b1. This action does not have any significant impact on other components of the circuit. Simultaneously, to validate the accuracy of the "c-d" path, the input "b" of the subsequent NOR gate must be set to the binary value 1'b0. This validation also ensures that the output "c" of the NAND gate remains at the binary value 1'b1, confirming the execution of the "b-c" path. This alignment allows both paths to be true simultaneously, leading to the confirmation of the overall "b-c-d" path as a true path.

In Figure 11.2(c), we encounter another scenario involving a false path "b-d-e-h". The "b-d" path can only be valid if the input "a" for the AND1 gate is set to the binary value 1'b1. Similarly, the "d-e" path attains validity when the input "g" is 1'b1 (the input "c" is 1'b0) for the AND2 gate. However, if both "a" and "g" are set to 1'b1, the output "f" generated by the AND3 gate assumes the value 1'b1, which then becomes a controlling factor for the subsequent OR gate. Given this condition of "f" being 1'b1, the final output "h" will also become 1'b1, rendering the "e-h" path false. Therefore, when the "b-d" and "d-e" paths are true, the following "e-h" path becomes false. In summary, the entire "b-d-e-h" path cannot be deemed a true path.

In Figure 11.2(d), we observe the "a-d-e-h" path, which is rightly classified as a true path. To activate the "a-d" path as valid, the input "b" should take on the binary value 1'b1, a change that doesn't impact other aspects. Similarly, for the "d-e" path to be true, the input "g" of the AND2 gate needs to be 1'b1, which doesn't immediately influence the output "f" from the AND3 gate. Consequently, the "e-h" path becomes true when the output "f" adopts the binary value 1'b0 in the OR gate. To achieve this condition for "f", the input "a" must be set to 1'b0, making "a-d" and "d-e" both true paths. As a result, the overall "a-d-e-h" path is accurately identified as a true path.

B. Other Examples of True/False Path

In Figure 11.3(a), the diagram illustrates a false path denoted as "a-b-d-e-h". This false path arises due to the possibility of executing either the combination of channel 0 in MUX1 and channel 1 in MUX2, or the combination of channel 1 in MUX1 and channel 0 in MUX2. This is made feasible by the opposite selection signals applied to the two multiplexers. Notably, the data path encompassing both channel 0 in MUX1 and MUX2 constitutes a false path.

Conversely, the data path labeled as "a-b-d-f-h", which involves the selected channel 0 in MUX1 and the selected channel 1 in MUX2, is classified as a true path, as highlighted in Figure 11.3(b).

In Figure 11.3(c), the diagram showcases the false path labeled as "b-d-e-f-h". Notably, the data path "b-d" holds no influence over other paths, rendering it a suitable candidate for a true path. However, for the data path "d-e" to be considered a true path, the input "c" must assume the value of binary 1'b1.

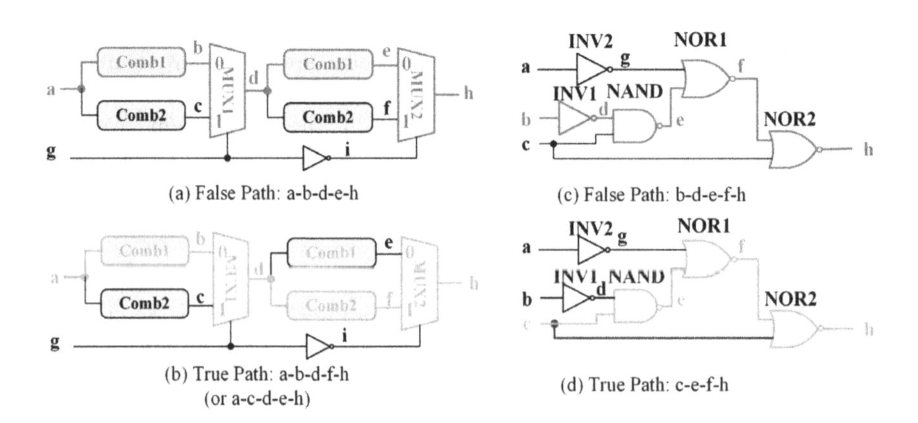

FIGURE 11.3
Other two examples of true/false path.

This setting ensures that the final output "g" becomes binary 1'b0, irrespective of the input "f" at the concluding NOR2 gate. Essentially, the value "c=1" transforms the data path "f-h" into a false path, thereby making the complete data path "b-d-e-f-h" unable to qualify as a true path.

Moving on to Figure 11.3(d), the depiction illustrates the second longest path, denoted as "c-e-f-h", within the same circuit. The data path "c-e" assumes the role of a true path when "d" is assigned the value of binary 1'b1. This condition can be executed alongside the input "b=0". For the subsequent data path "e-f" to be a true path, it should be ensured that "g" is set to binary 1'b0, all while not influencing any other path that functions as a false path. Finally, the ultimate data path involving the NOR2 gate, "f-h", qualifies as a true path when "c" takes the value of binary 1'b0. Consequently, if "c" is set to binary 1'b0, the output "e" of the NAND gate assumes a value of binary 1'b1 and the output "f" of the NOR1 gate assumes a vale of binary 1'b0, enabling the data path "c-e-f" to operate as a true path. As a result, the entirety of the data path "c-e-f-h" can be accurately executed as a true path.

11.1.3 Propagation/contamination delay and critical path

A. Propagation/Contamination Delay

Logic elements such as inverters, AND/NAND gates, and OR/NOR gates introduce physical delays. These delays signify the time required for the output of a logic element to update after an input change. There are two important delay metrics: contamination delay (referred to as "t_cd") and propagation delay (referred to as "t_pd"). Contamination delay represents the shortest path delay, while propagation delay signifies the longest path delay.

In an combinational circuit comprising multiple logic elements, the contamination delay is the aggregate of the contamination delays of all logic elements in the shortest true path. Similarly, the propagation delay is the sum of the propagation delays of all logic elements along the longest true path. It's crucial to emphasize that when analyzing propagation and contamination delays, only true paths are considered, as false paths are never executed.

B. Critical Path within Register-Transfer Design

The concept of a critical path holds great significance in register-transfer level designs. It pertains to the propagation delay that exists between two registers within the entire circuit. To illustrate, take into account the circuit depicted in Figure 11.4. The signal "a" originates from the initial register's output, traverses a combinational circuit, and ultimately manifests as the final output signal "f", which enters the second register. Conversely, the signal "g" emerges from the second register, navigates two additional logic elements, and transforms into the output "i", which enters the third register. The longer path between these two combinational circuits is "a-b-c-d-e-f", which encompasses the propagation delays introduced by the INV-AND1-AND2-AND3-OR1 logic

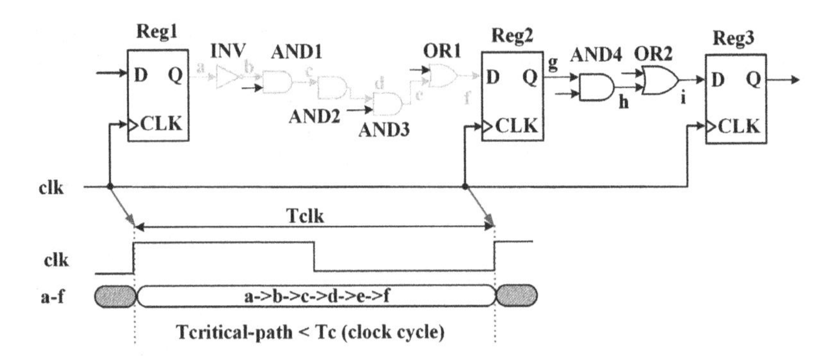

FIGURE 11.4
Critical path within register-transfer design circuit.

gates. Consequently, this path constitutes the critical path within the register-transfer circuit.

The critical path must be less than the clock period, represented as "$T_{critical\_path} < T_c$". Failing this condition, the computation of the combinational circuit will extend beyond the allocated clock cycle, resulting in timing violations and compromising the integrity of the entire register-transfer design. Consequently, the critical path holds immense significance in shaping both the clock frequency and the overall operational speed.

11.2 Timing Information

11.2.1 Timing constraints

In addition to the propagation delay and contamination delay encountered in combinational circuits, this section presents a comprehensive overview of various timing constraints and delays pertinent to sequential registers. These include the setup time (t_{setup}), hold time (t_{hold}), and clock-to-Q delay (t_{cq}). The t_{setup} and t_{hold} parameters serve as timing constraints for synchronous registers, whereas t_{cq} captures the duration between the active clock edge and the emergence of a valid output from the register. The following breakdown provides insight into these timing aspects.

- Setup Time (t_{setup}): This parameter signifies the duration for which the register input "D" must remain stable prior to the arrival of the active clock edge. Failure to satisfy the setup time requirement can lead to instability in the register's operation, potentially resulting in meta-stability.

- Hold Time (t_{hold}): This aspect refers to the duration during which the register input "D" must remain steady subsequent to the occurrence of the active clock edge. If the hold time condition is not met, the register's operation can become unstable, again inviting the possibility of meta-stability.
- Clock-to-Q Delay (t_{cq}): This delay corresponds to the time elapsed from the instant of the active clock edge to the point when a valid output "Q" displays from the register. Attempting to access the output prior to the expiration of this delay can yield inconsistent outcomes due to potential fluctuations in the register's output "Q".

Illustrated in Figure 11.5, two registers are synchronized by a common clock signal. The initial register yields an output "q1", which subsequently traverses a combinational circuit to produce the outcome "d2". This "d2" signal, in turn, becomes the input for the subsequent register. To ensure the accurate operation of the second register, it is imperative that the "d2" signal remains stable throughout both the setup and hold windows. This entails finalizing any updates to the signal prior to the initiation of the setup window and sustaining this stability until the conclusion of the hold window. Moreover, before attaining a dependable output "q1", a clock-to-Q delay is introduced to ascertain the stability of the register's output.

In conclusion, the timing attributes of a sequential circuit are concisely summarized in Table 11.1. Within these parameters, the t_{cq} delay and the t_{pd}/t_{cd} delay serve as timing delays, while t_{setup} and t_{hold} define timing constraints tailored to register operations. Emphasizing the significance of each of these timing parameters is crucial, as they collectively ensure the precise and reliable functionality of the register-transfer designs.

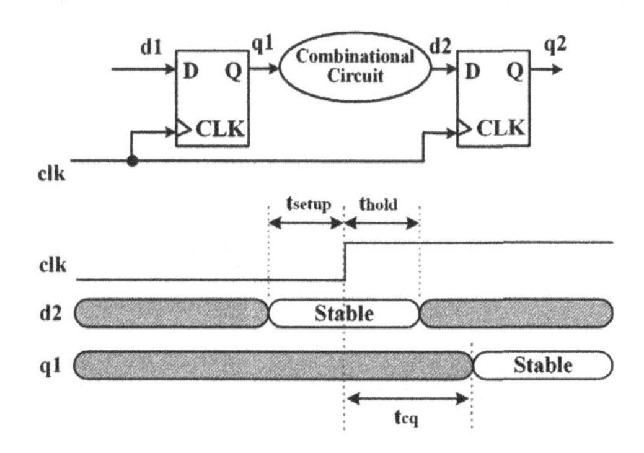

FIGURE 11.5
Setup time, hold time, and clock-to-Q delay.

TABLE 11.1
Timing attributes.

Timing Info	**Descriptions**
t_{pd}	Combinational Circuit Propagation Delay
t_{cd}	Combinational Circuit Contamination Delay
t_{cq}	Register Clock-to-Q Delay
t_{setup}	Register Setup Time Constraint
t_{hold}	Register Hold Time Constraint

11.2.2 Propagation/contamination delay examples

A. Timing Attributes

In this section, we circle back to the instances highlighted in Section 11.1 to showcase the process of deducing propagation delay and contamination delay within combinational circuits. For the sake of clarity in this demonstration, we consider that the delay attributed to wire connections can be overlooked, and the relevant cell delay data is furnished in Table 11.2.

B. Examples of Propagation/Contamination Delay

Regarding the circuit displayed in Figure 11.2(b), the critical path is discerned as "b-c-d". Consequently, the propagation delay across the entire circuit equals the cumulative propagation delays attributed to the NAND and NOR gates. This computation unfolds as follows:

$$t_{pd} = t_{pd}(NAND) + t_{pd}(NOR) = 1.5 + 1.5 = 3.0\,\text{ns}. \tag{11.1}$$

TABLE 11.2
Timing example for combinational circuits.

Circuits	t_{pd} (ns)	t_{cd} (ns)
AND/NAND	1.5	0.5
OR/NOR	1.5	0.5
XOR/XNOR	1.5	0.5
INV (Inverter)	1.0	0.5
BUF (Buffer)	1.5	1.0
Comparator	1.0	0.5
MUX (Multiplexer)	3.0	1.0
ADD (Adder)	2.5	1.5
SUB (Subtractor)	2.5	1.5
MUL (Multiplier)	3.5	2.0
SHF (Shifter)	0.5	0.5

Calculating the contamination delay of the circuit involves summing up the contamination delay relevant to the shortest true path. In the instance of Figure 11.2(b), the shortest true path is identified as "b-d". Consequently, the contamination delay is ascertained as follows:

$$t_{cd} = t_{cd}(NOR) = 0.5\,\text{ns}. \tag{11.2}$$

To compute the propagation delay for the circuit depicted in Figure 11.2(d), it involves adding up the propagation delay of the AND1, AND2, and OR gates along the critical path "a-d-e-h". This computation can be represented as follows:

$$t_{pd} = t_{pd}(AND1) + t_{pd}(AND2) + t_{pd}(OR) = 3 \times 1.5 = 4.5\,\text{ns}. \tag{11.3}$$

Determining the contamination delay hinges on the shortest true path present in the circuit. In the given example of Figure 11.2(d), the shortest true path is "a-f-h", and the contamination delay can be evaluated as follows:

$$t_{cd} = t_{cd}(AND3) + t_{cd}(OR) = 2 \times 0.5 = 1.0\,\text{ns}. \tag{11.4}$$

C. Other Examples of Propagation/Contamination Delay

Considering the example illustrated in Figure 11.3(b), let's presume that the critical path is represented by "a-b-d-f-h" (or alternatively "a-c-d-e-h"). For the combinational circuit #1 (denoted as "comb1" in the figure), a propagation delay of 10.0 ns is assumed, while combinational circuit #2 (denoted as "comb2" in the figure) exhibits a delay of 5.0 ns. Expanding upon this, the propagation delay for the entire circuit can be calculated by aggregating the propagation delay values for both combinational circuits, alongside the propagation delay for the two multiplexers. This calculation can be formulated as follows:

$$t_{pd} = t_{pd}(Comb1) + t_{pd}(Comb2) + 2 \times t_{pd}(MUX) = 10.0 + 5.0 + 2 \times 3.0 = 21.0\,\text{ns}. \tag{11.5}$$

Moving on to the contamination delay, it involves summing the delay pertinent to the shortest true path, which is identified as "g-i-h". Accordingly, the contamination delay can be computed using the following equation:

$$t_{cd} = t_{cd}(INV) + t_{cd}(MUX) = 0.5 + 1.0 = 1.5\,\text{ns}. \tag{11.6}$$

Regarding the example displayed in Figure 11.3(d), the critical path is deduced to be "c-e-f-h". As such, the calculation for the propagation delay in this circuit can be performed using the following formula:

$$t_{pd} = t_{pd}(NAND) + t_{pd}(NOR1) + t_{pd}(NOR2) = 1.5 + 2 \times 1.5 = 4.5\,\text{ns}. \tag{11.7}$$

Identifying the shortest true path as the "c-h" route via the NOR2 gate, the calculation for the contamination delay is as follows:

$$t_{cd} = t_{cd}(NOR2) = 0.5\,\text{ns}. \tag{11.8}$$

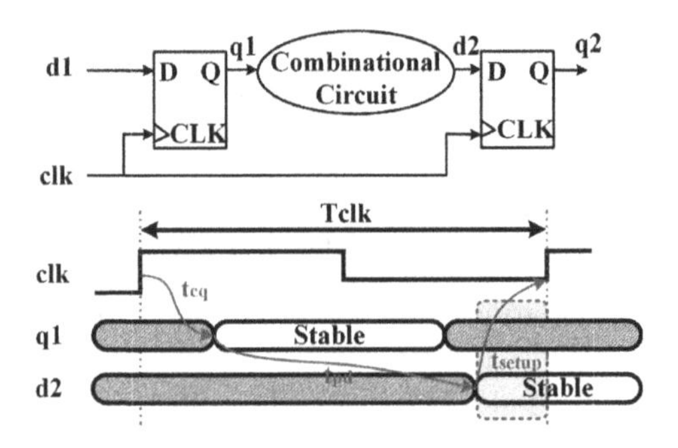

FIGURE 11.6
Maximum delay constraint.

11.3 Maximum Delay Constraints

11.3.1 Maximum delay constraint and setup time violation

A. Maximum Delay Constraint

The synchronization of timing delay and the setup time window must align within the bounds of the clock cycle. In Figure 11.6, should the summation of the clock-to-Q delay (t_{cq}) and the propagation delay (t_{pd}) across the combinational circuit prove excessively prolonged, it may render the input "d2" for the second register unstable within the setup window.

To provide an example, if $t_{cq} + t_{pd}$ stands at 10 ns and T_{clk} matches this at 10 ns, it would cause the output "d2" to update within the setup window, thereby introducing unstable data into the second register. This occurrence is known as a **setup time violation**. Expressing this mathematically, the clock period (T_{clk}) should surpass or equal the summation of the timing delay ($t_{cq} + t_{pd}$) and the setup requirement (t_{setup}), which is represented as:

$$T_{clk} \geq t_{cq} + t_{pd} + t_{setup}. \tag{11.9}$$

B. Solutions to Setup Time Violation

If a setup time violation occurs, it is possible that the data being transferred into the second register may enter a metastable state. This instability has the potential to disrupt the overall system functionality.

Addressing setup time violations can be approached in a couple of ways. One strategy involves extending the clock period, allowing sufficient time for data to be updated before the setup window closes. However, this adjustment would result in a reduction of the clock frequency. Alternatively, the solution can involve modifying the combinational circuit to minimize its propagation delay. In cases where the critical path is excessively lengthy, it can be partitioned into multiple shorter segments, with additional registers introduced along the data path.

11.3.2 Maximum operational frequency

In accordance with Equation 11.9, the minimum requisite clock period for the RTL design on sequential circuits, encompasses the clock-to-Q delay (t_{cq}), the propagation delay across the combinational circuit (t_{pd}), and the stipulated setup time (t_{setup}). This minimal clock period can thus be formulated as the amalgamation of these three temporal factors:

$$T_{clk(min)} = t_{cq} + t_{pd} + t_{setup}. \tag{11.10}$$

Subsequently, the inverse of the minimum clock period equates to the **MOF)** of the synchronous circuit, which is calculable as follows:

$$MOF = \frac{1}{T_{clk(min)}}. \tag{11.11}$$

To attain the utmost MOF, it is imperative to harness cutting-edge technology, characterized by reduced clock-to-Q delay and setup time restrictions. Furthermore, from RTL designers' vantage point, diminishing the critical path of the combinational circuit design can also contribute significantly to augmenting the MOF.

11.3.3 Setup time slack

The **setup time slack** delineates the available time window for the combinational circuit design between registers, and its computation is as follows:

$$t_{setup-slack} \leq T_{clk} - t_{cq} - t_{setup}. \tag{11.12}$$

Assuming the registers' timing details are provided in Table 11.3, let's consider an illustrative scenario where the clock frequency specification is set at or above 100MHz, consequently yielding a minimum clock period of 10 ns. From the data in the table, the clock-to-Q delay is 2.0 ns, and the setup time constraint is 1.0 ns. In order to satisfy the speed requirement, the maximum setup slack can be calculated as $t_{setup-slack} \leq 10 - t_{cq} - t_{setup} = 7.0$ ns. This implies that there is a 7.0 ns timeframe available to incorporate the

TABLE 11.3
Timing example for sequential registers.

Circuits	t_{cq} (ns)	t_{setup} (ns)	t_{hold} (ns)
Register	2.0	1.0	1.5

combinational functions between the registers. Should the critical path exceed this 7.0 ns window, it would necessitate the division of the combinational circuit into two or more shorter segments by inserting additional registers.

11.3.4 Design Example #1: binary counter MOF

In this section, consider the Verilog design of a binary counter provided below. This includes a conditional *assign* block in line 4 that executes the combinational circuit comprising a compactor and an adder. The *always* block in lines 5–11 denotes the 4-bit register responsible for storing the interim outputs produced by the combinational circuit.

```
1  module binary_counter (input              clk,
2                         input              rst,
3                         output reg [3:0]   cnt);
4  wire [3:0] nxt_cnt = (cnt==4'h9) ? 4'h0 : cnt+4'h1;
5  always @(posedge clk, negedge rst) begin
6    if (~rst) begin
7      cnt <= 4'h0   ;
8    end else begin
9      cnt <= nxt_cnt;
10   end
11 end
12 endmodule
```

The block diagram depicted in Figure 11.7 provides an overview of the analysis outcome for the binary counter design. This design encompasses a solitary register accompanied by a combinational circuit situated between the output and input of the register. The register's output, identified as "cnt", feeds into the combinational circuit composed of an adder, a multiplexer, and a compactor. The resulting output of this combinational circuit, labeled as "nxt_cnt", is then captured and stored by the register, thus forming a closed feedback loop.

The critical path of the combinational circuit involves the sum of the propagation delays of the adder and the multiplexer, as the adder contributing a longer delay than the comparator. Utilizing the timing particulars outlined in Table 11.2, the critical path delay for the combinational circuit computes as $t_{pd} = t_{pd}(ADD) + t_{pd}(MUX) = 2.5 + 3.0 = 5.5$ ns. This calculation subsequently allows the determination of the MOF for the counter design:

$$MOF = \frac{1}{t_{cq} + t_{pd} + t_{setup}} = \frac{1}{2.0 + 5.5 + 1.0} = 117.65\,\text{MHz} \tag{11.13}$$

where the variables t_{cq} and t_{setup} originate from the data provided in Table 11.3.

11.3.5 Design example #2: sequential circuit MOF

Here's another illustrative design example for the analysis of timing constraints. The circuit is comprised of three individual single-bit inputs, denoted as "a", "b", and "c", with the exceptions of "clk" and "rst". Additionally, there is a single-bit output labeled as "h". The definition of the combinational circuit is presented in lines 3–6, while the sequential circuit is elaborated upon in lines 8–22.

```
1  module (input clk, rst, a, b, c, output reg h);
2  reg  a1,b1,c1,c2,d2,e2;
3  wire d1 = a1 & b1 ;
4  wire e1 = a1 | b1 ;
5  wire f2 = c2 ? e2 : d2 ;
6  wire g2 = f2 & c2 ;
7
8  always @(posedge clk, negedge rst) begin
9    if (~rst) begin
10     a1<=1'b0; b1<=1'b0; c1<=1'b0;
11     c2<=1'b0; d2<=1'b0; e2<=1'b0;
12     h <=1'b0;
13   end else begin
14     a1 <= a ;
15     b1 <= b ;
16     c1 <= c ;
17     c2 <= c1;
18     d2 <= d1;
19     e2 <= e1;
20     h  <= g2;
21   end
22 end
23 endmodule
```

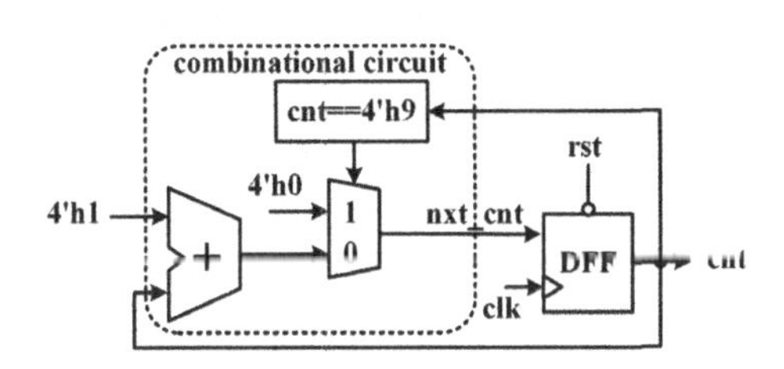

FIGURE 11.7
Design example #1: counter from hexadecimal 0 to 9.

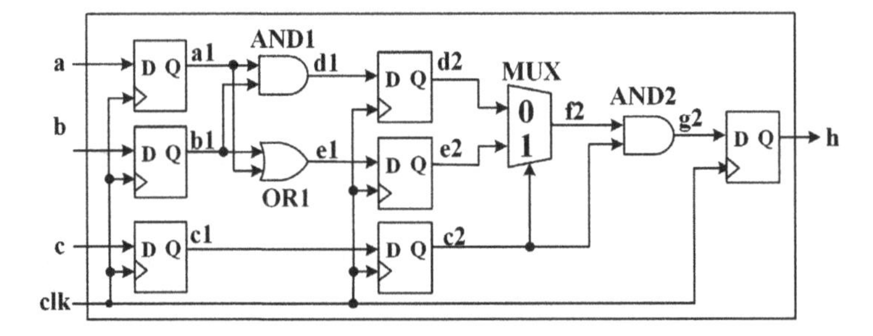

FIGURE 11.8
Design example #2: register-transfer design circuit.

The designed circuit is showcased within Figure 11.8, including two tiers of combinational circuits: one situated between the initial and second strata of registers, and another situated between the second and third strata of registers. In the initial stage, the concurrent execution of the AND1 and OR1 gates transpires, with the same route corresponding to the maximum propagation delay through "a1/b1-d1" or "a1/b1-e1". In accordance with the timing data provided in Table 11.2, both the AND and OR gates exhibit a propagation delay of 1.5 ns. The shortest route within this stage is embodied by the "c1" wire linking the output of the first-layer register to the input of the second-layer register.

Moving on to the subsequent stage, the lengthiest path "d2/e2/c2-f2-g2" encompasses the propagation delay of the multiplexer and the AND2 gate. Conversely, the shortest path "c2-g2" incorporates solely the delay introduced by the AND2 gate.

Derived from the analysis, the critical path within the complete circuit corresponds to the most extended true path named "d2/e2/c2-f2-g2". This path incurs a propagation delay of $t_{pd} = t_{pd}(MUX) + t_{pd}(AND2) = 3.0 + 1.5 = 4.5$ ns. Assuming a design directive defined by a 100 MHz clock frequency (a 10 ns clock period), the design adheres to the setup timing requisites due to the summation $t_{cq} + t_{pd} + t_{setup} = 1.0 + 4.5 + 2.0 = 7.5$ ns being shorter than the 10.0 ns clock period. This conveys that the data can be updated prior to the setup window, thereby achieving stability upon entry into the subsequent register.

Furthermore, the potential MOF can be mathematically computed as follows:

$$MOF = \frac{1}{t_{cq} + t_{pd} + t_{setup}} = \frac{1}{1.0 + 4.5 + 2.0} = 133.33\,\text{MHz}. \tag{11.14}$$

11.3.6 Design example #3: pipeline design MOF and latency

A. MOF Analysis of Design Example #3

For the assessment of design performance, encompassing computation latency and the MOF, this section presents an illustrative example showcased in Figure 11.9(a). It depicts an instance of a crucial pathway encompassing a multiplier, an adder, and a shifter. The aggregate propagation delay for this pivotal route is computed as $t_{pd} = t_{pd}(MUL) + t_{pd}(ADD) + t_{pd}(SHF) = 3.5 + 2.5 + 0.5 = 6.5$ ns. Subsequently, the MOF can be determined utilizing the following formula:

$$MOF = \frac{1}{t_{cq} + t_{pd} + t_{setup}} = \frac{1}{1.0 + 6.5 + 2.0} = 105.26\,\text{MHz}. \tag{11.15}$$

Utilizing this potential MOF value, it becomes evident that the time interval necessary for the "r/g/b" inputs to generate the "gs_data_out" output equates to 9.5 ns, precisely matching a single clock period.

The second design, depicted in Figure 11.9(b), executes the same computation as the first design but introduces intermediate registers between the multipliers and the subsequent adder. This division results in the computation spanning two clock cycles. The outputs from the multipliers, namely "rx38", "gx75", and "bx15", are stored in intermediate registers for the second-stage computation, which encompasses the adder and shifter. Therefore, the critical path for the entire circuit emerges as the summation of the maximum delays across distinct layers of registers. Between the initial and second strata, the propagation delay corresponds to the delay attributed to the multiplier, amounting to 3.5 ns. In contrast, between the second and third strata, the propagation delay aggregates the propagation delays of the adder and shifter, summing up to $2.5 + 0.5 = 3.0$ ns. As the multiplier incurs a lengthier latency, the critical path for the complete circuit stands at 3.5 ns.

Leveraging the timing particulars in Table 11.3, the MOF can be computed as follows:

$$MOF = \frac{1}{t_{cq} + t_{pd} + t_{setup}} = \frac{1}{1.0 + 3.5 + 2.0} = 153.85\,\text{MHz}. \tag{11.16}$$

This speed surpasses that of the first design. However, each "gs_data_out" output necessitates two clock cycles, represented as $2 \times \frac{1}{153.85} = 13$ ns, to be processed, indicating a longer duration compared to the first design.

B. Latency Analysis of Design Example #3

Let's consider a test scenario involving 256 data feed into the "r/g/b" inputs. In the context of the first design showcased in Figure 11.9(a), the computation journey spanning from inputs to outputs necessitates a solitary clock cycle per

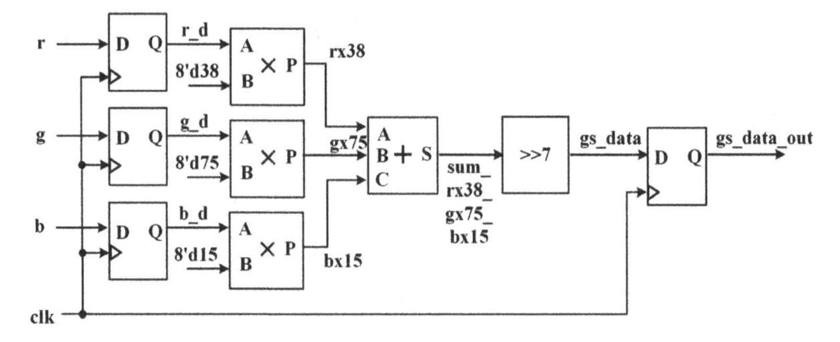

(a) Critical Path including Multipliers, an Adder, and a Shifter

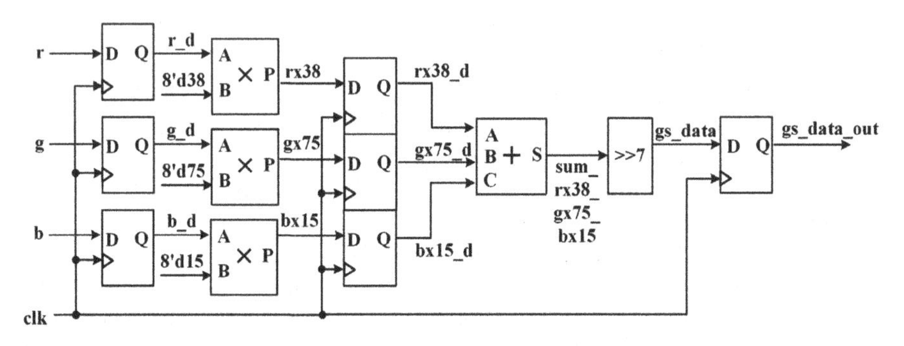

(b) Registers Insert within Critical Path

FIGURE 11.9
Design example #3: pipeline design with higher MOF.

data processing. This entails the execution of multiplications, addition, and data shifting operations. Consequently, the entirety of the 256 inputs will be processed within 256 clock cycles, as depicted in Figure 11.10(a). To simplify representation, in this figure, "MUL" denotes multiplications, and "SUM" signifies the amalgamation of addition and data shifting. The cumulative time consumption can be obtained by computing $256 \times \frac{1}{105.26} = 2,432$ ns.

Within the context of the second design illustrated in Figure 11.9(b), a division between multiplications and addition/data shifting takes place over two clock cycles. As showcased in Figure 11.10(b), the initial data computation spans clock cycles 1–2. During the initial clock cycle, the multiplication is executed, while the subsequent clock cycle involves the execution of addition and data shifting operation. The introduction of pipeline data processing further enables the initiation of the multiplication for the second data processing within the second clock cycle. This reduces the latency for the initial two data processing cycles to a total of three clock cycles. Consequently, the ultimate latency for processing 256 data inputs can be abbreviated to 257 clock cycles, equivalent to $257 \times \frac{1}{153.85} = 1,670.5$ ns.

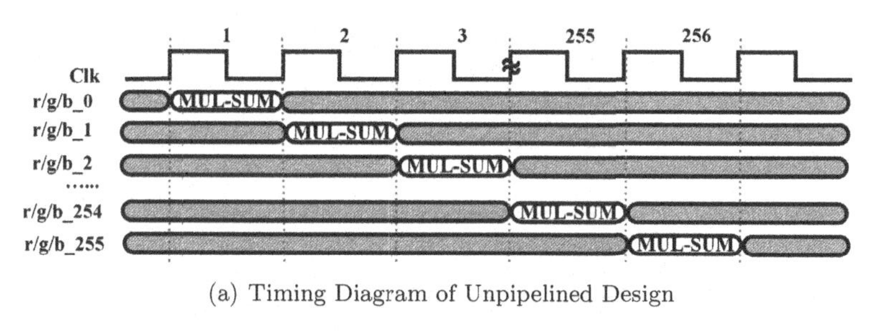

(a) Timing Diagram of Unpipelined Design

(b) Timing Diagram of Pipelined Design

FIGURE 11.10
Timing diagram of pipeline design with higher MOF.

While the pipeline design allocates two clock cycles to each data processing stage, it manages to diminish the latency for processing 256 data instances by approximately 30% through the utilization of pipeline data processing.

11.4 Minimum Delay Constraints

11.4.1 Minimum delay constraint and hold time violation

In this section, we explore the fundamental constraints governing minimum delay in register-transfer designs. In the context of Figure 11.11, our focus is on the time it takes for data to propagate from the initial register's clock edge to the subsequent register.

This journey encompasses both the clock-to-Q delay (t_{cq}) and the contamination delay specific to the combinational circuit (t_{cd}). When this interval is too short or if the hold time (t_{hold}) is too long, there is a risk that the data at the "d2" might be updated within the hold window, potentially leading to a **hold time violation**. To mitigate this scenario, it is imperative for the minimum delay to exceed the hold time (t_{hold}), and this principle can be expressed as:

$$t_{cq} + t_{cd} > t_{hold}. \tag{11.17}$$

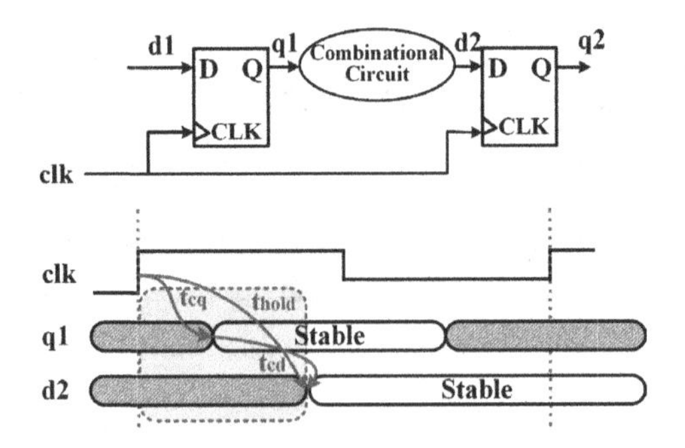

FIGURE 11.11
Minimum delay constraint.

Failure to prevent a hold time violation can result in unstable data entering the second register and corrupting the circuit's state. In such cases, circuit redesign may be necessary, such as adding buffers to delay the signal updating without changing the design functions. Designers should exercise caution to prevent such failures, as redesigning RTL code can be time-consuming, requiring regression testing and other procedures in the design flow.

11.4.2 Design example #1: hold timing analysis of binary counter

Let's take a look at the example demonstrated in Figure 11.7. During the period between the two successive rising clock edges, the shorter route travels through the comparator and the multiplexer. This configuration yields a contamination delay of $t_{cd} = t_{cd}(Comparator) + t_{cd}(MUX) = 0.5 + 1.0 = 1.5$ ns for the combinational circuit. As a result, the overall minimum delay, which combines the clock-to-Q delay and the contamination delay of the combinational circuit, sums up to $t_{cd} + t_{cq} = 1.5 + 2.0 = 3.5$ ns. This interval surpasses the hold time requirement ($t_{hold} = 1.5$ ns as per Table 11.3), thereby validating the condition $t_{cd} + t_{cq} > t_{hold}$. This successful outcome indicates that the timing check for the hold constraint is met.

11.4.3 Design example #2: hold timing constraint of sequential circuit

Let's revisit the example depicted in Figure 11.8. This example highlights the wire connection identified as "c1" between registers, signifying the most challenging scenario concerning the hold time requirement. Specifically, the minimal timing delay occurring between two successive active clock edges is determined by t_{cq}. This constraint introduces a condition to the second-layer register, necessitating that $t_{cq} > t_{hold}$. When consulting the timing data

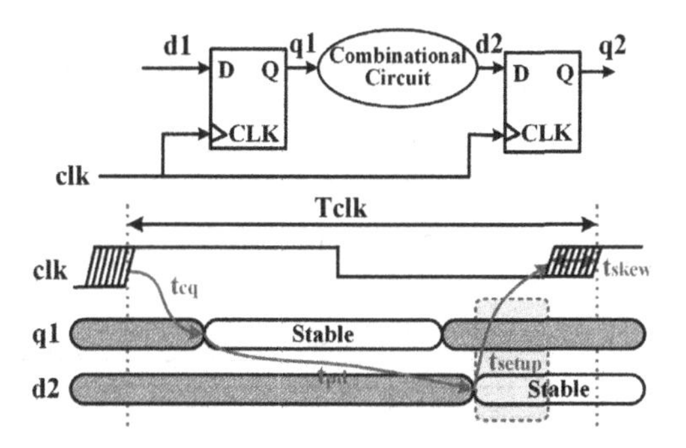

FIGURE 11.12
Maximum delay constraint with clock skew.

provided in Table 11.3, it becomes evident that $t_{cq} = 2.0$ ns and $t_{hold} = 1.5$ ns, and this alignment successfully meets the hold timing requirement.

11.5 Clock Skew

11.5.1 Maximum delay constraints with clock skew

Until now, the timing analysis has operated under the assumption of skew-free clocks. However, in reality, clocks arriving at different registers bring with them inherent uncertainties about their exact arrival times. This intrinsic unpredictability can lead to a reduced timeframe available for data processing. Figure 11.12 visualizes this by using hashed lines to indicate the range of potential clock arrival times, accounting for the presence of **clock skew**.

In the most unfavorable scenario involving the maximum delay constraint, the initial register encounters a delayed clock signal while the second register experiences an early clock signal, ultimately reducing the available timing window. Despite the steady clock period between ascending edges, clock skew elongates this period, which can be described by the equation:

$$T_{clk} \geq t_{cq} + t_{pd} + t_{setup} + t_{skew}. \tag{11.18}$$

In this scenario, introducing clock skew will lead to a decrease in the MOF of the synchronous circuit:

$$MOF = \frac{1}{t_{cq} + t_{pd} + t_{setup} + t_{skew}}. \tag{11.19}$$

With the available time for computation being diminished, the slack time for setup is adjusted as follows:

$$T_{setup-slack} \leq T_{clk} - t_{cq} - t_{setup} - t_{skew}. \tag{11.20}$$

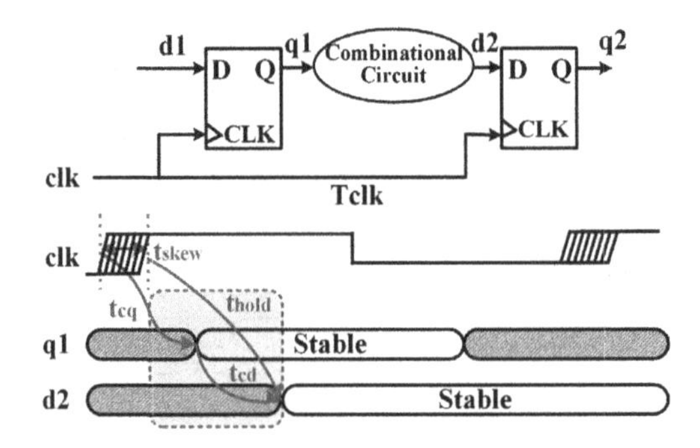

FIGURE 11.13
Minimum delay constraint with clock skew.

In conclusion, clock skew tends to elevate the maximum delay, resulting in a decrease in both the MOF and the available setup time slack.

11.5.2 Minimum delay constraints with clock skew

The most challenging situation concerning the minimum delay constraint arises when the first register receives its clock signal ahead of time, and the second register experiences a delayed clock, as demonstrated in Figure 11.13. In such instances, clock skew effectively extends the hold time requirement for the second register. This modified hold time requirement can be described as:

$$t_{cq} + t_{cd} > t_{hold} + t_{skew}. \quad (11.21)$$

If the minimum delay, encompassing both the clock-to-Q delay (t_{cq}) and the contamination delay (t_{cd}), falls short of the combined value of hold time (t_{hold}) and clock skew (t_{skew}), the input "d2" destined for the second register will be updated prematurely, resulting in a meta-stable condition within the hold window.

11.5.3 Design Example #1: clock skew analysis of binary counter

Let's reconsider the example portrayed in Figure 11.7. With a clock skew of 1.0 ns, the minimal clock period adjusts to $T_{min} = t_{cq} + t_{pd} + t_{setup} + t_{skew} = 2.0 + 5.5 + 1.0 + 1.0 = 9.5$ ns. Consequently, the MOF can be calculated as $MOF = \frac{1}{T_{min}} = 105.26$ MHz. In comparison to the outcome in Section 11.3.4 where a 117.65 MHz MOF was achieved, the introduction of clock skew reduces the MOF.

Regarding the hold time constraint, clock skew amplifies the needed timing for the hold time window to $t_{hold} + t_{skew} = 1.5 + 1.0 = 2.5$ ns. Since the summation of the clock-to-Q delay and the contamination delay, denoted as

$t_{cd} + t_{cq} = 1.5 + 2.0 = 3.5$ ns, surpasses the required timing, no hold violation occurs.

11.5.4 Design Example #2: clock skew analysis of sequential circuit

Let's recall the example presented in Figure 11.8. Considering the critical path, the minimum clock period can be computed as $t_{cq} + t_{pd} + t_{setup} + t_{skew} = 2.0 + 4.5 + 1.0 + 1.0 = 8.5$ ns. Consequently, the MOF can be determined using the formula:

$$MOF = \frac{1}{t_{cq} + t_{pd} + t_{setup} + t_{skew}} = \frac{1}{2.0 + 4.5 + 1.0 + 1.0} = 117.65\text{MHz}. \tag{11.22}$$

When compared to the results in Section 11.3.5, which yielded a 133.33 MHz MOF, it's evident that the presence of clock skew has led to a reduction in the MOF.

Taking the clock skew into consideration alongside the hold time requirement, the necessary delay should surpass $t_{hold} + t_{skew} = 1.5 + 1.0 = 2.5$ ns. However, with the clock-to-Q delay being only 2.0 ns (as per Table 11.3), a hold time violation will occur. To address this concern, a buffer can be introduced into the shortest path, denoted as the "c1" connection between the registers. The timing data in Table 11.2 indicates that a buffer's contamination delay is 1.0 ns. By incorporating this, the overall delay can be extended to $t_{cd}(BUF) + t_{cq} = 1.0 + 2.0 = 3.0$ ns, which satisfies the hold time requirement when accounting for clock skew.

11.6 Clock Domain Crossing

In RTL design, the dependable capture and storage of signal values within registers rely on the activation of clock edges. Nevertheless, signals that enter a metastable state during setup or hold windows have the potential to disrupt the exact timing requirements for registers, which can ultimately result in malfunctions within the circuit and system.

One of the most common scenarios prone to introducing signal race conditions is the RTL design of signals crossing between different clock domains. Timing violations associated with CDC (clock domain crossing) are elusive and extremely challenging to debug, underscoring the importance of correctly designing synchronization logic from the outset. In this section, we will delve into the fundamental design rules aimed at mitigating these challenges, using a simple design example as our focal point.

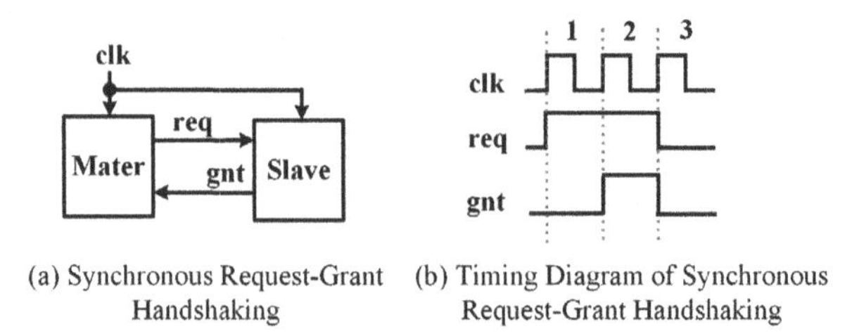

(a) Synchronous Request-Grant Handshaking (b) Timing Diagram of Synchronous Request-Grant Handshaking

FIGURE 11.14
Synchronous request-grant handshaking.

11.6.1 Request-grant handshaking

Let's revisit the handshaking protocol introduced in Chapter 8, focusing specifically on the single-master and single-slave structure, as depicted in Figure 11.14(a). In this context, the master can initiate bus requests, and the slave grants the requests if it is available, thereby completing the handshaking protocol. For the sake of simplicity, we assume that the slave is always prepared to respond, allowing the grant to be triggered whenever the request is active.

Further clarification is provided by the accompanying timing diagram displayed in Figure 11.14(b). In this timing diagram, the sequence begins with the master initiating a request using the "req" bus during the first clock cycle. Following this, the slave promptly detects the request and responds by asserting the grant signal through the "gnt" bus in the next clock cycle. This successful handshaking process is completed within the second clock cycle, coinciding with the deactivation of the "req" signal.

The example provided above is based on a synchronous design context, where both the master and the slave operate within the same clock domain. In contrast, Figure 11.15 illustrates an asynchronous design scenario involving communication between a master in clock domain A and a slave in clock domain B. This specific design configuration is commonly referred to as clock domain crossing, a situation that can introduce significant timing challenges. This is primarily because registered signals from asynchronous clock domains may potentially experience metastability when their data inputs fall within

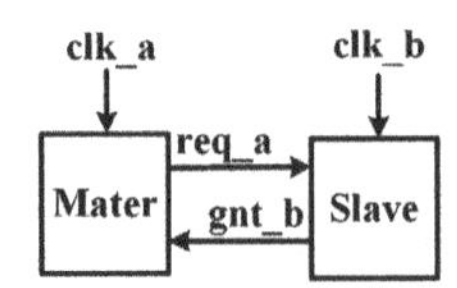

FIGURE 11.15
Asynchronous request-grant handshaking.

TABLE 11.4
Master IOs description.

Name	Direction	Bit Width	Description
clk_a	Input	1	Clock, rising edge, 33 MHz
rst_a	Input	1	Asynchronous reset, 0 valid
en_a	Input	1	Enable signal, 1 valid
req_a	Output	1	Request signal from clock domain A
gnt_b	Input	1	Grant signal from clock domain B

the setup or hold time window. As a result, the implementation of specific RTL design rules is crucial to ensure reliable data transfer across distinct clock domains.

11.6.2 Design specification of request-grant handshaking crossing asynchronous clock domains

A. Design IOs

As a case study, Table 11.4 provides a comprehensive summary of all the input and output signals originating from the master. It's important to note that signals ending with "_a" correspond to the clock A domain, while those ending with "_b" correspond to the clock B domain. The "en_a" input serves as a pulsed input responsible for enabling the master to initiate a request toward the slave. On the other hand, the "req_a" output signifies a level signal generated by the master within the clock A domain. Meanwhile, the "gnt_b" signal signifies the response generated by the slave within the clock B domain.

For the effective utilization of the "gnt_b" signal in the clock A domain, it's necessary to introduce a delay of two clock A cycles by employing a synchronization circuit, or namely a synchronizer. Subsequently, the "req_a" can be pulled down with the synchronized grant signal in the clock A domain, forming a "request-grant" handshake that spans across asynchronous clock domains.

Table 11.5 provides a concise overview of the input and output signals associated with the slave. The mechanism entails that upon receiving a request (designated as "req_a") from the master, the slave responds by issuing a grant (indicated as "gnt_b") to successfully complete the handshaking process. Given that the request and grant signals exist within distinct clock domains, it is imperative to introduce a delay of two clock B cycles to the "req_a" signal before its utilization in the clock B domain.

TABLE 11.5
Slave IOs description.

Name	Direction	Bit Width	Description
clk_b	Input	1	Clock, rising edge, 50 MHz
rst_b	Input	1	Asynchronous reset, 0 valid
gnt_b	Output	1	Grant signal from clock domain *B*
req_a	Input	1	Request signal from clock domain *A*

B. Timing Diagram for Clock Domain Crossing

Figure 11.16 provides an illustrative timing diagram depicting the interaction between the master and slave. In the initial cycle of "clk_a", the pulse signal "en_a" triggers the commencement of the request-grant handshake. Beginning in the subsequent "clk_a" cycle, the master transmits the request signal ("req_a") and awaits the response from the slave.

Moving to the slave side, the detection of the request signal "req_a" takes place during the second "clk_b" cycle. To prevent any potential race conditions, a delay of two "clk_b" cycles is introduced for the input "req_a", occurring across cycles 2 and 3 within the clock *B* domain. It is in the third cycle of "clk_b" that the output "gnt_b" materializes, utilizing the synchronized signal "req_b_d2".

In subsequent steps, the "gnt_b" input transitions into the clock *A* domain, demanding a delay spanning two "clk_a" cycles encompassing cycles 5 and 6. By the sixth cycle of "clk_a", the synchronized signal "gnt_a_d2" makes

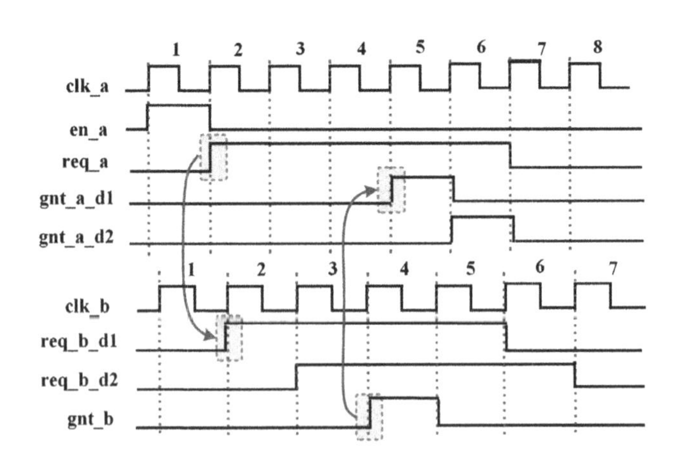

FIGURE 11.16
Timing diagram of request-grant handshaking crossing asynchronous clock domains.

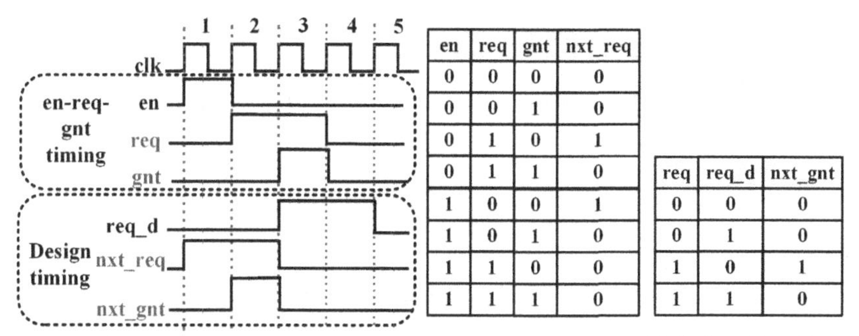

en	req	gnt	nxt_req
0	0	0	0
0	0	1	0
0	1	0	1
0	1	1	0
1	0	0	1
1	0	1	0
1	1	0	0
1	1	1	0

req	req_d	nxt_gnt
0	0	0
0	1	0
1	0	1
1	1	0

(a) Timing Diagram of en-req-gnt Design (b) Truth Table of Request (c) Truth Table of Grant

FIGURE 11.17
Timing diagram of request-grant handshaking.

its entrance, effectively pulling down the initial request "req_a" signal and formally concluding the handshaking procedure.

In summary, effectively coordinating handshaking operations across distinct clock domains requires the integration of two distinct design modules: a handshaking circuit and a specialized synchronizer designed for signals crossing different clock domains.

11.6.3 Design on request-grant handshaking

A handshaking circuit is a commonly used RTL design for coordinating commands and data exchanges between master and slave components. Figure 11.17(a) provides a timing diagram illustrating the logic behind the "req-gnt" handshaking process. This includes the enable signal "en" responsible for initiating the handshake, as well as the "nxt_req" and "nxt_gnt" signals that showcase the combinational logic used to determine the timing. Specifically, in the first clock cycle, the "en" input initiates a master request to the slave. Subsequently, in the second and third clock cycles, the master requests access to the bus ("req"), and the slave grants the bus request ("gnt"), respectively.

The "nxt_req" signal plays a crucial role in determining the timing of a request. It initiates when the "en" signal is activated and ceases when the "gnt" signal is detected. The truth table depicted in Figure 11.17(b) elucidates the design logic governing the computation of the "nxt_req" signal. It activates when "en" is triggered and both the current request and grant signals are not activated (as shown in the sixth row), marking the commencement of the next request state. Alternatively, it sustains the request when the current request is active but the grant signal has not yet been received (as shown in the fourth row). Similarly, the "nxt_gnt" signal plays a crucial role in determining the timing of the grant. It becomes active immediately upon the assertion of the request, assuming that the slave is always available for the handshake. As

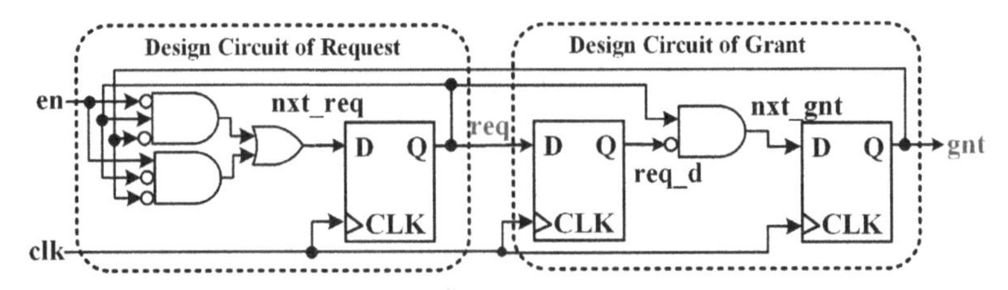

FIGURE 11.18
Block diagram of request-grant handshaking in synchronous clock domains.

depicted in Figure 11.17(c), the "req_d" signal is utilized to detect the first clock cycle when the request is initiated.

In what follows, Figure 11.18 illustrates the design circuit involving both the master and slave. It becomes evident that the combinational logic design for "nxt_req" comprises inverters, two 3-input AND gates, and one 2-input OR gate. On the other hand, the combinational logic design for "nxt_gnt" consists of one 2-input AND gate and one inverter. Additionally, two registers are employed to facilitate the updating of the "req" and "gnt" based on the next-state signals of "nxt_req" and "nxt_gnt". Furthermore, an additional register is required to introduce a one-clock-cycle delay to the "req" signal, thus triggering the bus grant.

Here is the Verilog description of the master circuit providing the "req" signal and the slave circuit for handshaking with the "gnt" signal. In the master circuit design from lines 1 to 15, line 7 describes the combinational logic illustrated in Figure 11.17(b), while lines 8 to 14 detail the design of the register output for "req". In the slave circuit from lines 17 to 32, line 23 describes the combinational logic depicted in Figure 11.17(c), while lines 24 to 32 detail the design of the register output for "gnt".

```
1  // master circuit for request
2  module master(input       clk,
3                input       rst,
4                input       en ,
5                input       gnt,
6                output reg  req);
7  wire nxt_req = (~en & req & ~gnt) | (en & ~req & ~gnt);
8  always @(posedge clk, negedge rst) begin
9    if (~rst) begin
10     req   <= 1'b0;
11   end else begin
12     req   <= nxt_req;
13   end
14 end
```

```
15  endmodule
16
17  // slave circuit for response
18  module slave (input        clk,
19                input        rst,
20                input        req,
21                output reg gnt);
22  reg req_d;
23  wire nxt_gnt  = req & ~req_d ;
24  always @(posedge clk, negedge rst) begin
25    if (~rst) begin
26      gnt   <= 1'b0;
27      req_d <= 1'b0;
28    end else begin
29      gnt   <= nxt_gnt;
30      req_d <= req     ;
31    end
32  end
33  endmodule
```

11.6.4 Synchronizer design

In the preceding section, we presented a handshaking circuit design in which both the request and grant signals, along with their associated logic, reside within synchronous clock domains. However, when these two design circuits operate in distinct clock domains – namely, with "req_a" in clock domain A and "gnt_b" in clock domain B – a synchronizer becomes essential to prevent timing violations.

In Figure 11.19(a), a generic design for the synchronizer is presented, depicting the flow of the "req_a" from clock domain A into clock domain B. Rather than directly employing the "req_a" signal within clock domain B, an arrangement is devised wherein a synchronizer comprising two registers is implemented. This configuration introduces a delay of two "clk_b" cycles to the input signal "req_a". The timing diagram depicted in Figure 11.19(b) further clarifies the process. Subsequently, the output signal "req_b_d2" originating from the second register can be safely utilized within the clock domain B. While the output emerging from the first register, denoted as "req_b_d1", may potentially enter a metastable state for a certain period, the wait time of two clock cycles serves to enhance the likelihood of its stabilization to a reliable level.

Below is the Verilog code illustrating the synchronizer design. The resulting signals are "req_b_d1", which is delayed by one "clk_b" cycle, and "req_b_d2", which is delayed by two "clk_b" cycles. Using a similar synchronizer design, it is essential to synchronize the "gnt_b" signal originating from the clock domain B before utilizing it within clock domain A.

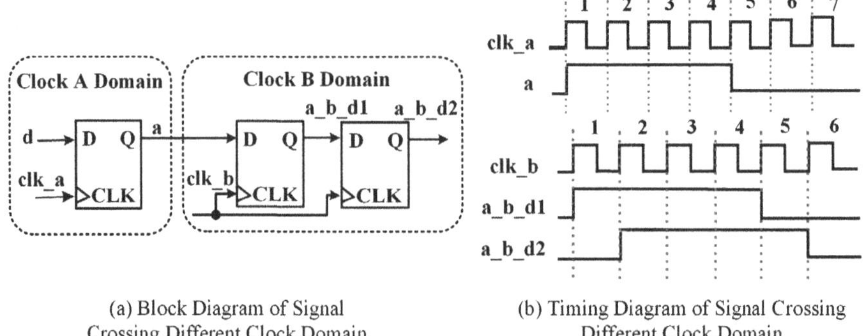

(a) Block Diagram of Signal Crossing Different Clock Domain

(b) Timing Diagram of Signal Crossing Different Clock Domain

FIGURE 11.19
Block diagram and timing diagram of synchronizer design.

```
always @(posedge clk, negedge rst) begin
  if (~rst) begin
    req_b_d1 <= 1'b0;
    req_b_d2 <= 1'b0;
  end else begin
    req_b_d1 <= req     ;
    req_b_d2 <= req_b_d1;
  end
end
```

Design Rules – Clock Domain Crossing

1. Clock domain crossing refers to the process of transferring data from one source clock domain to another destination clock domain. To ensure reliable transfer, a synchronizer is typically employed, comprising two series registers. Although the output from the first register may potentially enter a metastable state for a certain period, the two-clock-cycle delay helps increase the likelihood of stabilization to a reliable level.
2. It is important to note that a synchronizer is specifically designed for scenarios involving clock domain crossings where the signal can be consistently sampled by at least one clock edge of the destination clock. If the signal does not meet this criterion, alternative methods, such as utilizing asynchronous

FIFOs, should be considered to ensure reliable data transfer across clock domains.

11.6.5 System integration for request-grant handshaking and CDC

Concluding the discussion of the aforementioned design modules, Figure 11.20 presents the complete block diagram of the request-grant handshaking circuit. Returning to the timing diagram depicted in Figure 11.16, let's examine its components in detail.

- In the upper segment, a "req-gnt" handshaking circuit is used to convert the enable pulse input "en_a" into a level output "req_a", while also concluding the handshaking process in coordination with the synchronized "gnt_a_d2" input.
- Subsequently, in the lower section, the "req_a" signal is transferred to clock domain B. Following a 2-clock cycle delay of "clk_b", the request is granted when the bus request occurs.
- Finally, the "gnt_b" signal re-enters the upper section of clock domain A. To finalize the handshaking process, it is essential to introduce a 2-clock cycle delay of "clk_a" before utilizing the signal to de-assert the original request signal "req_a".

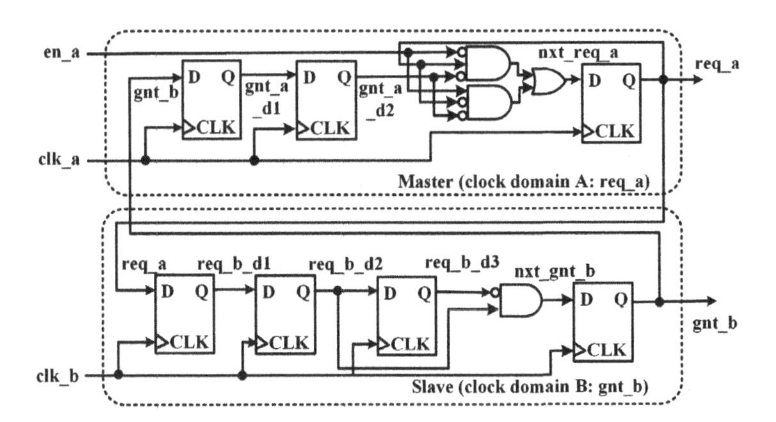

FIGURE 11.20
Block diagram of request-grant handshaking in asynchronous clock domains.

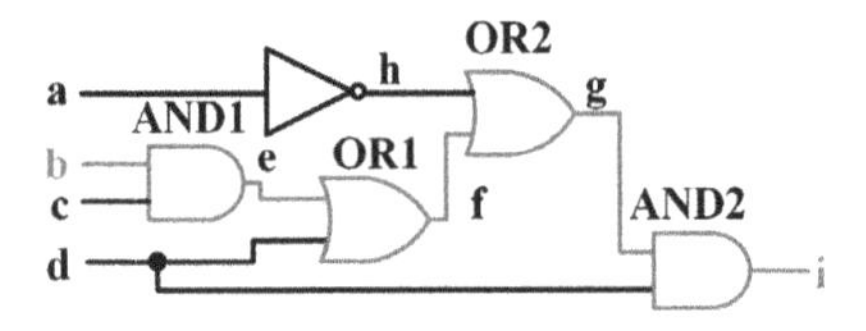

FIGURE 11.21
Design circuit #1 for true/false path.

Exercises

Please refer to Tables 11.2 and 11.3 for the timing parameters to be used in the following exercises. It is assumed that the clock skew is always 1.0 ns.

Problem 11.1. Referring to Figure 11.21, please answer the following questions:

1) Is the path "b-e-f-g-i" a true path or a false path? If it is a false path, what is the longest true path?

2) What is the shortest true path?

3) Assuming that the entire combinational circuit is between two register layers: all the signals "a", "b", "c", and "d" are outputs from the first register layer, and the signal "i" is the input to the second register layer. What is the MOF for the entire sequential circuit?

4) Does a hold time violation occur in the circuit?

Problem 11.2. Referring to Figure 11.22, please answer the following questions:

1) Determine whether the longest path "c-d-e-f-j" is a true path or a false path. If it is a false path, please identify the longest true path.

2) Identify the shortest true path in the circuit.

3) Assuming that the entire combinational circuit is between two register layers: all the signals "a", "b", and "c" are outputs from the first register layer, and the signal "j" is the input to the second register layer. What is the MOF for the entire sequential circuit?

4) Is there a hold time violation?

Problem 11.3.

1) Draw the circuit described with the Verilog code below.

2) Assuming a clock period of 12 ns, identify whether there is a potential setup violation if the logic is synthesized exactly as described. Provide a brief explanation for your answer.

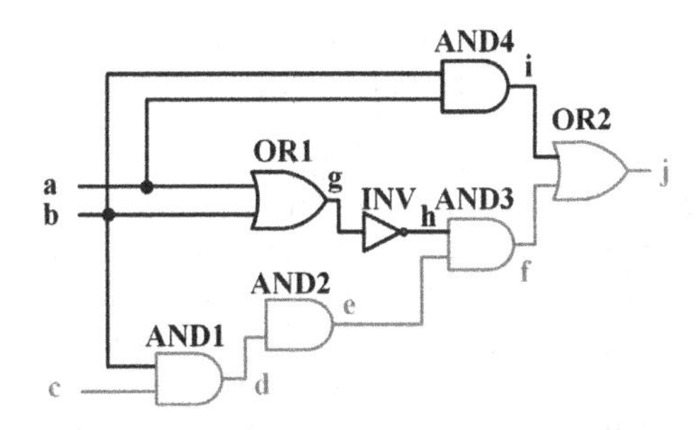

FIGURE 11.22
Design Circuit #2 for true/false path.

3) Identify whether there is a potential hold violation. Provide a brief explanation for your answer.

```
module (input a, b, c, d, e, f, output reg j);
reg g, h, i;

always @(posedge clk, negedge rst)
if (~rst) begin
    g<=0; h<=0; i<=0; j<=0;
end else begin
    g <= a & b                    ;
    h <= c | d                    ;
    i <= e ^ f                    ;
    j <= i ? ~|{g, h} : &{g, h};
end
endmodule
```

PBL 31: MOF Assessment of FP Numerical Hardware

In Chapter 2, we introduce an analytical approach to approximately estimate the maximum operational frequency (MOF) of register-transfer designs using the following equation:

$$MOF = \frac{1}{T_R - T_{WNS}}. \tag{11.23}$$

In this equation, T_R represents the reference clock period, and T_{WNS} represents the Worst Negative Slack, indicating the available time in each clock cycle.

1) Let's assume that the design specification requires a reference clock frequency of 100 MHz, which corresponds to a minimum clock period of $\frac{1}{100\text{ MHz}} = 10$ ns. Therefore, we can set T_R to 10 ns. In this scenario, determine the T_{WNS} for the circuit designed in PBL 18 in Chapter 8, which utilizes the single-clock-cycle FP operators. You can achieve this by configuring the constraint file and executing the Vivado synthesis and implementation tools. For detailed configuration information, please refer to Section 2.6.2 in Chapter 2.
2) Follow the same procedure for PBL 19 in Chapter 8, which utilizes the pipeline-designed FP operators.
3) Calculate the MOF for PBLs 18 and 19, and provide an explanation of the MOF differences.
4) Consider the performance disparities (latency and bandwidth) among PLBs 19, 20, and 21.

PBL 32: Request-Grant Handshaking and CDC

1) Design the handshaking circuit by integrating the master in clock domain A and the slave in clock domain B, following the block diagram depicted in Figure 11.20. The design's IO specifications are outlined in Tables 11.4 and 11.5. The "req_a" output from the master serves as the bus request input for the slave, while the "gnt_b" output from the slave facilitates the bus grant to the master.
2) Create a testbench to simulate the circuit and confirm its functionality by generating a pulse signal named "en_a". Monitor the timing behavior of all signals, with special attention to the interaction between the handshaking signals "req_a" and "gnt_b", which constitute the handshaking process.

PBL 33: Synchronous FIFO

1) Design a synchronous FIFO, encompassing the IOs detailed in Table 11.6. Notably, the FIFO incorporates an asynchronous reset signal denoted as "rst", which initializes the FIFO's memory and

TABLE 11.6
Synchronous FIFO IOs description.

Name	Direction	Bit Width	Description
clk	Input	1	Clock, rising edge, 50 MHz
rst	Input	1	Asynchronous reset, 0 valid
write_en	Input	1	FIFO write enable
read_en	Input	1	FIFO read enable
data_in	Input	parameterized	The data bus for FIFO writing
data_out	Output	parameterized	The data bus for FIFO reading
full	Output	1	FIFO full indicator
empty	Output	1	FIFO empty indicator

internal pointers upon activation. Operating in sync with the clock input labeled "clk", the FIFO facilitates data writing and reading via the "write_en" and "read_en" signals, respectively. Data is written into the FIFO using the "data_in" input, while data retrieval is achieved through the "data_out" output. The FIFO design further encompasses a "full" indicator to signify when the FIFO reaches capacity and can no longer accommodate additional data. Similarly, an "empty" indicator denotes that the FIFO is devoid of data.

The following are the design specifications and guiding principles:

- The FIFO design involves a circular buffer structure, realized by utilizing an array of registers, as depicted in line 13 within the provided Verilog code. It allows for a user-configurable depth and width, which can be established using *parameters*, as evident in line 9–10.
- The FIFO module should integrate two internal pointers: a write pointer (referred to as "write_ptr" in line 14) and a read pointer (identified as "read_ptr" in line 15). These pointers will serve to indicate the subsequent locations for data writing and reading, respectively. Upon executing a write operation, the "write_ptr" should increment by one. Similarly, during a read operation, the "read_ptr" should also increment by one. The bit width of these pointers should correspond to the FIFO depth, as indicated in line 11. For instance, in the provided Verilog code example, where the FIFO depth is parameterized as 64, a pointer width of 6 bits is needed.
- Additionally, the FIFO module should include a counter (labeled as "cnt" in line 16), which will keep track of the total number of data items stored within the buffer. Upon each write operation, the counter should increment by one. Conversely, during a read operation, the counter should decrement by one.

The state of the counter will dictate the status of the FIFO. When the counter reaches the value of $DEPTH$, the FIFO should be deemed full. Conversely, if the counter is at zero, the FIFO should be considered empty.

- A crucial aspect of the FIFO design is to offer indicators for its fullness or emptiness. In cases where the counter equals $DEPTH$, the FIFO must not accept further data even if the "write_en" signal is active. Likewise, when the counter is zero, the FIFO should not permit data extraction, even if the "read_en" signal is asserted.

```
1  module sync_fifo (input                   clk       ,
2                    input                   rst       ,
3                    input                   write_en  ,
4                    input                   read_en   ,
5                    input  [WIDTH-1:0]      data_in   ,
6                    output [WIDTH-1:0]      data_out  ,
7                    output                  full      ,
8                    output                  empty     );
9  parameter DEPTH     = 64;  // Depth of the FIFO
10 parameter WIDTH     = 8 ;  // Width of the FIFO
11 parameter PTR_WIDTH = 6;   // DEPTH = 1<<PTR_WIDTH;
12
13 reg [WIDTH-1:0]      mem [0:DEPTH-1];
14 reg [PTR_WIDTH-1:0]  write_ptr      ;
15 reg [PTR_WIDTH-1:0]  read_ptr       ;
16 reg [PTR_WIDTH-1:0]  cnt            ;
17
18 endmodule
```

2) Create a testbench to simulate the FIFO design, assessing its functionalities encompassing FIFO write and read operations, along with scenarios where the FIFO is either full or empty.

PBL 34: High-Speed Design on FP Matrix-Matrix Adder

1) Redesign the FP Matrix-Matrix Adder in Section 9.3 in Chapter 9, by leveraging the pipeline-designed FP operators provided in this book.

2) Create a testbench to facilitate the simulation of the design and validate its functionality using the same FP data input shown in Figure 9.11.
3) Comments on the difference between using single-clock-cycle IPs and pipeline-designed IPs.

PBL 35: High-Speed Design on FP AXPY Calculation

1) Redesign the FP AXPY calculation circuit in Section 9.4 in Chapter 9, by leveraging the pipeline-designed FP operators provided in this book.
2) Create a testbench to facilitate the simulation of the design and validate its functionality using the same FP data input shown in Figure 9.13.
3) Comments on the difference between using single-clock-cycle IPs and pipeline-designed IPs.

PBL 36: High-Speed Design on FP DDOT

1) Redesign the FP DDOT circuit in Section 9.5 in Chapter 9, by leveraging the pipeline-designed FP operators provided in this book.
2) Create a testbench to facilitate the simulation of the design and validate its functionality using the same FP data input shown in Figure 9.15.
3) Comments on the difference between using single-clock-cycle IPs and pipeline-designed IPs.

PBL 37: High-Speed Design on FP Matrix-Matrix Multiplier

1) Redesign the Multiplication-Addition Circuit (*MAC*) in PBL 25, by leveraging the pipeline-designed FP operators provided in this book.

2) Redesign the FP Matrix-Matrix Multiplier circuit in PBL 26, incorporating the high-speed MAC circuit as the design IP.
3) Create a testbench to facilitate the simulation of the design and validate its functionality using the same FP data input in PBL 26.
4) Comments on the difference between using single-clock-cycle IPs and pipeline-designed IPs.

PBL 38: High-Speed Streaming Design on DDOT

1) Redesign the Streaming Design on DDOT circuit in Section 10.2 in Chapter 10, by leveraging the pipeline-designed FP operators provided in this book.
2) Create a testbench to facilitate the simulation of the design and validate its functionality using the same FP data input shown in Figure 10.3.
3) Comments on the difference between using single-clock-cycle IPs and pipeline-designed IPs.

PBL 39: High-Speed Iterative Design with Streaming Width Four

1) Redesign the Iterative Design with Streaming Width Four in Section 10.3 in Chapter 10, by leveraging the pipeline-designed FP operators provided in this book.
2) Create a testbench to facilitate the simulation of the design and validate its functionality using the same FP data input shown in Figure 10.8.
3) Comments on the difference between using single-clock-cycle IPs and pipeline-designed IPs.

PBL 40: High-Speed Streaming Design on DFT2

1) Redesign the Streaming Design on DFT2 in PBL 29, by leveraging the pipeline-designed FP operators provided in this book.

2) Create a testbench to facilitate the simulation of the design and validate its functionality using the same FP data input shown in Figure 10.16(b).
3) Comments on the difference between using single-clock-cycle IPs and pipeline-designed IPs.

PBL 41: High-Speed Streaming Design on DFT4

1) Redesign the Streaming Design on DFT4 in PBL 30, by leveraging the pipeline-designed FP operators provided in this book.
2) Create a testbench to facilitate the simulation of the design and validate its functionality using the same FP data input as shown in Figure 10.18(b).
3) Comments on the difference between using single-clock-cycle IPs and pipeline-designed IPs.

12

SoC Design and Integration

Chapter 12 provides an introduction to system-on-chip (SoC) design and integration, with a focus on the widely used bus architecture prevalent in the IC design industry: AMBA (Advanced Microcontroller Bus Architecture) AXI (Advanced eXtensible Interface). Additionally, this chapter introduces commonly-used design components on SoCs, including direct access memory (DMA), image processing units, and neural networks. It also explores various design approaches for achieving high-performance computing, emphasizing the tradeoffs between optimizing results and efficiently managing hardware costs, along with a variety of design structures tailored to accommodate diverse specifications.

PBL 42–50 are provided in this chapter to showcase various design architectures (streaming and iterative designs) and design approaches (floating-to-fixed-point conversion, approximate design, arbitration, etc.), tailored for different design specifications concerning accuracy, hardware resource constraints, and latency bounds.

12.1 SoC Bus Architecture

SoC integration represents a cost-effective approach for incorporating a wide range of intellectual properties (IPs) into a system. These IP blocks encompass both in-house designed modules and third-party components, such as processing cores, DMA units, neural network engines, image/video/audio processing modules, industry-standard interfaces, and various other essential design elements required for SoC integration. To streamline the integration process for IPs originating from diverse sources, the inclusion of a universal bus standard becomes crucial to guarantee compatibility among different IP blocks.

Within the industry, various bus protocols are available, including options such as AMBA AHB (Advanced High-performance Bus)[11] and AMBA AXI[12] from ARM Holdings, Wishbone from Silicore Corporation [15], OCP (Open Core Protocol) from OCP-IP (OCP International Partnership) [14], and others. Among these, AMBA AXI stands out due to its widespread adoption and implementation across the semiconductor sector. It enjoys support from commercial IP products, FPGA development, and electronic design

DOI: 10.1201/9781003187080-12

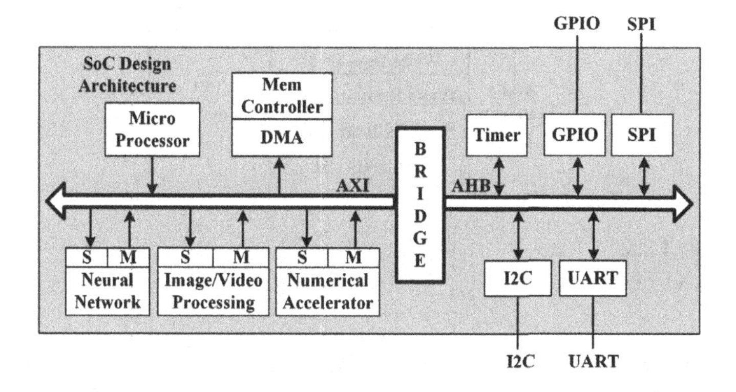

FIGURE 12.1
SoC architecture with AMBA AXI and AHB.

automation (EDA) tools. Consequently, this chapter is dedicated to providing a concise introduction to the AMBA AXI bus standard.

12.1.1 AMBA AXI bus architecture

Figure 12.1 illustrates a standard SoC architecture based on the AMBA framework. Within this architecture, the microprocessor assumes the role of the central **manager** positioned on the high-speed AMBA AXI bus. It exercises control over a multitude of interconnected IPs. These IPs, which encompass elements like the neural network engines, image/video processing units, and the floating-point (FP) hardware accelerators, operate as **subordinate** entities through the control bus, denoted by the "S" interface. This manager-subordinate connection constitutes the fundamental bus communication of the AMBA AXI protocol.

Additionally, each of these IPs can also behavior as AXI managers, denoted as the "M" interface, allowing them to access memory via both the DMA and the memory controller. Notably, within this configuration, the DMA and memory controller themselves function as subordinate entities within the manager-subordinate structure.

AMBA AHB serves as the earlier version of the AMBA bus protocol and is primarily utilized for low-speed bus transfers. Bridging the high-speed AXI bus to the lower-speed AHB bus becomes essential. On the AHB side, peripherals like I2C, SPI, UART, GPIO, etc., are treated as **subordinates** to the microprocessor. In this setup, the microprocessor serves as the **manager** on the AHB bus, exerting control through the bridge.

In conclusion, AMBA AHB is recognized for its simpler protocol, involving fewer signaling lines. This simplicity enhances implementation ease and reduces SoC overhead. Conversely, the advanced iteration of the AMBA bus framework, AMBA AXI, features a more intricate protocol with diverse

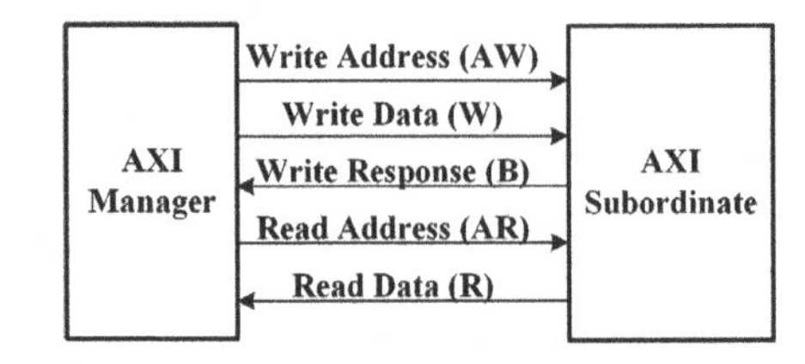

FIGURE 12.2
AMBA AXI channels.

channels for data transactions, rendering it more fitting for intricate and expansive SoC designs. Given its widespread adoption within the IC design field, this chapter predominantly concentrates on the advanced AXI bus standard.

12.1.2 AMBA AXI channels

The AXI protocol establishes a point-to-point connection between a manager and a subordinate, as illustrated in Figure 12.2. This protocol facilitates full-duplex communication, allowing simultaneous read and write operations to take place on distinct channels. The interaction between these two interfaces involves five distinct channels: Write Address (AW), Write Data (W), and Write Response (B) dedicated to write operations, and Read Address (AR) and Read Data (R) designed for read operations.

Both Write Address and Read Address channels carry essential control information pertaining to the nature of the data transfer. In the context of a write operation, the manager transmits an address through the Write Address channel and concurrently transfers data via the Write Data channel to the subordinate entity. Subsequently, the subordinate writes the data to the specified address. Following the completion of the write process, the subordinate transmits a response message to the manager through the Write Response channel.

During a read operation, the manager initiates the process by transmitting the targeted address via the Read Address channel. The subordinate entity, in response, retrieves the data associated with the provided address and transmits it back to the manager through the Read Data channel. In the event of an error, such as encountering an invalid address or lacking the necessary security permissions, the subordinate has the capability to communicate this error message through the Read Data channel.

12.1.3 AMBA AXI bus valid-ready handshaking

The AXI protocol functions through a valid-ready handshaking mechanism, ensuring communication synchronization between the source and destination of each channel. This mechanism is depicted in Figure 12.3. The source entity

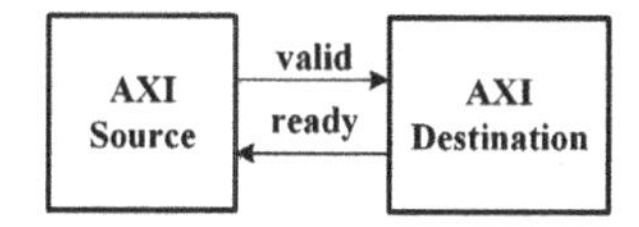

FIGURE 12.3
AMBA AXI valid-ready handshaking.

asserts the "valid" signal when it possesses valid information for transmission, while the destination entity indicates its readiness to receive information by asserting the "ready" signal.

It's important to note that the roles of source and destination can interchange depending on the specific channel in use. For instance, in the Write Address, Read Address, and Write Data channels, the manager takes on the role of the source, while the subordinate assumes the role of the destination. Conversely, in the Write Response and Read Data channels, the roles are reversed, with the subordinate serving as the source and the manager as the destination.

Figure 12.4 illustrates two instances of the valid-ready handshaking mechanism between a source and a destination. Each example showcases a transaction involving four data transfers from the source to the destination. It's important to note that the nature of the data can vary; it might represent addresses in the Write/Read Address channels, actual data in the Write/Read Data channels, or responses in the Write Response channel, contingent upon the channels in use.

In Figure 12.4(a), the data transfers between the source and destination span multiple clock cycles. The handshaking process is elaborated as follows:

- In clock cycle 2, the source asserts the "valid" signal to transmit the first data item "d0" to the destination. However, in this cycle, the destination isn't yet prepared to receive "d0". Consequently, the source retains the "valid" signal and data "d0" until the subsequent clock cycle.

- In clock cycle 3, the destination asserts its "ready" signal, signifying its readiness to accept "d0".

- In clock cycle 4, the source updates the data to "d1" and asserts the "valid" signal. As the destination's "ready" signal is asserted, it becomes capable of receiving "d1".

- In clock cycle 5, the source updates the data to "d2" and asserts the "valid" signal. Since the destination isn't prepared to receive "d2", the source holds both the "d2" data and the "valid" signal until the following clock cycle.

- In clock cycle 6, the destination is now prepared to receive "d2". In the subsequent clock cycle 7, the source updates the data to "d3", and the destination promptly receives "d3" within the same cycle.

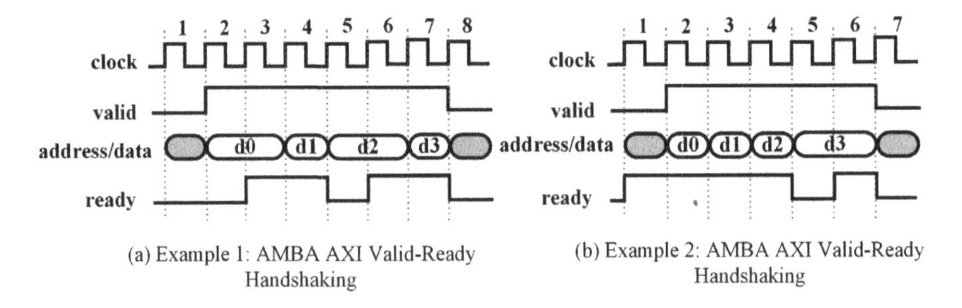

(a) Example 1: AMBA AXI Valid-Ready Handshaking

(b) Example 2: AMBA AXI Valid-Ready Handshaking

FIGURE 12.4
Examples of AMBA AXI valid-ready handshaking.

Figure 12.4(b) offers another illustrative instance of the valid-ready handshaking process. In this specific scenario, the "ready" signal is asserted prior to the "valid" signal. This signifies that the destination is prepared to accept the data even before detecting an asserted "valid" signal. The sequence of handshaking across clock cycles is clarified as follows:

- The "ready" signal is asserted from cycles 1 to 4, signaling the destination's readiness to receive data. In response, the source sequentially sends addresses/data, namely "d0", "d1", and "d2", during cycles 2 to 4. These items are promptly received by the destination.
- In cycle 5, the destination isn't equipped to accept further data, hence the "ready" signal remains unasserted. During this cycle, the source retains the "valid" signal and the associated "d3" data.
- In cycle 6, the destination becomes capable of receiving data once more. Consequently, the last data item, "d3", is received by the destination. This successfully concludes all four addresses/data transfers.

12.1.4 AMBA AXI write and read operations

The AXI bus operates on a burst-based protocol that encompasses five distinct transaction channels, allowing the simultaneous transfer of multiple data items within a single transaction. Figure 12.5(a) illustrates a straightforward write transaction that involves four data transfers. This transaction encompasses the operations of the Write Address channel, Write Data channel, and the Response channel.

- The Write Address channel initiates the process as the manager dispatches the write address (denoted as "a0" on the "awaddress" bus), the write burst length (represented as 8'h3 on the "awlen" bus), and other transaction-related details such as burst type and size (not shown in this figure). The interaction between the manager and the subordinate within the Write Address channel culminates in clock cycle 3.

- Subsequent to the initial handshake, the manager proceeds to transmit four data to the subordinate via the Write Data channel. Precisely, the manager transmits the sequence of data: "d0", "d1", "d2", and "d3", on the write data bus "wdata", while concurrently asserting "wvalid" during clock cycles 4, 5, 7, and 8, respectively. The subordinate, in turn, enables data reception in the corresponding clock cycles. It's important to note that during clock cycle 6, the subordinate should disregard the data on the "wdata" bus since the "valid" signal is not asserted and the data is deemed invalid. The handshake within the Write Data channel concludes by clock cycle 8.

- The Response channel comes into play as the subordinate employs it to validate the completion of the write transaction after receiving all write data. In clock cycle 9, the subordinate asserts "bvalid" while driving "bresp" (in this instance, the response is "OKAY") to denote the outcome – either the success or failure – of the write transaction. Concurrently, the manager signals its readiness to receive the response by asserting "bready" during the same cycle to finalize the handshake within the Response channel.

Figure 12.5(b) provides insight into a read transaction that encompasses four data transfers, involving the operations of two key channels: the Read Address channel and the Read Data channel.

- Initially, the Read Address channel comes into play as the manager sends the read address (referred to as "a0" on the "araddress" bus) and the read burst length (represented as "8'h3" on the "arlen" bus). Alongside these details, additional commands such as burst size and type (not shown in the figure) are sent to the subordinate. The handshake within the Read Address channel concludes by clock cycle 3.

- Subsequently, the subordinate leverages the Read Data channel to transmit four data transfers back to the manager. Specifically, the subordinate asserts "rvalid" and concurrently transmits the corresponding read data "d0" and "d1" during clock cycles 4 and 5, as well as "d2" and "d3" in cycles 7 and 8. The manager is primed to receive the data, as evident from the asserted "rready" bus within clock cycles 4–8. During clock cycle 6, the manager should disregard the data on the "rdata" bus, as it is flagged as invalid by the subordinate.

  Notably, the read transaction adopts the Read Data channel to convey the status of each read operation. As depicted in the figure, the subordinate imparts "OKAY" responses on the "rresp" bus and asserts the "rlast" signal during the final cycle, thereby indicating the conclusion of the read transaction. It's crucial to emphasize that if an error arises during any of the transfers, the transaction must still continue until the designated burst length is

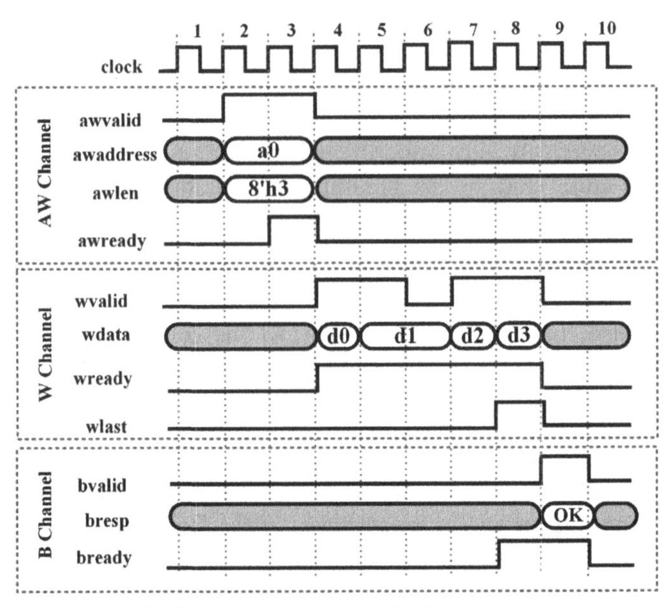

(a) An Example of AMBA AXI Write Operations

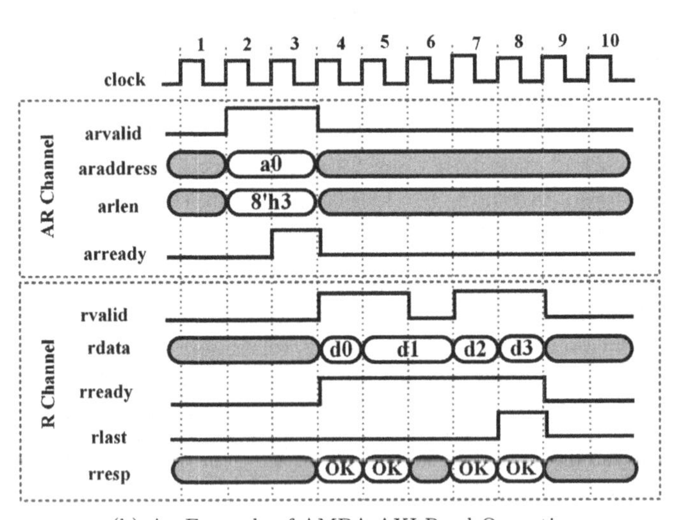

(b) An Example of AMBA AXI Read Operations

FIGURE 12.5
Examples of AMBA AXI transactions.

achieved. The protocol does not incorporate a mechanism for prematurely terminating the burst operation.

12.1.5 AMBA AXI burst length and burst size

As previously mentioned, the AXI protocol defines the burst length for write and/or read operations, indicating the number of data transfers in a single bus transaction. For read transactions, the burst length is denoted by the "arlen[7:0]" bus, while for write transactions, it is indicated by the "awlen[7:0]" bus. In a more general sense, "axlen" encompasses both "arlen[7:0]" and "awlen[7:0]".

The burst length can be determined using the formula:

$$burst\_length = axlen + 1. \tag{12.1}$$

This calculation permits a maximum burst length of 256 for the 8-bit "axlen" bus in the AXI4 protocol. In contrast, the AXI3 protocol accommodates burst lengths ranging from 1 to 16, given that the "axlen" is specified as a 4-bit bus. Throughout this book, we predominantly illustrate examples based on the AXI4 bus protocol.

The AXI protocol also defines the burst size, which pertains to the width of the data bus. Specifically, the "arsize[2:0]" bus signifies the burst size for read transactions, while the "awsize[2:0]" bus pertains to write transactions. In a broader context, "axsize" encompasses both "arsize[2:0]" and "awsize[2:0]".

A comprehensive listing of burst sizes designated by the AXI protocol is presented in Table 12.1, encompassing values from 1 byte up to 128 bytes. To ascertain the total data size within a transaction, the following formula can be employed:

$$trans\_size = burst\_length \times burst\_size. \tag{12.2}$$

For instance, consider a scenario where a manager dispatches a command with "axlen=8'h3" (yielding $burst\_length = axlen + 1 = 4$ bursts) and "axsize=3'b010" (as shown in the Table, corresponding to $burst\_width = 4$ bytes within each burst). In this context, the data size within the bus transaction can be computed as $trans\_size = burst\_length \times burst\_size = 4 \times 4 = 16$ bytes.

12.1.6 AMBA AXI burst types

The AXI bus protocol introduces three distinct burst types: FIXED, INCR, and WRAP. As detailed in Table 12.2, the burst type is specified by the "arburst[1:0]" bus for read transactions and the "awburst[1:0]" bus for write transactions. Broadly, the "axburst[1:0]" bus encompasses both "arburst[1:0]" and "awburst[1:0]", encapsulating burst type determination.

TABLE 12.1
AXI burst size.

axsize[2:0]	Bytes in Each Burst
3'b000	1
3'b001	2
3'b010	4
3'b011	8
3'b100	16
3'b101	32
3'b110	64
3'b111	128

During each bus transaction, the manager solely initializes the address for the initial data transfer. Subsequently, as the series of data transfers unfold, the subordinate must deduce the addresses pertaining to subsequent data transfers, employing the distinct burst types. An important point to highlight is that a data burst is constrained from traversing a 4KB address boundary.

A. AMBA AXI FIXED Burst Type

In a FIXED transaction, the address remains consistent for all data transfers encompassed within the transaction. To illustrate, consider a four-beat FIXED transaction exemplified in Figure 12.6. At the initiation, the address is designated as "axaddr=32'h14", the burst length is denoted by "axlen=8'h3" (equivalent to four data bursts), and the burst size is indicated as "axsize=3'b10" (representing word-sized bursts).

On the receiving end shown on the "mem addr" bus, all four memory accesses derive from the same "32'h14" address due to the FIXED burst type. This particular burst type proves useful for situations requiring repetitive accesses to a specific memory location, as observed when loading or clearing a FIFO.

TABLE 12.2
AXI burst types.

axburst[1:0]	Burst Types
2'b00	FIXED
2'b01	INCR
2'b10	WRAP
2'b11	Reserved

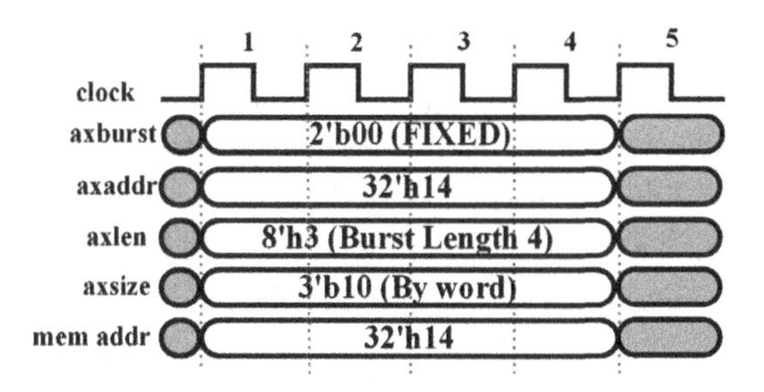

FIGURE 12.6
An example of AMBA AXI FIXED burst type.

B. AMBA AXI INCR Burst Type

In an increment transaction, denoted as INCR, the address associated with each data transfer within a transaction experiences a systematic increase by a predetermined value. The extent of this increment is contingent upon the burst size parameter.

As portrayed in Figure 12.7(a), a four-beat INCR transaction features an initial address designated as "axaddr=32'h14", a burst length of "axlen=8'h3" (equivalent to four data bursts), and a burst size defined as "axsize=3'b10" (indicating word-sized bursts).

Given the burst size of four bytes, the address for each data transfer advances by four units. This progression yields a sequence of memory addresses: 32'h14, 32'h18, 32'h1c, and 32'h20, as visually represented on the "mem addr" bus in Figure 12.7(a) and the memory access illustrated in Figure 12.7(b). This burst type finds common application in scenarios necessitating access to sequentially ordered memory locations.

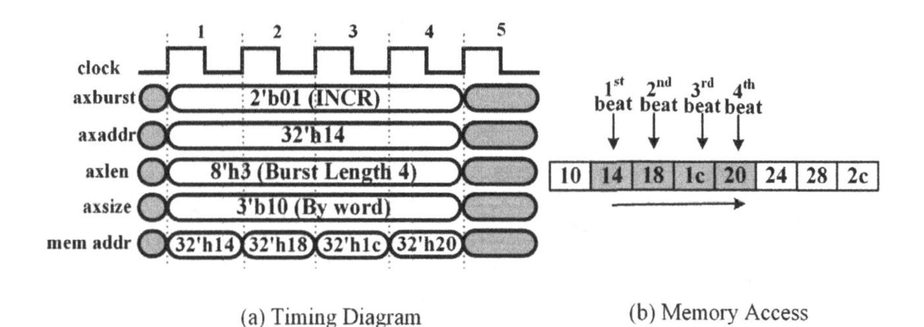

FIGURE 12.7
An example of AMBA AXI INCR burst type.

C. AMBA AXI WRAP Burst Type

Within a WRAP transaction, the operational principle mirrors that of an increment burst, yet with an additional feature: the address cycle returns to the foundational address when the higher address value coincides with the wrap boundary. To compute the base address, refer to the equation:

$$base\_addr = \{axaddr[MSB - 1 : N], N'b0\}. \tag{12.3}$$

In this context, N signifies the bit width of the data transaction size, denoted as $2^N = trans\_size$. Consequently, $N'b0$ is appended to the least significant bits of the initial address. Subsequently, the wrap boundary can be further articulated as:

$$wrap\_boundary = base\_address + trans\_size. \tag{12.4}$$

For example, let's take the case of a four-beat WRAP transaction depicted in Figure 12.8(a). At the outset, the address is "axaddr=32'h14", the burst length is "axlen=8'h3", corresponding to four data bursts, and the burst size is "axsize=3'b10", indicating word-sized bursts. Consequently, the transaction size can be computed as $burst\_size \times burst\_length = 16$ bytes (in hexadecimal 32'h10), leading to the bit width of the data transaction size: $N = 4$ (due to $2^4 = 16$). Therefore, substituting the four least significant bits with zeros yields the base address as 32'h10, as per Equation 12.3. Furthermore, the wrap boundary is established as $base\_address + trans\_size = 32'h10 + 32'h10 = 32'h20$.

Functioning as a WRAP transaction, therefore, the memory address cyclically returns to the base address 32'h10 when it reaches 32'h20, as demonstrated in Figures 12.8(a) and 12.8(b). The memory access takes the form of a loop, commencing from the starting address (32'h14), progressing until the wrap boundary (32'h20), and finally circling back to the base address (32'h10).

Within the AXI bus protocol, it's important to note that the WRAP burst length is constrained to values of 2, 4, 8, or 16. For all other scenarios, the methodology for calculating the base address is succinctly outlined in Table 12.3.

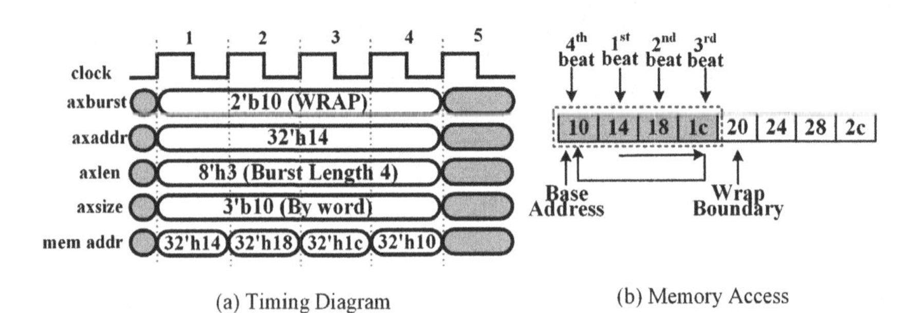

FIGURE 12.8
An example of AMBA AXI WRAP burst type.

TABLE 12.3
Base address for wrap burst type.

Burst Length	Burst Size	N	Base Address
2	1	1	{axaddr[MSB-1:1], 1'b0}
	2	2	{axaddr[MSB-1:2], 2'b0}
	4	3	{axaddr[MSB-1:3], 3'b0}
	8	4	{axaddr[MSB-1:4], 4'b0}
	16	5	{axaddr[MSB-1:5], 5'b0}
	32	6	{axaddr[MSB-1:6], 6'b0}
	64	7	{axaddr[MSB-1:7], 7'b0}
	128	8	{axaddr[MSB-1:8], 8'b0}
4	1	2	{axaddr[MSB-1:2], 2'b0}
	2	3	{axaddr[MSB-1:3], 3'b0}
	4	4	{axaddr[MSB-1:4], 4'b0}
	8	5	{axaddr[MSB-1:5], 5'b0}
	16	6	{axaddr[MSB-1:6], 6'b0}
	32	7	{axaddr[MSB-1:7], 7'b0}
	64	8	{axaddr[MSB-1:8], 8'b0}
	128	9	{axaddr[MSB-1:9], 9'b0}
8	1	3	{axaddr[MSB-1:3], 3'b0}
	2	4	{axaddr[MSB-1:4], 4'b0}
	4	5	{axaddr[MSB-1:5], 5'b0}
	8	6	{axaddr[MSB-1:6], 6'b0}
	16	7	{axaddr[MSB-1:7], 7'b0}
	32	8	{axaddr[MSB-1:8], 8'b0}
	64	9	{axaddr[MSB-1:9], 9'b0}
	128	10	{axaddr[MSB-1:10], 10'b0}
16	1	4	{axaddr[MSB-1:4], 4'b0}
	2	5	{axaddr[MSB-1:5], 5'b0}
	4	6	{axaddr[MSB-1:6], 6'b0}
	8	7	{axaddr[MSB-1:7], 7'b0}
	16	8	{axaddr[MSB-1:8], 8'b0}
	32	9	{axaddr[MSB-1:9], 9'b0}
	64	10	{axaddr[MSB-1:10], 10'b0}
	128	11	{axaddr[MSB-1:11], 11'b0}

12.1.7 AMBA AXI address alignment

Memory accessed via the AXI bus necessitate address alignment in accordance with the designated burst sizes. For instance, when the burst size is specified as "axsize=3'b010", signifying four bytes, it mandates that the bus address maintains alignment with four bytes for each burst. In more precise terms, the two least significant bits of memory addresses must be set to binary 2'b00.

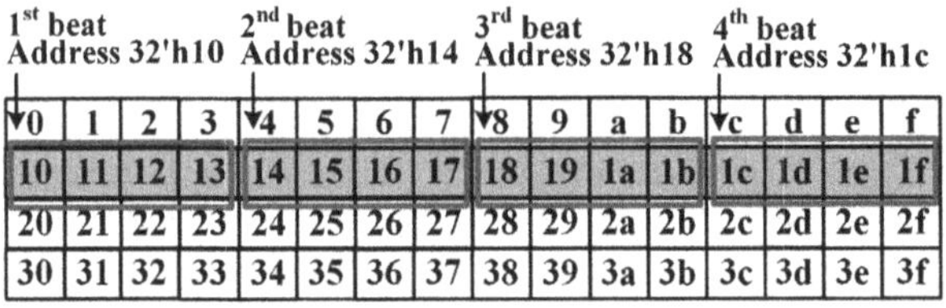

1st beat Address 32'h10 | 2nd beat Address 32'h14 | 3rd beat Address 32'h18 | 4th beat Address 32'h1c

0	1	2	3	4	5	6	7	8	9	a	b	c	d	e	f
10	11	12	13	14	15	16	17	18	19	1a	1b	1c	1d	1e	1f
20	21	22	23	24	25	26	27	28	29	2a	2b	2c	2d	2e	2f
30	31	32	33	34	35	36	37	38	39	3a	3b	3c	3d	3e	3f

FIGURE 12.9
An example of address alignment.

Similarly, in the case of a burst size defined as "axsize=3'b011", or double-word-sized, alignment necessitates the three least significant bits of memory addresses to assume the value 3'b000.

In a broader context, the formula for calculating the aligned address is given by:

$$aligned\_addr = INT(\frac{start\_addr}{burst\_size}) \times burst\_size. \tag{12.5}$$

where the function $INT(x)$ is utilized to round down x to the nearest integer value.

For instance, a manager issues a command with transfer type "axburst=2'b01" (indicating an INCR transaction), a burst length of "axlen=8'h3" ($burst\_length = 4$), and a burst size of "axsize=3'b10" ($burst\_size = 4$). In the scenario where the start address is erroneously set to $0x11$ by software, leading to misalignment with the burst size, hardware will rectify the address to 32'h10 using the equation $aligned\_addr = INT\left(\frac{17}{4}\right) \times 4 = 16$, in hexadecimal 32'h10. In this context, the value 17 arises from the software-configured address $0x11$, while the value 4 corresponds to the $burst\_size$. As a consequence, Figure 12.9 illustrates the correction of the first memory address to 32'h10, followed sequentially by the second, third, and fourth addresses – 32'h14, 32'h18, and 32'h1c – on the "axaddr" address bus.

12.2 Direct Memory Access (DMA)

DMA plays a crucial role in facilitating data movement within the SoC design. It enables individual masters on the data bus to interact directly with memory, bypassing the need for central processing unit (CPU) involvement. As a result, it leads to a notable improvement in the efficiency of data movement between bus masters and memory, promoting rapid exchange of substantial data blocks.

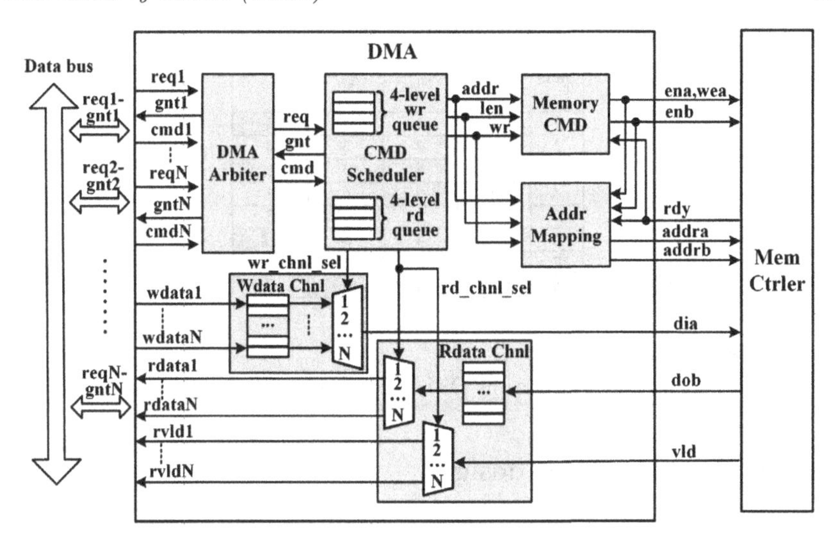

FIGURE 12.10
DMA design architecture.

In this section, we introduce a fundamental DMA design that revolves around a straightforward bus protocol tailored to handle incremental data bursts.

12.2.1 DMA design architecture

Figure 12.10 depicts the block diagram of the DMA design, encompassing several key components: the Bus Arbiter, Command Scheduler, Memory Command Converter, and Address Mapping module. Additionally, the Write Data and Read Data channels coordinate data write and read operations between the bus masters and the memory. The data bus accommodates multiple bus masters, each primed for memory access. When a master seeks memory interaction, it sends a request to the Bus Arbiter. Determined by priority assignments, the arbiter resolves access allocation, permitting engagement with a single master at a time. Further insights into the arbiter's design particulars are unveiled in Section 12.2.2.

In what follows, commands initiated by the authorized master are directed into the Command Scheduler, which consists of separate write and read queues. In this example, each of these queues has the capacity to hold a maximum of four commands individually. This mechanism efficiently frees up the bus promptly, preventing unnecessary delays caused by the immobilization of the master's requests. The Memory Command and Address Mapping modules work together to translate bus-specific commands into memory accesses. This translation process is designed to be compatible with memory interfaces, ensuring the optimization of data transfer efficiency. Further details regarding memory access intricacies and data processing can be found in Section 12.2.3.

Time Slots	M1 Priority No.	M2 Priority No.	M3 Priority No.	M4 Priority No.
1	1	2	3	4
2	2	4	6	4
3	3	6	3	8
4	4	8	6	4
5	5	2	9	8
6	6	4	3	12
7	7	6	6	4
8	1	8	9	8

FIGURE 12.11
Dynamic priority mechanism on DMA arbitration.

12.2.2 DMA arbiter design

A. DMA Arbitration

In systems featuring multiple masters, the DMA Arbiter assumes a pivotal role by determining the master permitted to govern the bus and access memory. In cases where multiple masters concurrently submit bus requests, authorization is extended to the master holding the highest priority, with subsequent masters being granted access in descending order of priority. As an illustrative instance, this section introduces an arbiter design that implements dynamic priority adjustments. Depicted in Figure 12.11, four distinct masters – labeled M1, M2, M3, and M4 – emerge, each endowed with an initial priority value: one, two, three, and four correspondingly. Notably, these initial priority values can be configured by the CPU.

During the initial time slot, M4 emerges with the highest priority value, promptly securing permission – an evident representation in the shaded box within the second row. Subsequent to this, in the succeeding time slot, priority values assigned to the remaining three masters undergo augmentation by their respective initial values. This augmentation is designed to bolster their likelihood of bus access, given the lack of authorization in the prior time slot. As demonstrated in the second time slot, M3's entreaty obtains consent due to its elevated priority value of six.

Advancing to the third time slot, M3's priority value reverts to its original setting of three, whereas M1, M2, and M4 experience priority elevation. This coordinated adjustment culminates in M4's request prevailing during the third time slot, facilitated by its commanding priority value of eight. Analogously, time slots 4 through 8 observe a sequential procession wherein M2, M3, M4, M1, and M3 are consecutively sanctioned. Each instance witnesses the highest-priority master attaining bus control.

TABLE 12.4
DMA Arbiter IOs.

Names	Width	IOs	Description
req1~req4	1	Input	Bus requests from M1~M4
cmd1~cmd4	24	Input	Bus commands from M1~M4
gnt1~gn4	1	Output	Bus grants for M1~M4
req	1	Output	Command scheduler request
cmd	24	Output	Command scheduler command
gnt	1	Input	Command scheduler grant

B. DMA Arbiter IOs

Table 12.4 offers an overview of the DMA Arbiter's IO assignments. The four inputs, labeled "req1-req4", represent bus requests originating from masters M1 through M4. In parallel, the corresponding set of outputs, designated as "gnt1-gnt4", indicates the bus grants issued by the DMA Arbiter. The allocation of bus access depends on the priority associated with each master's request, and the corresponding commands are conveyed through the "cmd1-cmd4" buses.

At the same time, the "req-gnt" arbitration mechanism becomes significant in managing interactions between the DMA Arbiter and the Command Scheduler. The arbiter's request can be granted as long as the command queue has available capacity, which is a condition different from being completely full. Additionally, the granted DMA Commands carried on the "cmd" bus can be pushed into the queue for further processing.

The DMA command encapsulates vital particulars concerning the memory access, encompassing attributes such as the access type (write or read), the length of the data burst, and the initial address for memory interaction. This configuration is visually depicted in Figure 12.12. Notably, the most significant bit specifies the memory access type – 1'b1 for write operations and 1'b0 for read operations. Meanwhile, the bits spanning from 22 to 16 signify the extent of data transfer, while the lower half-word portion serves to designate the commencement address.

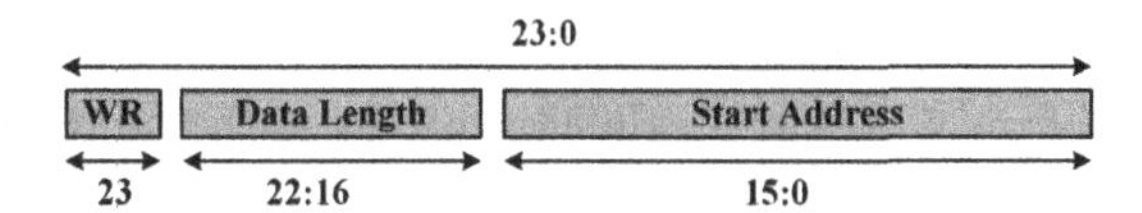

FIGURE 12.12
DMA Command.

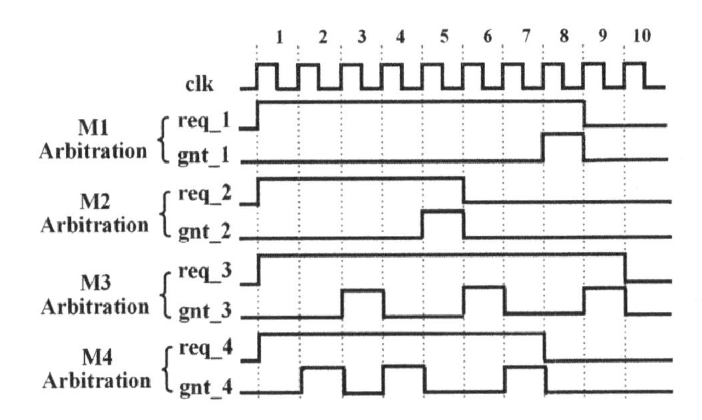

FIGURE 12.13
Timing diagram of DMA arbitration.

C. Timing Diagram of DMA Arbiter

The timing diagram illustrating bus arbitration is presented in Figure 12.13. In the initial clock cycle, all four masters initiate requests to access the bus. The first and second masters (M1 and M2) transmit a single command each, while the third and fourth masters (M3 and M4) concurrently submit three commands apiece. Moving to clock cycle 2, M4's request (denoted as "req_4") is granted by virtue of its superior priority, and the request signal remains unaltered since two more commands/requests are queued within its pipeline.

Upon entering the third clock cycle, the arbiter grants M3's request labeled as "req_3", maintaining the request signal throughout the fourth clock cycle. This is due to the presence of two pending commands/requests in M3's pipeline. Moving to the fourth clock cycle, the second request from M4 receives approval. In the fifth clock cycle, M2's request is granted, and its request signal, denoted as "req_2", is deactivated, indicating the completion of its single request.

Advancing to the sixth clock cycle, the second request from M3 obtains consent, with the final request from M4 being authorized during the subsequent seventh clock cycle. In the seventh clock cycle, M4's request signal is deactivated, signaling the fulfillment of its three requests. The eighth clock cycle confirms the approval of M1's sole request, referred to as "req_1", and the last request from M3 is given the green light in the ninth clock cycle.

12.2.3 DMA Command and address mapping

A. Memory Command and Address Mapping IOs

Table 12.5 presents the IOs for the Memory Command and Address Mapping design modules. Both modules are fed with disassembled DMA commands,

TABLE 12.5
Memory Command and Address Mapping IOs.

Names	Width	IOs	Description
wr	1	Input	DMA write/read indicator: 1'b1 for write, 1'b0 for read
len	7	Input	DMA write data length
addr	16	Input	DMA write address
ena	1	Output	Memory port "a" enable
wea	4	Output	Memory write enable to each data byte
enb	1	Output	Memory port "b" enable
rdy	2	Input	Memory ready response: MSB for write, LSB for read
addra	16	Output	Memory port "a" address
addrb	16	Output	Memory port "b" address

which encompass the write/read indication labeled as "wr" (binary 1'b1 for write and 1'b0 for read), the data burst length marked as "len", and the address referred to as "addr". These inputs necessitate conversion into memory commands to facilitate memory access.

In this specific design, a simple dual-port register array is employed, utilizing the following memory inputs and outputs: For port "a", three signals are involved, which include the enable signal "ena", the write enable signal "wea", and the address "addra". Simultaneously, for port "b", the components consist of an enable signal "enb" and an address "addrb". It's important to note that the memory enable signal and the write enable signal are generated as outputs from the Memory Command module, while the addresses originate from the Address Mapping module.

Additionally, both the Memory Command and Address Mapping components necessitate an essential input signal designated as "rdy[1:0]". The most significant bit, denoted as "rdy[1]", assumes relevance within write operations by signaling the memory's readiness to accept incoming data through the "dia" bus. Conversely, the least significant bit, labeled "rdy[0]", pertains to read operations, indicating the presence of valid data in the memory data bus "dout".

B. Timing Diagram of Memory Command and Address Mapping

The data processing mechanisms within the DMA are graphically outlined in Figure 12.14. Spanning clock cycles 2 to 5, four commands undergo insertion into the command queue, followed by prompt retrieval as the memory readies itself for access.

CMD Scheduler Channel: Starting with the initial command on the bus (in the second clock cycle), represented as hexadecimal 24'h840000, it indicates a

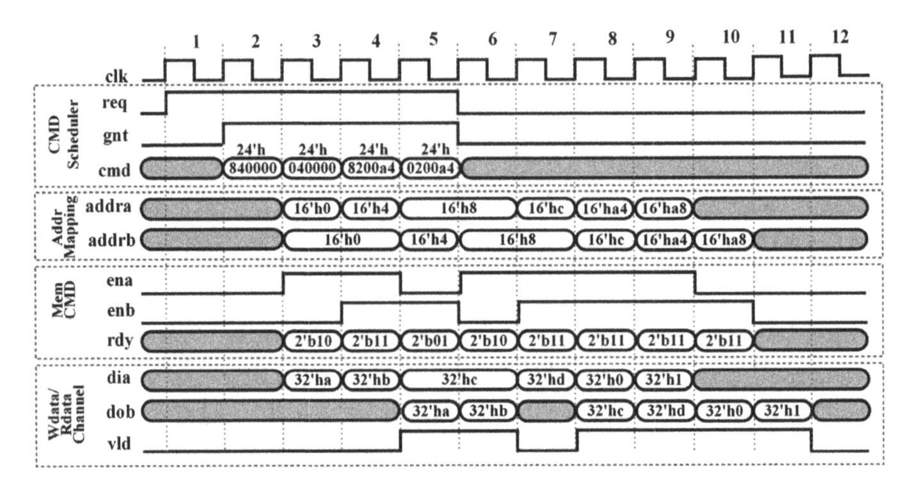

FIGURE 12.14
Timing diagram of Memory Command and Address Mapping.

write operation, as indicated by cmd[23]=1'b1. This write operation involves 4-beat data transfers, denoted by cmd[22:16]=7'h04. The operation is directed at an initial address of hexadecimal 16'h0, captured within cmd[15:0]=16'h0. Subsequently, the second command emerges (in the third clock cycle), which is a 4-beat read operation, as indicated by cmd[23]=1'b0 and cmd[22:16]=7'h04. It targets the same addresses as the previous command. The sequence continues with the third and fourth commands (in clock cycles 4–5), which encompass write and read operations and are related to the memory address hexadecimal 16'ha4, as reflected by cmd[15:0]=16'ha4. These commands entail data packet lengths encoded as hexadecimal 7'h2 within cmd[22:16]=7'h2.
Memory Command and Address Mapping Channels: The commands extracted from the command queue find their way to the Memory Command and Address Mapping modules. Assuming a data bus size of four bytes, the Address Mapping module calculates memory addresses using the initial address and the data package length. As an illustration, for the first write command, the starting address stands at hexadecimal 16'h0, accompanied by a data package length of hexadecimal 7'h4. Consequently, the memory write addresses – 16'h0, 16'h4, 16'h8, and 16'hc – are presented on the "addra" bus.

However, these memory addresses are updated solely when the ready signal (indicated as "rdy[1]") is asserted, denoting the memory's readiness to accept data at the corresponding address. In situations where "rdy[1]" is unasserted, the address bus "addra" preserves its current address status until a valid "rdy[1]" signal is restored. An example of this transpires in clock cycle 5, wherein "rdy[1]" is deasserted, and the address "addra=16'h8" remains suspended. It's only in clock cycle 6, marked by the assertion of "rdy[1]", that the address resumes propagation.

For read operations, the least significant bit of "rdy[0]" assumes a role in suspending read address updates when the memory's timely response to the read command is hindered. For instance, the first read command triggers address updates beginning in clock cycle 4 and concluding by clock cycle 8 – unveiling values 16'h0, 16'h4, 16'h8, and 16'hc. Yet, clock cycle 6 encounters a scenario where the memory controller cannot promptly address hexadecimal 16'h8. Consequently, the update of the address is postponed until clock cycle 7, allowing for data retrieval to take place.

The final pair of commands, encompassing write and read operations, play out in clock cycles 8 and 9, bearing addresses "addra=16'ha4" and "addra=16'ha8", along with clock cycles 9 and 10, characterized by "addrb=16'ha4" and "addrb=16'ha8", respectively.

Write and Read Data Channels: The Write and Read Data Channels play a crucial role in facilitating data transfers between the data bus and the memory controller. To enhance bus efficiency, a write data FIFO is integrated into the write data channel. When a bus command is granted and queued, the corresponding write data packages are placed into the write data FIFO. Subsequently, the memory controller retrieves and processes the data through the memory write bus ("dia").

It's important to note that data validity is indicated by the "vld" signal. Validity is established only when the "vld" signal is asserted, signifying that the data on the read data bus ("dob") can be relied upon. Conversely, when the "vld" signal is not asserted, data on the "dob" output must be disregarded, as demonstrated in clock cycle 7.

12.3 RGB-to-Grayscale Converter

Modern ICs are facing growing demands for minimized power consumption, reduced chip area, and robust high-performance computing capabilities. This emphasizes the importance of incorporating cost-effective design strategies, including techniques such as floating-to-integer conversion and approximate implementations. These methodologies will be exemplified through a practical demonstration using an RGB-to-grayscale converter, providing an illustration of their real-world applications.

12.3.1 Floating-point design

A. Grayscale and Color Images

An image pixel, the smallest unit composing an image, derives its name from the fusion of the terms "picture" and "element". In the case of grayscale images, pixels convey only intensity information, ranging from black (weakest intensity) to white (strongest intensity). A black pixel can be represented by the 8-bit value 8'h0, while a white pixel corresponds to the 8-bit value 8'hff. To define the image grid, we simply multiply the width by the height of the pixel matrix. This grid is commonly referred to as the image's resolution. For instance, when we say "640 × 480", it indicates an image resolution with 640 columns and 480 rows.

In images that use color, each pixel may consist of three or four components, such as red, green, and blue in the RGB model, or cyan, magenta, yellow, and black in the CMYK model. For example, in RGB, various colors are created by combining red, green, and blue light components. Each of these light components represents a portion of the original image and can be described using multiple bits. In a format like RGB888, a single color pixel (red, green, or blue) is composed of eight bits of data, with all three components utilizing eight bits each. Therefore, a color pixel can be represented by 24 bits of data, ranging from 24'h0 to 24'hffffff. For example, {8'h0, 8'h0, 8'hff} or 24'hff signifies a blue pixel, with blue set to 8'hff while red and green are set to 8'h0.

Comparing this to a one-dimensional grayscale image, a color image using the RGB888 format can be visualized as a three-dimensional grid or array, often referred to as a "tensor". The three dimensions of this tensor represent the width, height, and the image's color channels.

B. RGB-to-Grayscale Conversion Algorithm

Grayscale images, unlike RGB images that require three data dimensions, simplify the field of image and video processing. The RGB-to-grayscale converter plays a pivotal role in this context by converting RGB images into grayscale versions. This transformation is significant as it can lead to benefits such as reduced computational demands and memory usage.

To visually demonstrate the RGB-to-grayscale conversion for specific images, Figure 12.15 illustrates the process. This conversion can be achieved using the formula:

$$grayscale = 0.2989 \times r + 0.587 \times g + 0.114 \times b \tag{12.6}$$

where r denotes red pixels, g signifies green pixels, and b represents blue pixels. In general, the formula involves multiplying each color pixel by its corresponding weight: 0.2989 for red pixels, 0.587 for green pixels, and 0.114 for blue pixels.

From a hardware design perspective, the most straightforward approach for realizing the color-to-grayscale converter involves the utilization of three FP multipliers and two FP adders, as depicted in Figure 12.16. This technique entails taking each red, green, and blue pixel and subjecting them to multiplication by their respective weights (0.2989 for red, 0.587 for green, and 0.114 for blue) using the FP multipliers. Subsequently, the outcomes are combined through addition to generate the final grayscale output.

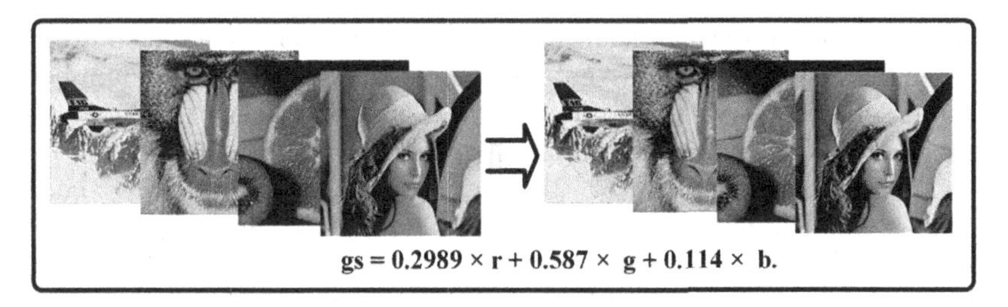

FIGURE 12.15
RGB-to-grayscale conversion.

It's worth noting that the precision of both the FP adders and multipliers is variable, spanning from byte, half-word, single-precision word, double-word, and beyond. Opting for modules with higher precision contributes to enhanced accuracy in the conversion process.

12.3.2 Floating-to-integer conversion and approximate designs

The FP design necessitates more resources and consumes greater power than equivalent integer designs, regardless of whether the design is intended for an FPGA or an ASIC platform. This disparity arises because integer arithmetic operations are less intricate and computationally demanding than FP arithmetic operations. Consequently, RTL designers are tasked with weighing the trade-offs when transitioning to integer data types and comprehending the advantages this shift offers. To illustrate the trade-off between the quality of results and hardware expenditure, this section introduces several integer implementations of the RGB-to-grayscale converter.

A. Floating-to-integer Conversion

The integer representation is primarily employed to portray integer values, distinct from FP numbers. As an illustrative example, this section delves

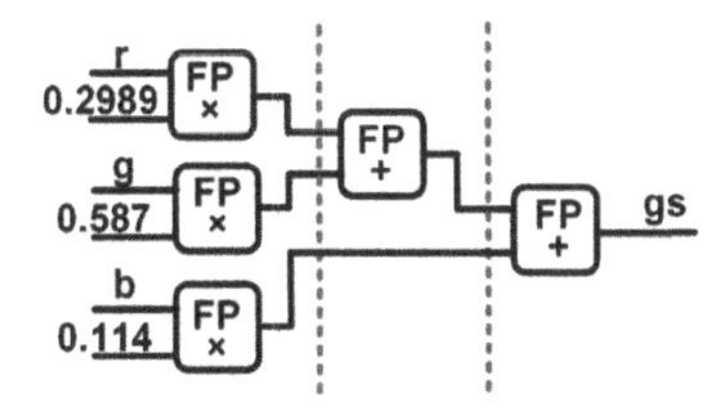

FIGURE 12.16
Floating-point design on RGB-to-grayscale converter.

into a common hardware design technique that converts FP data into fixed-point numbers. This process entails scaling a single precision number by a power of two to yield an integer value approximating the original FP number. For instance, to represent a single precision number like 3.14159, the number can be scaled by 2^{16} (decimal 65536) to generate the integer value 205887, which is mathematically represented and approximated as $3.14159 \times 65536 = 205887.67264 \approx 205887$. Subsequently, this integer value is encoded using multiple bits of binary data to depict the FP number 3.14159.

Nevertheless, it's imperative to recognize that this technique involves a compromise in precision in order to render the FP number as an integer. Notably, not all decimal places are retained in the integer representation. The selection between integer and floating-point design hinges on the particular requisites of the application at hand. While integer designs tend to exhibit enhanced power efficiency and cost-effectiveness for certain applications, they might not deliver the exact accuracy and quality of results required. Designers must meticulously assess the design specifications and weigh the trade-offs before determining the most suitable implementation approach.

B. Integer Design #1: Floating-to-integer Conversion

In Figure 12.17(a), we recall the FP implementation of the RGB-to-grayscale converter. In what follows, the initial integer design example is expressed mathematically using the following equation, considering that multiplying the weights by 2^7 and then dividing the final sum by 2^7 doesn't alter the ultimate grayscale result:

$$\begin{aligned} gs &= 0.2989 \times r + 0.5870 \times g + 0.1140 \times b \\ &= [2^7 \times (0.2989 \times r + 0.5870 \times g + 0.1140 \times b)]/2^7 \\ &\approx (38 \times r + 75 \times g + 15 \times b)/2^7 \end{aligned} \tag{12.7}$$

In this equation, the products of 0.2989×2^7, 0.5870×2^7, and 0.1140×2^7 are approximated as integers 38, 75, and 15 respectively. Consequently, Integer Design #1 can be constructed as shown in Figure 12.17(b). This design solely employs integer multipliers and adders, along with a 7-bit right shifter to accomplish the mathematical division ($/2^7$).

C. Integer Design #2: Integer Adder Based Design

Assuming that integer adders consume fewer hardware resources than multipliers, Figure 12.17(c) showcases an alternative design option where multipliers are replaced with adders and shifters. This substitution is possible due to the constancy of the weights assigned to the red, green, and blue pixels. To execute the shifting, zeros can be appended either on the left or the right side. For example, $38 \times r$ can be expressed as $(32 + 4 + 2) \times r$ or $(2^5 + 2^2 + 2^1) \times r$. In Verilog notation, the representation for $2^5 \times r$ is $\{r, 5'b0\}$, for $2^2 \times r$ it is

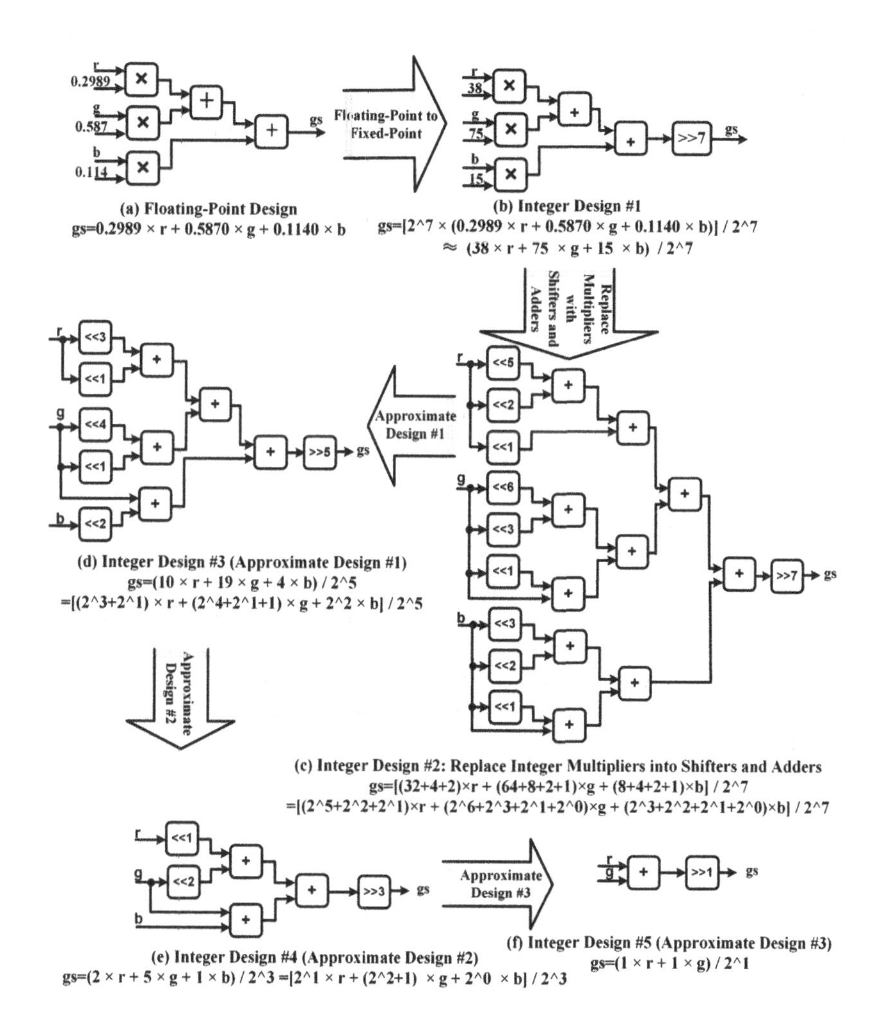

FIGURE 12.17
Floating-to-integer designs of RGB-to-grayscale converter.

$\{r, 2'b0\}$, and for $2^1 \times r$ it is $\{r, 1'b0\}$. The summation of these components results in $\{r, 5'b0\} + \{r, 2'b0\} + \{r, 1'b0\}$.

Likewise, the calculation of $75 \times g$ can be deconstructed into $(64+8+2+1) \times g$ or $(2^6 + 2^3 + 2^1 + 2^0) \times g$. In Verilog notation, this becomes $\{g, 6'b0\}$ for $2^6 \times g$, $\{g, 3'b0\}$ for $2^3 \times g$, and $\{g, 1'b0\}$ for $2^1 \times g$. These results are combined to yield $\{g, 6'b0\} + \{g, 3'b0\} + \{g, 1'b0\} + g$. Similarly, the expression $15 \times b$ can be expanded as $(8 + 4 + 2 + 1) \times b$ or $(2^3 + 2^2 + 2^1 + 2^0) \times b$. The Verilog representation for $2^3 \times b$ is $\{b, 3'b0\}$, for $2^2 \times b$ it is $\{b, 2'b0\}$, and for $2^1 \times b$ it is $\{b, 1'b0\}$. The summation of these components culminates in $\{b, 3'b0\} + \{b, 2'b0\} + \{b, 1'b0\} + b$.

The design of the integer hardware circuit, referred to as #2, is portrayed in Figure 12.17(c), exclusively utilizing integer adders. To assess its performance in terms of resource utilization and computation speed, we assume that each addition operation is completed within a single cycle, while ignoring the delay from shifters and wire connections. Consequently, the circuit's critical path is determined by the propagation delay of the four consecutive adders.

D. Integer Design #3–5: Approximate Designs

While integer implementation #1, as depicted in Figure 12.17(c), relies solely on adders for their execution, they necessitate ten adders, which is twice the number of adders and multipliers utilized in integer design #1 from Figure 12.17(b). To mitigate this, alternative design structures are proposed in Figure 12.7(d)-(f) that minimize the adders' utilization.

For instance, the formula for the conversion from FP to integer in Equation 12.7 can be approximated by multiplying the weights with the integer 32 (or 2^5), rounding up the resulting products, and subsequently dividing the final summations by the integer 32. The mathematical expression for integer design #3 is detailed below:

$$\begin{aligned} gs &= 0.2989 \times r + 0.5870 \times g + 0.1140 \times b \\ &= 2^5 \times (0.2989 \times r + 0.5870 \times g + 0.1140 \times b)/2^5 \\ &\approx (10 \times r + 19 \times g + 4 \times b)/2^5 \\ &= (2^3 \times r + 2 \times r) + (2^4 \times g + 2 \times g + g) + (2^2 \times b) \end{aligned} \tag{12.8}$$

In this hardware design, five integer adders are employed, as illustrated in Figure 12.17(d).

Similarly, the process of converting from FP to integer can be approximated by multiplying the weights with the integer 8 (equivalent to 2^3) and subsequently dividing the final sum by the integer 8. This gives rise to the mathematical formulation of integer design #4:

$$\begin{aligned} gs &= 0.2989 \times r + 0.5870 \times g + 0.1140 \times b \\ &= 2^3 \times (0.2989 \times r + 0.5870 \times g + 0.1140 \times b)/2^3 \\ &\approx (2 \times r + 5 \times g + 1 \times b)/2^3 \\ &= (2 \times r) + (2^2 \times g + g) + b \end{aligned} \tag{12.9}$$

In this hardware design, the implementation necessitates merely three integer adders, as depicted in Figure 12.17(e).

The most conservative accuracy scenario is demonstrated in Figure 12.17(f). The formulation for design #5 is provided below:

$$\begin{aligned} gs &= 0.2989 \times r + 0.5870 \times g + 0.1140 \times b \\ &= 2^1 \times (0.2989 \times r + 0.5870 \times g + 0.1140 \times b)/2^1 \\ &\approx 1 \times r + 1 \times g + 0 \times b/2^1 \\ &= (r + g)/2 \end{aligned} \tag{12.10}$$

In this context, the process of converting color to grayscale necessitates the employment of just one integer adder. Howcvcr, it's evident that this approach heavily depends on design approximations, which can result in a noticeable reduction in the quality of the results.

12.3.3 Hardware utilization vs. speed estimation

Table 12.6 provides an overview of the resource utilization for the integer designs #1 to #5. It is assumed that all the inputs "r", "g", and "b" are registered-in, and the output "gs" is registered-out. Notably, designs #5 exhibits the shortest critical path, involving a single adder. This attribute theoretically contributes to its higher MOF value when evaluating the performance of the sequential circuit.

Among the designs, #2, #4, #5, and #6 feature an increased degree of approximation, leading to a reduction in the quality of the color-to-grayscale conversion outcomes. However, this trade-off results in a decreased hardware cost due to a reduced utilization of integer adders. Striking a balance between quality of results and hardware efficiency is a pivotal consideration in RTL design.

TABLE 12.6
Resource utilization vs. speed of fixed point designs.

Designs	**Hardware Cost**	**Critical Path**
Design#1	3 Multiplier+2 Adders	1 Multiplier + 2 Adders
Design#2	10 Adders	4 Adders
Design#3	5 Adders	3 Adders
Design#4	3 Adders	2 Adders
Design#5	1 Adder	1 Adder

12.4 Neural Networks

The increasing demand for machine learning applications underscores the necessity for specialized hardware acceleration tailored to diverse neural networks. This demand arises from the substantial computational needs inherent in complex neural networks, as well as the requirement for parallel processing and energy efficiency to effectively address these demands. To illustrate, this section delves into various design frameworks applicable to a single-layer perceptron (SLP) neural network. The objective is to introduce a design methodology encompassing the creation of sigmoid neurons and the construction of a basic network architecture.

12.4.1 Single-layer perceptron neural network

A. Design Structure of SLP Neural Network

A basic SLP neural network can be constructed with an input layer, a hidden layer, and an output layer. Increasing the number of neurons in the hidden layer can enhance the network's potential accuracy. For instance, Figure 12.18 illustrates the structure of an SLP network with 12 sigmoid neurons. The input layer requires 784 neurons to accommodate a 28×28 pixel matrix, while the output layer consists of ten neurons. These output neurons produce inference results relevant to distinct classification tasks. As a case in point, in the context of handwritten digit recognition, the results would represent the recognition percentages for digits ranging from zero to nine. The digit associated with the highest percentage among the outputs serves as the final classification result.

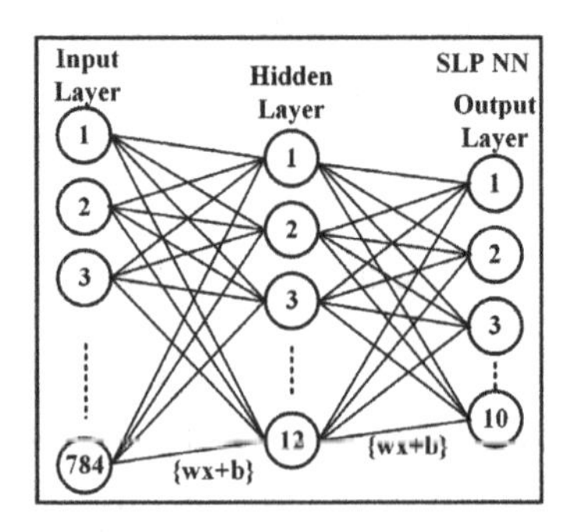

FIGURE 12.18
Design structure of SLP neural network.

B. SLP Sigmoid Neuron

The fundamental building block in constructing the network is the sigmoid neuron. Mathematically, a sigmoid neuron within the hidden layer (hn_i) can be formalized through the following equation:

$$hn_i = \frac{1}{1 + exp(-\sum_{j=1}^{784} wh_{ij} \times x_j - bh_i)} \tag{12.11}$$

In this equation, i denotes the index of sigmoid neurons within the hidden layer, spanning from 1 to 12, while j signifies the index of input elements, ranging from 1 to 784. The variable wh_{ij} corresponds to the weights pertaining to the connections in the hidden layer, and bh_i signifies the bias for the specific hidden neuron. To facilitate this calculation, training results encompassing 784×12 weights and 12 biases are required.

The computation begins by multiplying the 784 input elements x_j by their respective weights wh_{ij} and summing the products. Next, the negative of this summation is calculated and subtracted from the corresponding bias bh_i. Finally, the output of each hidden-layer sigmoid neuron is determined through a sequence of operations, which include applying the exponential function, adding 1.0, and applying the reciprocal function.

Similarly, the expression for the sigmoid neuron within the output layer (on_i) is as follows:

$$on_i = \frac{1}{1 + exp(-\sum_{j=1}^{12} wo_{ij} \times x_j - bo_i)} \tag{12.12}$$

In this context, the index i ranges from 1 to 10, representing the ten output neurons, while the index j ranges from 1 to 12, corresponding to the 12 hidden layer neuron outputs. The computational process for the output neurons mirrors that of the hidden layer neurons. It involves multiplying the 12 hidden neuron outputs x_j by their corresponding output weights wo_{ij}, summing the products, applying the negative summation, subtracting the bias bo_i, applying the exponential function, adding 1.0, and finally taking the reciprocal to obtain the output for each output neuron. To facilitate this calculation, training results involving 12×10 weights and 10 biases are required.

12.4.2 Streaming design on SLP neural network

12.4.2.1 Streaming design on sigmoid neurons

From a hardware design standpoint, the implementation of a sigmoid neuron necessitates multiple FP operators, encompassing FP adders, multipliers, subtractors, dividers, and exponential operators. An illustrative depiction of the streaming design configuration for a hidden-layer sigmoid neuron can be

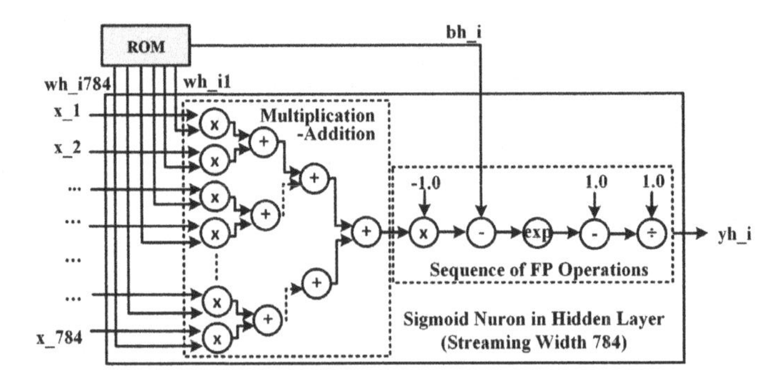

FIGURE 12.19
Streaming design structure of hidden-layer neuron.

found in Figure 12.19. This design can be conceptually divided into two distinct segments.

The initial segment concerns itself with the computation of $\sum_{j=1}^{784} wh_{ij} \times x_j$, as exemplified in the multiplication-addition segment within the figure. This process employs 784 FP multipliers concurrently to derive the products, followed by an accumulation of these 784 products using 783 FP adders. The inputs "x_1 ~ x_784" are sourced from the inputs, while the FP weights "wh_i1 ~ wh_i784" are pre-stored within a Read-Only Memory (ROM), to be accessed when required. The index i spans from 1 to 12, representing the number of neurons within the hidden layer. Consequently, a total of 784×12 weights are implicated across all 12 hidden-layer neurons.

Assuming each FP operator consumes a single clock cycle, the latency associated with the multiplication-addition phase can be computed as $1 + \lceil \log_2 784 \rceil = 11$ clock cycles. Here, the initial clock cycle is allocated to procure the 784 products in parallel, while $\lceil \log_2 784 \rceil = 10$ cycles are expended for the sequential accumulation of additions.

The subsequent segment encompasses a series of FP operations, namely FP multiplication ($x \times -1.0$), subtraction ($x - bh_i$), exponential function ($exp(x)$), addition ($x + 1.0$), and the final reciprocal operation ($1.0/x$). Due to the assumption that each FP operator necessitates a single clock cycle, the sequence of five operations entails an additional five clock cycles after the multiplication-addition phase. Consequently, the latency associated with each sigmoid neuron can be represented as $1 + \lceil \log_2 784 \rceil + 5 = 16$ clock cycles. This encompassing latency includes the multiplication-addition operation as well as the sequence of five subsequent FP operations.

It's important to recognize that the figure exclusively illustrates the design structure of a single sigmoid neuron. When incorporating 12 neurons in the hidden layer, the number of FP operators increases twelvefold. Additionally, the input data quantity surges to $784 \times 12 \times 32$ bits, resulting in a significant

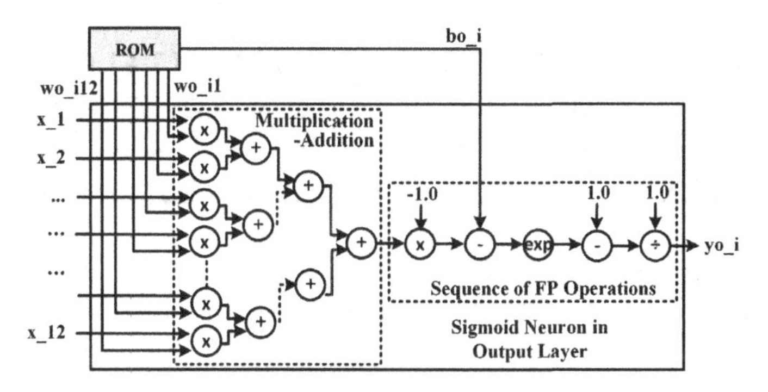

FIGURE 12.20
Streaming design structure of output-layer neuron.

increase in data flow. To mitigate the hardware utilization, one potential solution involves reusing a single sigmoid neuron for the computations related to all 12 neurons within the hidden layer. To coordinate the timing of sigmoid executions in a pipelined and parallel manner, a timing controller is necessary.

Shifting our focus to Figure 12.20, it provides insight into the design configuration of the output-layer sigmoid neurons. The inputs, labeled as "x_1 $\sim$ x_12", are derived from the outputs of the hidden-layer neurons. Each sigmoid neuron accesses 12 FP weights from the ROM. The computational process carried out within this layer closely resembles that of the hidden layer neurons, involving FP multiplications and additions, followed by a sequence of five FP operations. The latency can be calculated as $1 + \lceil \log_2 12 \rceil + 5 = 10$ clock cycles. Within this period, one cycle is allocated for the simultaneous computation of the 12 multiplications, four cycles are dedicated to the sequential additions, and five cycles are accounted for in the subsequent sequence of FP operations.

12.4.2.2 Timing diagram of streaming design on SLP neural network

Figure 12.21 provides a timing diagram of the streaming design, illustrating the activation of the 12 hidden-layer neurons (labeled as HNx on the left-hand side) occurring smoothly over clock cycles 1 to 27. Sequentially, the computation of the 10 output-layer neurons (designated as ONx on the right-hand side) takes place during clock cycles 28 to 46.

Delving into specifics, the inception of the computational process, occurring in the initial clock cycle, involves the utilization of 784 FP multipliers (indicated by "M" in the figure) to facilitate the multiplications of the 784 inputs by their corresponding weights. Over the subsequent ten clock cycles, these intermediate products are aggregated progressively (illustrated as "A0-A9"). The culmination of this iterative summation is witnessed in a coherent sequence spanning clock cycles 12 to 16. This sequence encompasses

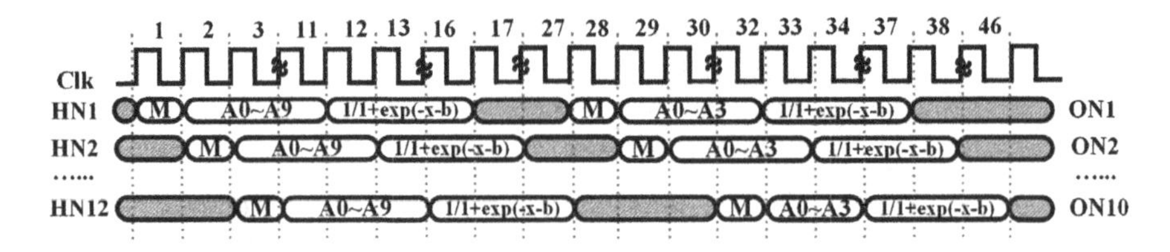

FIGURE 12.21
Timing diagram of steaming design on SLP neural network.

transformative steps such as negation ($-x$), bias subtraction ($-x - b$), exponential calculation ($exp(-x - b)$), addition to 1.0 ($1 + exp(-x - b)$), and culminates with the reciprocal calculation ($1/[1+exp(-x-b)]$). Consequently, the computational traversal for the first hidden-layer neuron (HN1) encompasses a span of 16 clock cycles.

The initiation of the second hidden-layer neuron's calculation coincides with the advent of the second clock cycle and culminates in clock cycle 17. This pattern continues, with subsequent computations proceeding in parallel to the execution of the first hidden-layer neuron. Likewise, the remaining ten hidden-layer neurons seamlessly unfold within a pipeline framework, spanning clock cycles 3 to 27.

Upon the culmination of the hidden-layer neuron computations, the activation of the output-layer neurons commences as of clock cycle 28. Each output-layer neuron is engaged with 12 FP inputs, resulting in an aggregate latency of ten clock cycles. This latency encompasses a single cycle devoted to the simultaneous execution of 12 FP multiplications, four cycles dedicated to the accumulation of the resultant products, and an additional five cycles encompassing operations such as negation, subtraction, exponential calculation, addition, and reciprocal calculation.

Subsequently, the remaining nine output-layer neurons progress through their calculations in a streamlined pipeline progression, requiring an additional nine clock cycles to produce the ten resultant outputs. As a culmination, the comprehensive computational process for all ten output-layer neurons is accomplished within a total of 19 clock cycles. Conclusively, the streaming design framework mandates 46 clock cycles, including the latency of both hidden-layer and output-layer computations, for the complete classification of a 28×28 matrix.

12.4.3 Iterative design on SLP neural network

The streaming design efficiently utilizes a significant number of FP operators and IOs operations to construct each hidden-layer neuron. In this section,

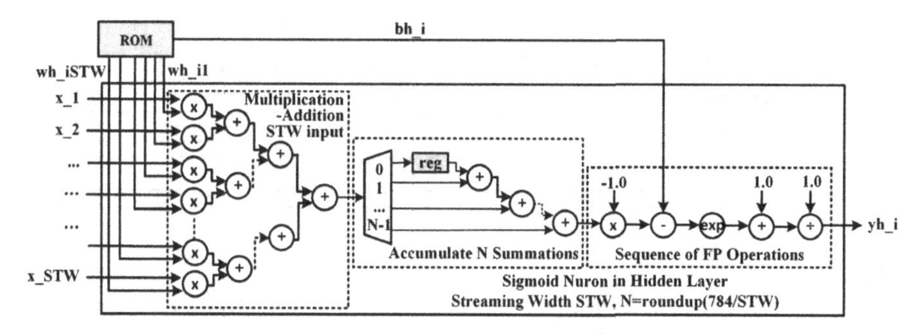

FIGURE 12.22
Iterative design structure of hidden-layer neuron.

we explore some iterative design variations that are characterized by narrower streaming widths. These alternatives aim to reduce the number of FP modules and IO operations. However, it's crucial to acknowledge that this approach inevitably results in increased latency due to the repeated utilization of FP operators.

12.4.3.1 Iterative design on sigmoid neurons

Figure 12.22 presents a detailed iterative design structure for a hidden-layer neuron. The parameter STW defines the streaming width of the iterative design, which is less than the total number of 784 inputs. To accommodate all 784 FP inputs, this approach divides the computation into a total of $N = \lceil 784/STW \rceil$ batches of FP multiplication-addition operations, with each batch handling $STW\times$ inputs. In comparison to the conventional streaming design structure, this iterative approach results in a reduction in FP multipliers and adders by a factor of N.

As depicted in the figure, the iterative design comprises two additional components: the sequential accumulation of multiplication-addition results within a pipeline and a subsequent sequence of five FP operations. Accumulation plays a critical role in the iterative design. The outputs from the pipeline, totaling $N\times$ outputs, are gradually aggregated to obtain the final multiplication-addition result across the entire array of 784 FP inputs. This process requires introducing a one-clock cycle delay for the result of the initial batch, after which it is sequentially added to the result of the second batch. This iterative addition process continues until the final multiplication-addition result is obtained in the $(N-1)^{th}$ iteration.

Ultimately, the final step involves a sequence of operations, specifically $x \times -1.0$, $x - bh_i$, $exp(x)$, $x + 1.0$, and $1.0/x$, which are executed over an additional five clock cycles.

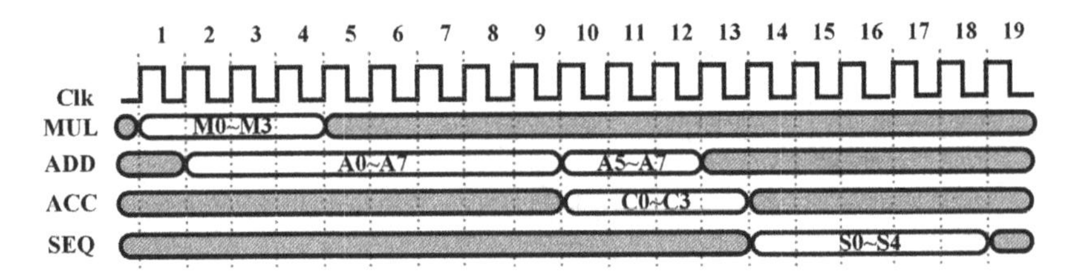

FIGURE 12.23
Timing diagram of hidden-layer sigmoid neuron within iterative design.

12.4.3.2 Latency assessment of hidden-layer sigmoid neurons in iterative design

In accordance with the aforementioned discussion, the computational latency encompassing all hidden-layer neurons can be computed as follows:

$$HL = \lceil 784/STW \rceil + \lceil \log_2 STW \rceil + 1 + 5 + 11 \times \lceil 784/STW \rceil \qquad (12.13)$$

Here, the initial term $\lceil 784/STW \rceil + \lceil \log_2 STW \rceil + 1 + 5$ calculates the computation latency for the first hidden-layer neuron. The subsequent term $11 \times \lceil 784/STW \rceil$ calculates the latency for the remaining 11 neurons in a pipeline configuration.

To provide a detailed illustration of the latency equation, a timing diagram showing the computation of a hidden-layer neuron is presented in Figure 12.23. In this specific design instance, a streaming width of 196 is employed, meaning that 196 FP inputs can be executed during each multiplication-addition phase. The series of operations performed by each neuron includes multiplications (indicated as “MUL”), additions (represented as “ADD”), accumulation (denoted as “ACC”), and ultimately concludes with a final sequence (termed as “SEQ”).

- **Multiplication:** Within the multiplication phase (“MUL”), a span of $\lceil 784/196 \rceil = 4$ clock cycles (“M0-M3”) is expended to effectuate the multiplication of 784 FP inputs with their corresponding weights. This process entails 196 parallel multiplications conducted during each cycle.

- **Addition:** Concurrently, the additions phase (“ADD”) initiates in the second clock cycle, proceeding in parallel with the multiplications. The “ADD” stage mandates $\lceil \log_2 196 \rceil = 8$ clock cycles (“A0-A7”) for the creation of the initial multiplication-addition result, while an additional three clock cycles (“A5-A7” spanning clock cycles 10 to 12) are dedicated to crafting the residual three multiplication-addition outcomes. Consequently, the cumulative latency for the four multiplication-addition operations amounts to $\lceil 784/196 \rceil + \lceil \log_2 196 \rceil = 12$ clock cycles.

- **Accumulation:** Spanning clock cycles 9 to 12, the four summations are fashioned, followed by their subsequent accumulation during clock cycles 10 to 13 ("C0-C3"). Given that accumulation overlaps with the additions stage, the accumulation phase ("ACC") merely requires a supplementary cycle.

- **Sequence of FP Operations:** Subsequently, the final sequence stage ("SEQ") mandates a span of five clock cycles ("S0-S4") to meticulously complete the sequence of FP operations. Conclusively, the combined duration of the accumulation and sequence stages encompasses $1 + 5 = 6$ clock cycles.

As a result, the computation of an individual hidden-layer neuron requires 18 clock cycles. In contrast, the output-layer neuron only involves the computation of 12 FP inputs. Therefore, the same design structure as the streaming paradigm is still applicable. As demonstrated in Figure 12.21, the latency associated with calculating the output-layer neuron amounts to 19 clock cycles.

12.4.3.3 Timing diagrams of iterative designs on SLP hidden-layer neural network

In this section, we present four different design examples crafted to assess the performance of various iterative designs. Our discussion revolves around the timing diagrams for each of these examples. These diagrams play a crucial role in illustrating the delicate balance between latency and hardware utilization that distinguishes these iterative design variations.

A. Iterative Design with Streaming Width 196

Figure 12.24 provides a detailed timing diagram that illustrates the coordination of 12 hidden-layer neurons in an iterative design with a streaming width of 196. This design approach notably results in a significant reduction in IO operations, reducing them by a factor of four compared to the streaming design structure shown in Figure 12.19. Additionally, the iterative design optimizes the multiplication-addition circuit within the hidden layer neuron construction, where both the count of FP multipliers and adders is efficiently reduced to 196 and 195 units, respectively.

Based on the previous analysis, generating the output of each hidden-layer neuron requires 18 clock cycles. Subsequently, the calculation of the remaining 11 hidden-layer neurons can be efficiently executed in a pipelined config-

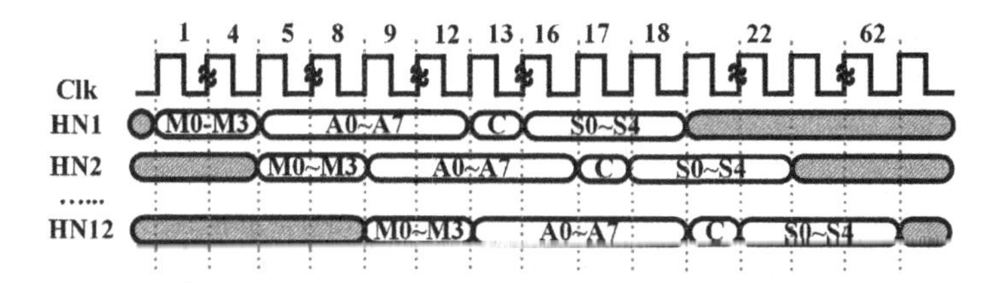

FIGURE 12.24
Timing diagram of iterative design (streaming width 196) on hidden-layer sigmoid neurons.

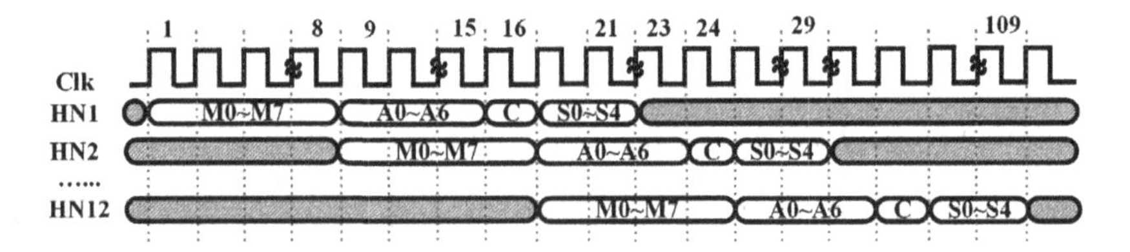

FIGURE 12.25
Timing diagram of iterative design (streaming width 98) on hidden-layer sigmoid neurons.

uration. As the multiplication for one hidden-layer neuron is completed, the multiplication for the next neuron begins, creating a seamless progression. This results in an additional 44 clock cycles to account for the 11 remaining hidden-layer neurons. Therefore, the computations for all 12 hidden-layer neurons add up to $18 + 44 = 62$ clock cycles in total.

B. Iterative Design with Streaming Width 98

The timing diagram, illustrating an iterative design with a streaming width of 98, is presented in Figure 12.25. This particular design approach introduces a significant reduction in streaming IO operations – specifically, a substantial reduction by a factor of eight when compared to the streaming design. Furthermore, the design efficiently reduces the number of FP multipliers and adders within the multiplication-addition circuit to 98 and 97 units, respectively.

Calculating the multiplication-addition latency reveals a value of $\lceil 784/98 \rceil + \lceil \log_2 98 \rceil = 15$ cycles. Here, $\lceil 784/98 \rceil = 8$ represents the delay associated with the 784 FP multiplications, and $\lceil \log_2 98 \rceil = 7$ denotes the delay for cascading additions. Consequently, the computation of the first hidden-layer neuron requires 21 clock cycles. This includes an additional cycle for accumulation (labeled as "C") and a sequence of five consecutive FP operations (labeled as "S0-S4"). The subsequent computations for the remaining 11 hidden-layer neurons follow in a pipeline fashion. The multiplication phase of each neuron begins every eight clock cycles after the previous one. Therefore, the total latency amounts to $21 + 8 \times 11 = 109$ cycles. This encompasses the 21 cycles needed for the initial sigmoid neuron's completion, along with an additional $8 \times 11 = 88$ cycles for the pipeline-based multiplications involving the remaining 11 neurons.

C. Iterative Designs with Streaming Widths 49 and 28

In this section, we introduce two additional iterative designs. The timing diagram of the iterative design with a streaming width of 49 is depicted in

Figure 12.26(a), while the one with a streaming width of 28 is shown in Figure 12.26(b). In contrast to the streaming design, the former design structure reduces the streaming IO operations by a factor of 16, while the latter reduces it by a factor of 28. Within the multiplication-addition circuits, these two designs involve 49 and 28 multipliers, as well as 48 and 27 adders, respectively.

Reducing the streaming width to 49 and 28 alters the latency for the 784 multiplications to $\lceil 784/49 \rceil = 16$ and $\lceil 784/28 \rceil = 28$ clock cycles for the respective designs. In the design with a streaming width of 49, it takes a total of $\lceil 784/49 \rceil + \lceil \log_2 49 \rceil = 22$ clock cycles to obtain the final multiplication-addition result. For the design with a streaming width of 28, the latency can be reduced to $\lceil 784/28 \rceil + \lceil \log_2 28 \rceil = 33$ clock cycles. Additionally, an extra six clock cycles are needed for the subsequent accumulation and sequence of operations in both designs. Consequently, the overall latency for each sigmoid neuron computation is $22 + 6 = 28$ for the design with a streaming width of 49 and $33 + 6 = 39$ for the design with a streaming width of 28.

The latency for the remaining 11 sigmoid neurons is determined by the initiation of multiplications. In the design with a streaming width of 49, each neuron's computation begins with a delay of 16 cycles, resulting in a total of 204 clock cycles for the execution of the hidden-layer neurons $(28 + 11 \times 16)$. Similarly, the design with a streaming width of 28 requires 347 clock cycles $(39 + 11 \times 28)$, with each neuron's computation commencing after a delay of 28 cycles.

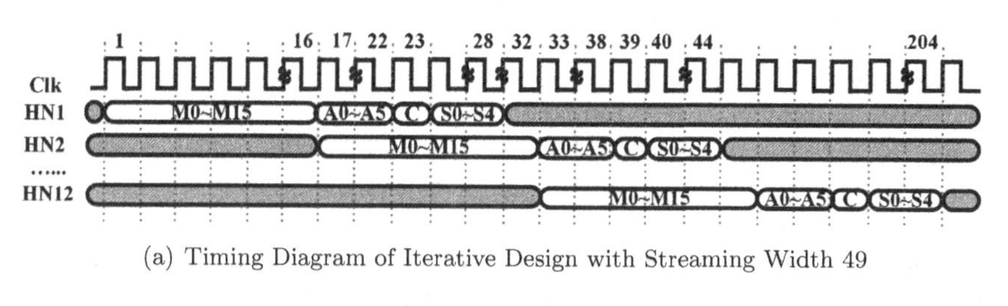

(a) Timing Diagram of Iterative Design with Streaming Width 49

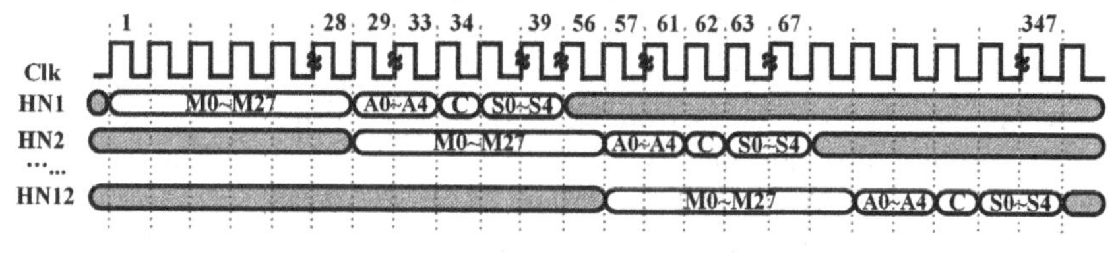

(b) Timing Diagram of Iterative Design with Streaming Width 28

FIGURE 12.26
Timing diagram of iterative design (streaming widths 49 and 28) on hidden-layer sigmoid neurons.

TABLE 12.7
Resource cost vs. latency of hidden-layer neurons.

NNs-STW	Hardware Cost					Latency /(Cycles)
	MUL	ADD	SUB	EXP	REC	
784	785	783	2	1	1	27
196	197	198	2	1	1	62
98	99	104	2	1	1	109
49	50	63	2	1	1	204
28	29	54	2	1	1	347

12.4.4 Hardware utilization vs. latency estimation

A. Design Performance of Hidden-Layer Neurons

Table 12.7 offers a comprehensive overview of hardware resource costs and latency associated with the computation of 12 hidden-layer neurons. The design employing a streaming width of 784 is referred to as "streaming design" or "STW 784" in this section, while the other designs are collectively referred to as "iterative designs" or "STW 196 – STW 28", as outlined in Table 12.7.

Specifically, in the "STW 784" design, 784 FP multipliers (referred to as "MUL") work in parallel for each hidden-layer neuron, necessitating 783 FP adders (labeled as "ADD") for the cumulative sum of the 784 products. It's worth noting that within the sequence, an additional FP multiplier and FP adder are utilized for computing $-1.0 \times x$ and $1.0 + x$, respectively. This results in the total number of FP multipliers and adders in the streaming design being 785 and 784, respectively. For each sigmoid neuron, sequence operations require one FP subtractor ("SUB"), one exponential operation ("EXP"), and one reciprocal calculation ("REC").

For the iterative designs, the structures "STW 196", "STW 98", "STW 49", and "STW 28" employ 197, 99, 50, and 29 FP multipliers, respectively. To calculate the multiplication-addition results, 195, 97, 48, and 27 FP adders are needed for cascaded additions. Additionally, $N - 1$ FP adders are required to accumulate the outcomes from N batches of multiplication-addition results, therefore, an additional 3, 7, 15, and 27 FP adders are necessary for the "STW 196", "STW 98", "STW 49", and "STW 28" designs, respectively. Including one additional adder in the sequence, the total number of FP adders for the iterative designs is presented in the third column as 199, 105, 64, and 55, respectively. The number of other operators remains consistent across all designs, as indicated in columns 4–6.

The latency for both the streaming and iterative designs is presented in the final column. Remarkably, the streaming design achieves the highest speed, while the iterative design with the smallest streaming width of 28 incurs the highest latency. This highlights that increased resource costs often result in

TABLE 12.8
Resource cost vs. latency of SLP neural networks.

NNs-STW	Hardware Cost					Latency
	MUL	ADD	SUB	EXP	REC	/(Cycles)
784	798	796	2	2	2	46
196	210	211	2	2	2	81
98	112	117	2	2	2	128
49	63	76	2	2	2	223
28	42	67	2	2	2	366

faster performance, but these trade-offs are typically constrained by factors such as IO limitations, chip size, and power consumption.

B. Design Performance of SLP Neural Networks

The utilization of FP operators and the overall latency of the SLP network are succinctly presented in Table 12.8. The output-layer neuron necessitates 13 FP multipliers and 11 FP adders for cascading product summations. For the various design structures, namely "STW 784", "STW 196", "STW 98", "STW 49", and "STW 28", the corresponding FP multiplier counts are 798, 210, 112, 63, and 42, respectively. Likewise, the FP adder counts are 796, 211, 117, 76, and 67, respectively. The quantities of FP subtractors, exponential functions, and reciprocal operators are doubled. As each output-layer neuron demands 19 clock cycles for execution, the cumulative latency is presented in the final column of the table.

Exercises

Problem 12.1. Utilizing the AMBA AXI protocol, a manager initiates a command to transfer a block of 16 bytes to an associated subordinate memory controller.

1) If the burst size is set to bytes (axsize=3'b000), what value represents the burst length ("axlen")?

2) Given an initial address of "axaddr=32'h14" and a burst type of $FIXED$ ("axburst=2'b00"), what sequence of memory addresses pertains to the remaining data transfers?

3) In the case of a burst type set to INCR ("axburst=2'b01"), which memory addresses correspond to the subsequent data transfers?

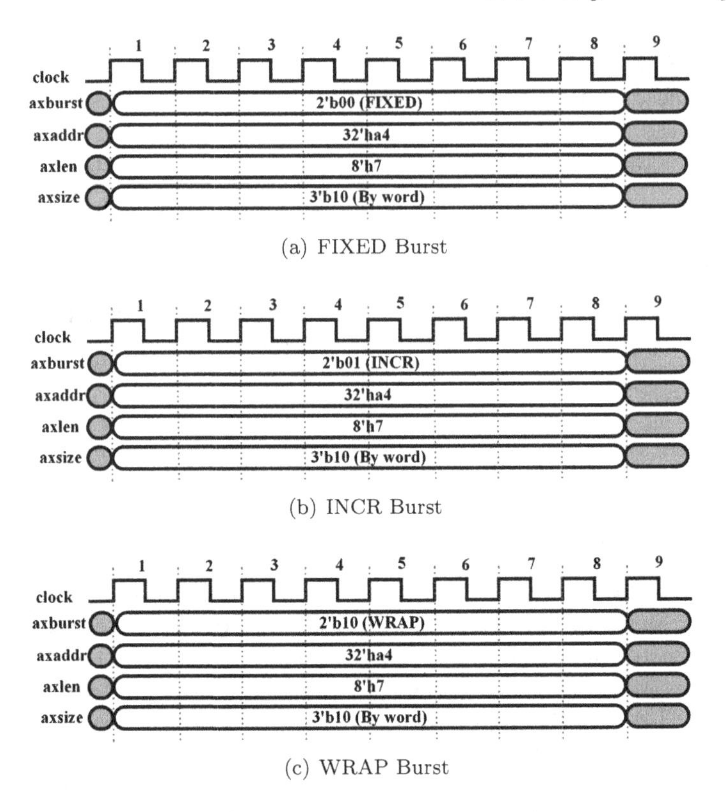

(a) FIXED Burst

(b) INCR Burst

(c) WRAP Burst

FIGURE 12.27
AXI bus transaction examples.

4) When the burst size is defined in words ("axsize=3'b010"), what burst length ("axlen") value should be used? Additionally, please revisit questions 2) and 3) accordingly.

Problem 12.2. Figure 12.27 represents three distinct bus transactions involving a manager and a subordinate. These transactions encompass a FIXED burst, as depicted in Figure 12.27(a), an INCR burst, visualized in Figure 12.27(b), and finally, a WRAP burst illustrated in Figure 12.27(c).

In each of the three bus transactions, the initial memory address is specified as "axaddr=32'ha4". What are the subsequent memory addresses in the FIXED burst? What are the subsequent memory addresses in the INCR burst? What are the subsequent memory addresses in the WRAP burst?

Problem 12.3. In an AXI bus transaction, the size of each data transfer in bytes is determined by the "axsize". Assuming the CPU configures the memory address as 32'h17 through the AXI bus, the memory address on the hardware side can be aligned with the transfer size as follows:

- If the "axsize" is set to 3'h000, resulting in one byte transferred in each transaction, what memory address on the hardware side corresponds to the configured memory address 32'h17?
- If the "axsize" is configured as binary 3'b001, indicating two bytes transferred in each transaction, what memory address on the hardware side aligns with the configured memory address 32'h17?
- When the "axsize" is set to binary 3'b010, leading to four bytes transferred in each transaction, what memory address on the hardware side aligns with the configured memory address 32'h17?
- Assuming the "axsize" is defined as binary 3'b011, causing eight bytes to be transferred in each transaction, what memory address on the hardware side aligns with the configured memory address 32'h17?
- In the scenario where the "axsize" is specified as binary 3'b100, resulting in 16 bytes transferred in each transaction, what memory address on the hardware side corresponds to the configured memory address 32'h17?
- Given that the "axsize" is set to binary 3'b101, leading to 32 bytes transferred in each transaction, what memory address on the hardware side corresponds to the configured memory address 32'h17?
- If the "axsize" is configured as binary 3'b110, resulting in 64 bytes transferred in each transaction, what memory address on the hardware side corresponds to the configured memory address 32'h17?
- Assuming the "axsize" is set to binary 3'b111, resulting in 128 bytes transferred in each transaction, what memory address on the hardware side corresponds to the configured memory address 32'h17?

Problem 12.4. Figure 12.28 presents two instances of the AXI bus handshaking between an AXI source and an AXI destination. These examples showcase bus transactions involving eight data transfers ("d0-d7"). In Figure 12.28(a), the "ready" signal spans 13 clock cycles, with the initial data transfer ("d0") commencing during the second clock cycle where the "valid" signal is asserted. By clock cycle 13, the "valid" signal is negated, concluding the entire bus transaction. Depict the signal trace for the "valid" signal and the data ("d1" to "d7" inclusive) within clock cycles 3 to 12.

In Figure 12.28(b), the "valid" signal is presented alongside the eight data transfers ("d0-d7") across 13 clock cycles. Additionally, the "ready" signal is indicated during the first and last clock cycles. Illustrate the "ready" signal's profile within the clock cycles spanning from 2 to 12.

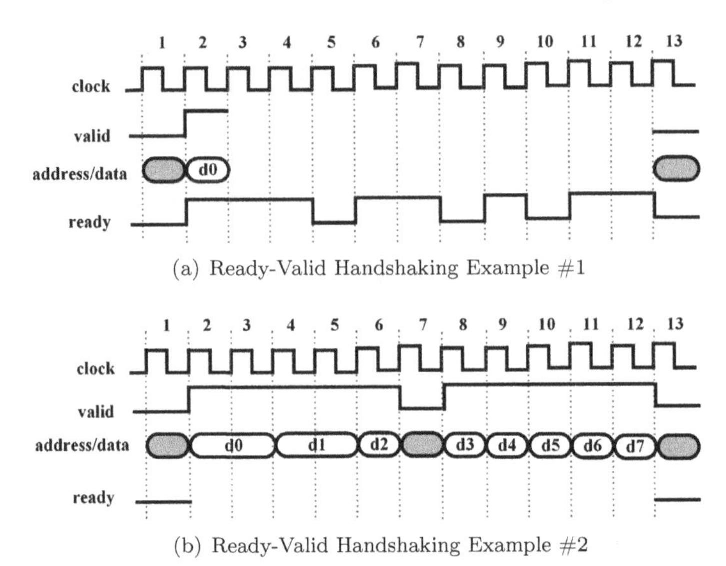

FIGURE 12.28
AXI ready-valid handshaking examples.

PBL 42: DMA Arbiter Design

1) Design the DMA Arbiter in accordance with the dynamic priority mechanism illustrated in Figure 12.11. Assume that four masters, denoted as M1, M2, M3, and M4, concurrently initiate requests for bus access, each initially assigned a priority value of one, two, three, and four, respectively. The IO specifications are detailed in Table 12.9, encompassing the request inputs from the four bus masters (labeled as "req1", "req2", "req3", and "req4"), as well as the corresponding grant outputs from the DMA Arbiter, serving as the bus slave for interfacing (designated as "gnt1", "gnt2", "gnt3", and "gnt4").

TABLE 12.9
DMA Arbiter IOs description.

Name	Direction	Bit Width	Description
clk	Input	1	Clock, rising edge, 100 MHz
rst	Input	1	Asynchronous reset, 0 valid
req1–req4	Input	1	M1–M4 requests
gnt1–gnt4	Output	1	M1–M4 grants

2) Create a testbench for conducting simulations of the DMA Arbiter design. The bus functional model, emulating the behavior of the bus master, should initiate requests from M1 to M4, adhering to the timing diagram depicted in Figure 12.13. To be specific, M1 and M2 are expected to transmit a single command each, while M3 and M4 should send three consecutive commands. For the sake of simplicity, the command inputs are not included in this project. After receiving all the grants from the DMA Arbiter, the bus masters should deactivate their requests to complete the request-grant handshake.

PBL 43: FP RGB-to-Grayscale Converter

1) Design the RGB-to-Grayscale Converter depicted in Figure 12.17(a), by leveraging the single-clock-cycle FP adder and multiplier provided in this book.
2) Create a testbench to facilitate the simulation of the converter design and validate its functionality using 4×4 matrix of RGB inputs.

PBL 44: High-Speed Design on FP RGB-to-Grayscale Converter

1) Redesign the circuit in PBL 43 by leveraging the pipeline-designed FP operators provided in this book.
2) Create a testbench to facilitate the simulation of the converter design and validate its functionality using the same 4×4 matrix of RGB inputs in PBL 43.
3) Comments on the difference between PBL 43 using single-clock-cycle IPs and PBL 44 using pipeline-designed IPs.

PBL 45: Integer Designs on RGB-to-Grayscale Converter

1) Design the RGB-to-Grayscale Converter illustrated in Figure 12.17(b)–(f), utilizing Verilog operators "+" and "*" for the binary operators of the adder and multiplier.

TABLE 12.10
Performance evaluation of hardware cost, MOF, and power consumption.

Designs		Resource Cost		MOF (MHz)	Power (mW)
		LUTs	FFs		
Floating-Point	PBL 43: Single-Cycle FP Operators				
	PBL 44: Pipelined FP Operators				
PBL 45: Fixed-Point	Design #1				
	Design #2				
	Design #3				
	Design #4				
	Design #5				
	Design #6				

2) Create a testbench to facilitate the simulation of the converter designs and verify their functionality using the identical 4×4 matrix of RGB inputs in PBL 43. It's important to note that the conversion from the floating-point to integer format is a critical step for data input. You can refer to Section 12.3.2 for insights into the methods of converting FP numbers into integer numbers.

3) Execute synthesis and implementation for all designs, selecting the same FPGA part in Vivado. Configure DSP utilization to 0 and assess the resource utilization in terms of LUTs and FFs. You can find the resource utilization data in the generated Resource Utilization Report. Populate the hardware cost in the third and fourth columns in Table 12.10. Additionally, provide comments on the disparities for different designs in resource utilization.

4) Assuming the design specification requires a reference clock frequency of 100 MHz, approximate the achievable MOF for all designs. Populate the MOFs in the fifth column in Table 12.10. Provide comments on the differences in MOF.

5) Following the synthesis and implementation phases, evaluate the total power dissipation, which includes both static power and dynamic power. You can locate power consumption data in the generated Power Utilization Report. Fill in the power consumption values in the sixth column in Table 12.10. Provide comments on the differences in power estimation.

6) Conclude the performance evaluation, encompassing the hardware resource cost, power dissipation, and design speed in MOF.

PBL 46: Streaming Design on Sigmoid Neuron

1) Design a sigmoid neuron using the streaming design structure referred in Figure 12.19. Incorporate the single-clock-cycle FP operators provided in this textbook. Assume that the input matrix has a size of 4×4 FP data, and the streaming width is set to 16. The corresponding weights, wh_1 - wh_{16}, are FP 1.0, −1.0, 2.0, −2.0, until 8.0 and −8.0.
2) Create a testbench to perform simulations of the streaming design. The bus functional model should input 16 FP data every clock cycle, x_1 - x_{16}, referring the timing diagram illustrated in Figure 12.21.
3) Conclude the performance evaluation, which includes the computational latency measured in clock cycles (as observed in the simulation waveform) for each 4×4 FP matrix input, and the utilization of FP operators (as counted in your design). Populate the results in the second row of Table 12.11.

PBL 47: High-Speed Streaming Design on Sigmoid Neuron

1) Redesign the circuit in PBL 46 by leveraging the pipeline-designed FP operators provided in this book.
2) Create a testbench to facilitate the simulation of the sigmoid neuron design and validate its functionality using the same data inputs in PBL 46.
3) Comments on the difference between using single-clock-cycle IPs and the pipeline-designed IPs. Populate the results in the third row in Table 12.11.

TABLE 12.11
Performance evaluation of computational latency and FP operator utilization.

Designs		Latency (Clock Cycles)	Number of FP Operators
Streaming Design (Streaming Width: 16)	PBL 46: Single-Cycle FP Operators		
	PBL 47: Pipelined FP Operators		
Iterative Design (Streaming Width: 8)	PBL 48: Single-Cycle FP Operators		
	PBL 50: Pipelined FP Operators		
Iterative Design (Streaming Width: 4)	PBL 49: Single-Cycle FP Operators		
	PBL 50: Pipelined FP Operators		

PBL 48: Iterative Design on Sigmoid Neuron with Streaming Width Eight

1) Design a sigmoid neuron using the iterative design structure referred in Figure 12.22. Incorporate the single-clock-cycle FP operators provided in this textbook. Assume that the input matrix has a size of 4×4 FP data, and the streaming width is set to eight. Utilize the same weights in PBL 46.
2) Create a testbench to perform simulations of the iterative design with a streaming width of eight. The bus functional model should input eight FP data every clock cycle, $x_1 - x_8$, and it takes two clock cycles to feed in each data set containing 16 FP data. Refer to the timing diagram illustrated in Figure 12.23 for guidance.
3) Conclude the performance evaluation, which includes the computational latency measured in clock cycles (as observed in the simulation waveform) for each 4×4 FP matrix input, and the utilization

of FP operators (as counted in your design). Populate the results in the fourth row in Table 12.11.

PBL 49: Iterative Design on Sigmoid Neuron with Streaming Width Four

1) Design a sigmoid neuron using the iterative design structure referred in Figure 12.22. Incorporate the single-clock-cycle FP operators provided in this textbook. Assume that the input matrix has a size of 4×4 FP data, and the streaming width is set to four. Utilize the same weights in PBL 46.
2) Create a testbench to perform simulations of the iterative design with a streaming width of four. The bus functional model should input four FP data every clock cycle, $x_1 - x_4$, and it takes four clock cycles to feed in each data set containing 16 FP data. Refer to the timing diagram illustrated in Figure 12.23 for guidance.
3) Conclude the performance evaluation, which includes the computational latency measured in clock cycles (as observed in the simulation waveform) for each 4×4 FP matrix input, and the utilization of FP operators (as counted in your design). Populate the results in the sixth row in Table 12.11.

PBL 50: High-Speed Iterative Designs on Sigmoid Neuron with Streaming Width Eight and Four

1) Redesign the circuits in PBL 48 and PBL 49 by leveraging the pipeline-designed FP operators provided in this book.
2) Create a testbench to facilitate the simulation of the two design circuits and validate their functionality using the same FP data input.
3) Comments on the difference between using single-clock-cycle design IPs and the pipeline-designed IPs. Populate the results in the fifth and seventh rows in Table 12.11.

Bibliography

[1] OpenROAD. https://theopenroadproject.org/. Online; accessed 12 January 2024.

[2] In *Xilinx 7 Series FPGAs Configurable Logic Block - User Guide*, page 20, September 27, 2016.

[3] Chips Alliance. Chisel – Software-defined hardware. https://www.chisel-lang.org/. Online; accessed 12 January 2024.

[4] John Clow, Georgios Tzimpragos, Deeksha Dangwal, Sammy Guo, Joseph McMahan, and Timothy Sherwood. A Pythonic Approach for Rapid Hardware Prototyping and Instrumentation. In *2017 27th International Conference on Field Programmable Logic and Applications (FPL)*, pages 1–7. IEEE, 2017.

[5] Wikipedia Contributors. Apple A11-A17. https://en.wikipedia.org/wiki/Artificial_intelligence, 2017. Online; accessed 12 January 2024.

[6] J. W. Cooley and J. W. Tukey. An algorithm for the machine calculation of complex Fourier series. *Mathematics of Computation*, 19(90):297–301, 1965.

[7] IRDS. 2020 International Roadmap For Devices and Systems. https://irds.ieee.org/editions/2020/executive-summary, 2020.

[8] ITRS. 2015 International Technology Roadmap for Semiconductors. https://www.semiconductors.org/resources/2015-international-technology-roadmap-for-semiconductors-itrs/, 2015.

[9] N. P. Jouppi, C. Young, N. Patil, and D. Patterson. In-datacenter performance analysis of a tensor processing unit. *in Proceedings of the 44th Annual International Symposium on Computer Architecture (ISCA 2017)*, pages 1–12, 2017.

[10] David Koeplinger, Matthew Feldman, Raghu Prabhakar, Yaqi Zhang, Stefan Hadjis, Ruben Fiszel, Tian Zhao, Luigi Nardi, Ardavan Pedram, Christos Kozyrakis, and Kunle Olukotun. Spatial: A Language and Compiler for Application Accelerators. In *Proceedings of the 39th ACM SIGPLAN Conference on Programming Language Design and Implementation*, pages 1–16, 2018.

[11] ARM Limited. AMBA AHB Specification. Technical report, Sunnyvale, CA, USA, 1999.

[12] ARM Limited. AMBA AXI Protocol Specification. Technical report, Sunnyvale, CA, USA, 2010.

[13] Derek Lockhart, Gary Zibrat, and Christopher Batten. PyMTL: A Unified Framework for Vertically Integrated Computer Architecture Research. In *Proceedings of the 47th Annual IEEE/ACM International Symposium on Microarchitecture*, pages 37–49. IEEE, 2014.

[14] OCP International Partnership (OCP-IP). OCP 2.2 Specification. `https://www.ocpip.org/download/ocp-2.2.pdf`, 2009. Online; accessed 12 January 2024.

[15] SiliconSapiens. Wishbone System-on-Chip (SoC) Interconnection Architecture. `http://cdn.opencores.org/downloads/wbspec_b4.pdf`, 2003. Online; accessed 12 January 2024.

[16] Emil Talpes, Douglas Williams, and Debjit Das Sarma. Dojo: The microarchitecture of tesla's exa-scale computer. In *2022 IEEE Hot Chips 34 Symposium (HCS)*, pages 1–28, 2022.

[17] Mario Vega, Xiaokun Yang, John Shalf, and Doru Thom Popovici. Towards a flexible hardware implementation for mixed-radix fourier transforms. In *2023 IEEE High Performance Extreme Computing Conference (HPEC)*, pages 1–7, 2023.

Index

Printed in the United States
by Baker & Taylor Publisher Services